LIGHTWEIGHT SANDWICH CONSTRUCTION

LIGHTWEIGHT SANDWICH CONSTRUCTION

Edited by
J.M. Davies
DSc, PhD, FREng, FICE, FIStructE
Professor of Structural Engineering,
The University of Manchester, UK

Published on behalf of CIB Working Commission
W056 Sandwich Panels (joint CIB-ECCS commission)

Blackwell
Science

Editorial Offices:
Osney Mead, Oxford OX2 0EL
25 John Street, London WC1N 2BS
23 Ainslie Place, Edinburgh EH3 6AJ
350 Main Street, Malden
MA 02148 5018, USA
54 University Street, Carlton
Victoria 3053, Australia
10, rue Casimir Delavigne
75006 Paris, France

Other Editorial Offices:

Blackwell Wissenschafts-Verlag GmbH
Kurfürstendamm 57
10707 Berlin, Germany

Blackwell Science KK
MG Kodenmacho Building
7–10 Kodenmacho Nihombashi
Chuo-ku, Tokyo 104, Japan

Iowa State University Press
A Blackwell Science Company
2121 S. State Avenue
Ames, Iowa 50014-8300, USA

First published 2001

Set in 9.5/12pt Times
by DP Photosetting, Aylesbury, Bucks
Printed and bound in Great Britain by
MPG Books Ltd, Bodmin, Cornwall

DISTRIBUTORS

Marston Book Services Ltd
PO Box 269
Abingdon
Oxon OX14 4YN
(*Orders:* Tel: 01235 465500
Fax: 01235 465555)

USA
Blackwell Science, Inc.
Commerce Place
350 Main Street
Malden, MA 02148 5018
(*Orders:* Tel: 800 759 6102
781 388 8250
Fax: 781 388 8255)

Canada
Login Brothers Book Company
324 Saulteaux Crescent
Winnipeg, Manitoba R3J 3T2
(*Orders:* Tel: 204 837 2987
Fax: 204 837 3116)

Australia
Blackwell Science Pty Ltd
54 University Street
Carlton, Victoria 3053
(*Orders:* Tel: 03 9347 0300
Fax: 03 9347 5001)

A catalogue record for this title is available from the British Library

ISBN 0-632-04027-0

Library of Congress
Cataloging-in-Publication Data
Lightweight sandwich construction/edited
by J.M. Davies.
p. cm.
Includes bibliographical references and index.
ISBN 0-632-04027-0
1. Sandwich construction. 2. Lightweight
construction. I. Davies, J.M.
TA660.S3 L55 2001
624.1′779–dc21 00-052912

For further information on
Blackwell Science, visit our website:
www.blackwell-science.com

Contents

List of Contributors

Professor Dr K. Berner Professor of Structural Engineering, FH Mainz, University of Applied Sciences, Germany. *(Chapters 7 and 14).*

Dr G.M.E. Cooke International Fire Safety Consultant, London, UK. *(Chapter 5).*

Professor J.M. Davies Professor of Structural Engineering, The University of Manchester, UK. Chairman of ECCS Committee TC7 and Joint Chairman of the ECCS/CIB Committee responsible for updating the European Recommendations for Sandwich Panels. *(Chapters 1, 5, 8, 9 and 13).*

Mr J. Gustafsson Manager, Paroc Ab, Sweden. *(Chapters 3 and 4).*

Mr P. Hassinen Research Scientist and Laboratory Manager, Helsinki University of Technology, Finland. *(Chapters 10 and 11).*

Mr A. Helenius Research Scientist, Technical Research Centre of Finland. Secretary of the ECCS/CIB Committee responsible for updating the European Recommendations for Sandwich Panels. *(Chapters 2 and 13).*

Mr L. Heselius Technical Director, Paroc Oy Ab Panel Systems, Parainen, Finland. CIB Joint Chairman of the ECCS/CIB Committee responsible for updating the European Recommendations for Sandwich Panels. *(Chapter 6).*

Dr T. Kellner Appointed expert for light gauge steel constructions, IFBS, Düsseldorf, Germany.

In addition, the following each contributed to one of the chapters, although they did not attend any of the meetings:

Mr D. Stemmann Former Technical Director, Hoesch, Germany, Former Chairman of ECCS Technical Working Group TWG 7.4 responsible for the original Preliminary European Recommendations. *(Chapter 1).*

Dr A.J. Brookes Brookes, Stacey, Ramsden, Fursdon, Architects, London, UK. *(Chapter 8).*

Various authors are responsible for Chapter 12.

Preface

Sandwich panels are being extensively and increasingly used in single storey and multi-storey building construction because they are light in weight, energy efficient, aesthetically attractive and can be easily handled and erected. An authoritative book which considers all aspects of their design may, therefore, be considered to be overdue.

This book has had a long period of gestation. The authors have all been closely involved in the evolution of the definitive documents for the design and application of sandwich panels within Europe and it was in the context of this committee work that the need for a comprehensive book, covering all of the important considerations which arise in the design, manufacture and use of sandwich panels, was first seen.

The *Preliminary European Recommendations for Sandwich Panels* were published by the European Convention for Constructional Steelwork (ECCS) in two parts. *Part I: Design* was published as ECCS Publication No. 66, in 1991. Its companion volume, *Part II: Good Practice* was published as ECCS Publication No. 62 in 1990. These Recommendations were drafted by ECCS Technical Working Group TWG 7.4 and concentrated on sandwich panels with thin metal faces and plastic foam cores. Two years later, the Recommendations for design were extended by CIB Commission W56 to cover sandwich panels with cores made of mineral wool lamellas and other slabstock materials. The work of the CIB Commission was published in CIB Publication No. 148, originally in 1993 and as a reprint again in 1995.

This work of producing these design recommendations has brought together experts from all over Europe on a regular basis and has led to a sharing of knowledge and experience. Furthermore, a great deal of additional experience of the design and use of sandwich panels with new material combinations and new areas of application has been gained since the Recommendations were first drafted. The work of harmonisation of the design rules and the standards for loads, materials and structures within Europe under the auspices of CEN has also emphasised the need to revise the European Recommendations for sandwich panels. The updating work on *Part I: Design* has again been carried out by a joint ECCS and CIB Committee and is now complete. The updated Recommendations will be published at about the same time as this book.

The authors of this book have been intimately involved in the above scenario and the book represents their efforts to bring together in one volume the sum of their accumulated knowledge and experience. It follows that the book has been drafted in an unusual way because it has been largely written in committee under the auspices of CIB Commission W56 'Lightweight Constructions'. Practical considerations have necessitated that certain members have taken responsibility for drafting and updating individual chapters in accordance with decisions taken at meetings but the members have accepted a joint responsibility for the balance and content of the individual chapters and the book as a whole. Harmony of style has been ensured by the efforts of the Editor.

The technology of sandwich construction continues to advance and the work of developing the guidelines for design, testing and use will continue long after this book is published. The authors, together with the other members of committees ECCS TC7 and CIB W56 will, therefore, welcome critical comments and proposals to improve the various design documents for which they have responsibility.

J.M. Davies
Editor

Chapter 1
Introduction

1.1 THE ORIGINS OF SANDWICH TECHNOLOGY

The first successful landing of a space ship on the moon on 20 July 1969 was the result of the successful application of a number of new technologies including rocketry, computers and sandwich construction. Although public interest centred on rocketry and computer technology, it was only with the help of sandwich technology that a shell of the spacecraft could be constructed that was light in weight and yet strong enough to sustain the stresses of acceleration and landing. Figure 1.1 shows the wall construction of the Apollo capsule which consisted of two interconnected sandwich shells. Figure 1.2 shows details of the outer shell, which comprised two thin steel facings and a honeycomb core. The inherent advantage of sandwich construction is immediately apparent, namely, high strength and rigidity at low weight.

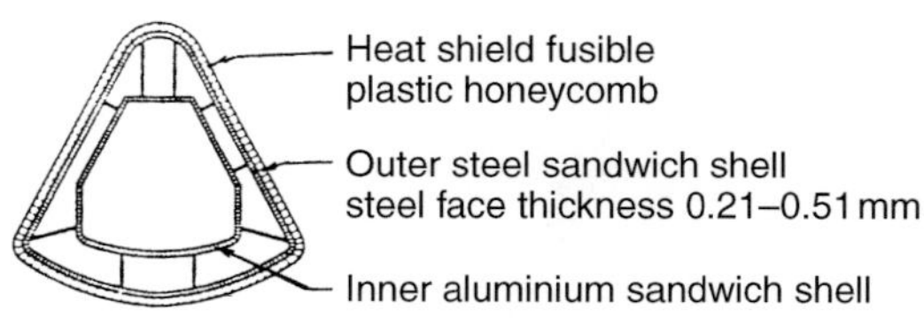

Fig. 1.1 Sandwich construction of the Apollo capsule.

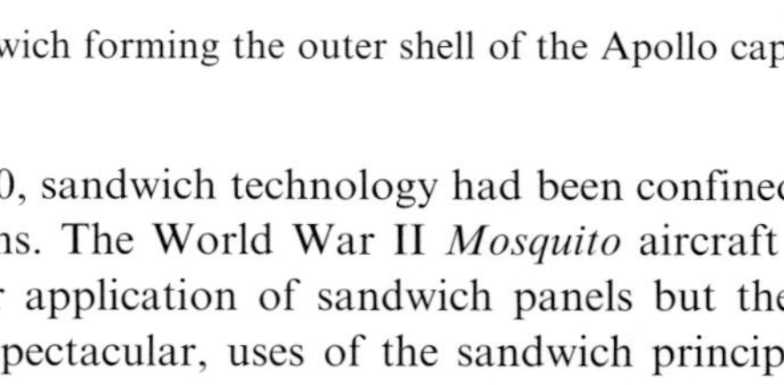

Fig. 1.2 Cellular sandwich forming the outer shell of the Apollo capsule.

Prior to about 1960, sandwich technology had been confined almost entirely to aerospace applications. The World War II *Mosquito* aircraft is often quoted as being the first major application of sandwich panels but there were numerous earlier, though less spectacular, uses of the sandwich principle. By about 1960, increasing numbers of alternative uses were being discovered, such as in the building, refrigerated storage, automobile and shipbuilding industries. This period was also the beginning of a worldwide boom in prefabricated building elements for diverse applications.

1.2 PRINCIPLES OF SANDWICH CONSTRUCTION

The structure of sandwich panels always follows the same basic pattern. Two facings, which are relatively thin and of high strength, enclose a core which is relatively thick and light and which has adequate stiffness in a direction normal to the faces of the panel. A great many alternative forms of sandwich construction may be obtained by combining different facing and core materials. The facings may be steel, aluminium, wood, fibre-reinforced plastic or even concrete. The core may be made of cork, balsa wood, rubber, solid plastic material (polyethylene), rigid foam material (polyurethane, polystyrene, phenolic foam), mineral wool slabs or from honeycombs of metal or even paper. Figure 1.3 shows three examples of structural sandwich elements, namely:

- panel with a polyurethane or polystyrene core
- panel with a metal or paper honeycomb core
- panel with a mineral wool core

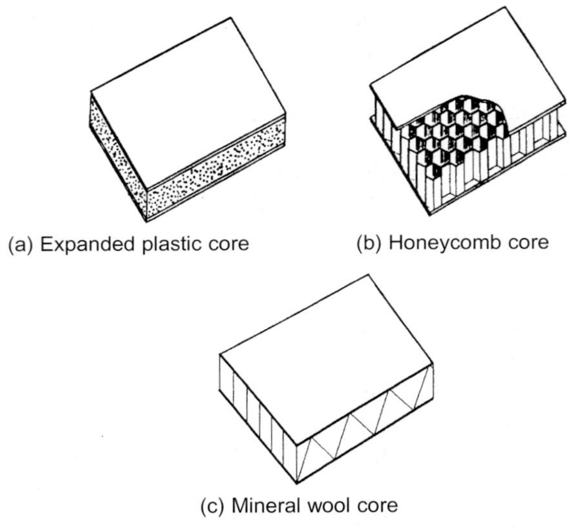

Fig. 1.3 Examples of structural sandwich elements.

In each case, the faces may be colour coated steel, or aluminium, or they may be non-metallic, such as plywood particle board or glass reinforced plastic.

This possibility of combining materials to form composite panels enables optimum designs to be produced for particular applications. In composite panels, the positive properties of the individual materials can be combined and the negative properties eliminated. For example, the good thermal insulating properties of plastic foam materials or mineral wool may only be used if they are protected against moisture by raintight and diffusion-proof facings; whereas the strength and stiffness of thin-walled metal cladding can only be fully utilised if it is stiffened against buckling under compressive force by the presence of the core material.

This book concentrates upon a particularly favourable combination of materials that is finding wide and increasing application as the cladding of buildings. Sandwich panels with thin steel or aluminium facings with low density plastic or mineral wool cores have a particular combination of properties that make them ideal for use as walls and roofs. They combine the positive properties of metal facings, namely their load-bearing capacity, protection of the insulation against mechanical damage, weather protection and vapour barrier with the complementary positive core properties, namely thermal and acoustic insulation and corrosion protection on the inside.

The resulting composite panel owes its success to the following favourable properties:

- high load-bearing capacity at low weight
- excellent and durable thermal insulation
- absolute water and vapour barrier
- excellent airtightness
- surface finished facings providing resistance to weather and aggressive environments
- capacity for rapid erection without lifting equipment; easier installation in hostile weather conditions
- easy repair or replacement in case of damage
- economical mass production to pre-cut lengths of components of uniform high quality
- long life at low maintenance cost
- behaviour in fire of panels with mineral wool cores

Naturally, sandwich elements have some less favourable properties and the following may also be identified:

- behaviour in fire of elements with rigid plastic foam cores
- deformation when one side is exposed to heat, e.g. strong sunlight
- creep under sustained load of elements with rigid foam cores
- low thermal capacity
- although sandwich elements give good sound insulation compared with alternative wall and roof construction of similar weight, the insulation level achieved is characteristic of lightweight construction

These negative properties and means of overcoming them will be discussed in detail in later sections.

On the basis of the above general discussion, it is now appropriate to define more precisely the sandwich panels to which the remainder of this book primarily refers. Thus, sandwich panels are wall or roof units in which both the inner and outer faces are formed of flat or profiled metal sheets that act compositely with a relatively low strength core which has suitable insulating and stiffening properties. The components of this sandwich must be bonded together in such a way that they act as a composite load-bearing unit for the expected working life. This bonding may be achieved by a line foaming process, by use of separate adhesives or by mechanical fastenings.

Notwithstanding the above limitations, the basic principles expounded in this book are generally of much wider application and can be used for other material combinations. For this reason, Chapter 2 *Materials*, in particular, gives brief consideration to some of the other materials which may be used in sandwich building panels.

1.3 REQUIREMENTS FOR SANDWICH CLADDING PANELS

In contrast to aerospace and automobile applications, where other considerations may prevail, in the building industry the ratio of cost to performance ranks above all other considerations because of the many competing types of construction. Bearing this in mind, the following list of functional requirements may not be complete but includes all the essentials under the four headings of safety, serviceability, durability and aesthetic requirements.

- *Safety requirements* concern the security of the work and its constituent elements:
  - stability under stresses during fabrication, transport, erection and use
  - capacity to accommodate the strains resulting from temperature influences, particularly from the exposure of one face to sunlight
  - adequate resistance to fire
- *Serviceability requirements* concern the performance of the panels in use:
  - resistance to the penetration of water, snow, air and dust at both the surfaces and joints
  - sufficient thermal insulation capacity
  - adequate thermal storage capacity
  - adequate protection from the results of moisture condensation
  - good sound insulation
- *Durability requirements* serve to ensure that the panels retain their performance for the specified lifetime:
  - long-term resistance to weather
  - resistance to corrosion from aggressive environments
  - resistance to degradation of the core material and its bond to the faces
  - stability of any colour coating
- *Aesthetic requirements* concern the visual quality of the completed construction:
  - geometrical tolerances, including flatness of face areas
  - colour variations

In addition, the following requirements with regard to construction must also be met:

- the cross-sections must be suitable for the materials being used
- the joints between elements must be designed for rapid and easy erection
- fastenings to the supporting structure must be secure and visually acceptable

- the number of different panel types must be as few as possible
- transport requirements must not be complicated and no lifting equipment should be required during construction
- handing and finishing at site should be easy to carry out with the usual tools

These requirements will all be considered in more detail in the subsequent chapters of this book.

Evidently, it may not be possible to meet all of these requirements simultaneously and a compromise must be achieved in which the most desirable properties are achieved at an acceptable cost.

1.4 WALL PANELS

Figure 1.4 shows two typical cross-sections of wall panels. The metal facings are very thin. Steel sheets have a minimum thickness of about 0.5 mm and aluminium of about 0.7 mm. With these thicknesses, the facings are sufficiently robust to allow safe handling during fabrication and erection. Element widths are usually in the range 600 mm–1200 mm.

As shown in Fig. 1.4(a), the metal surfaces are often reinforced by light profiling in the form of rolled-in longitudinal stiffeners usually between 1 mm and 4 mm deep. Such lightly profiled panels are also sometimes known as 'lined' or, when there are a large number of relatively small rolled-in profiles, 'micro-profiled'. There are several reasons for this provision:

- the rigidity of the faces is increased and consequently so is the resistance to local buckling; this means that the load-carrying capacity is enhanced
- potentially uneven sheets have a better surface appearance (it is very difficult to produce flat surfaces without lining)
- special architectural effects may be achieved
- at the edges of the panel, longitudinal folds are necessary parts of the sidelaps between adjacent panels

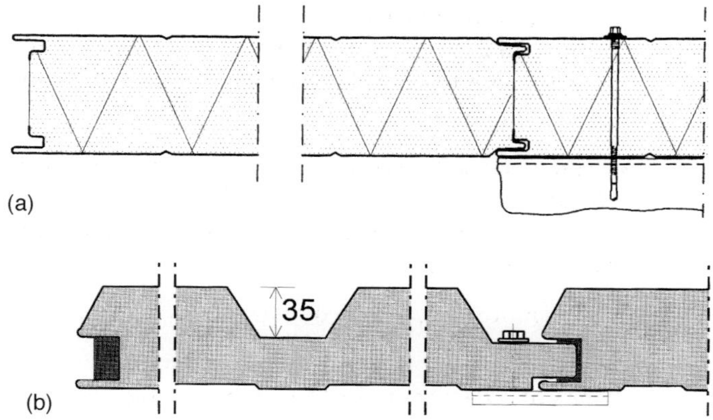

Fig. 1.4 Typical cross-sections of wall panels.

As an alternative to providing longitudinal stiffeners, a surface embossment may be applied. This is more frequently used with aluminium facings and has not been found to result in any significant increase in load-carrying capacity.

The practical performance of the various types of panel is considerably influenced by the design of the sidelaps which must form tight connections between adjacent elements. Tongue-and-groove joints, as shown in Fig. 1.4(a), are the most common. In this example, the edges of the metal sheets forming the facing members are folded back and into the core which, in this case, is mineral wool. In this way, a thermal bridge between the metal sheets is avoided. A soft sealant is applied to the groove and this is compressed when the tongue of the neighbouring element is inserted during erection, thus rendering the assembly adequately waterproof and air-tight. Self-tapping screws pass through both facings in order to connect the panel to the supporting structure.

Figure 1.4(b) shows a variation of the first example in which the core is rigid polyurethane foam and one of the two facings has a deeper profile.

In each of the examples shown in Fig. 1.4, tight connection with neighbouring panels is achieved by special edge detailing. No sidelap fasteners are required, the panels being connected together by the screws or clips providing connection to the supporting structure. In addition, sealing strips are also required.

Figure 1.4(a) shows an example of a sandwich panel in which the core is made of mineral wool in order to satisfy the requirements of fire resistance. In order to increase the strength of the core in a direction normal to the faces, the mineral fibre slabs have their fibres running transverse to the faces and are glued to these faces. When used in this way, the mineral wool elements are known as lamellas. Recently, a number of manufacturers have set up sophisticated facilities for the continuous lamination of panels with mineral wool lamellas as the core material.

An 80-mm thick element with this construction has a 90-minute fire resistance and this makes it particularly suitable for use in multi-storey construction where there is a requirement for a high degree of fire protection.

1.5 PANELS FOR COLD STORES

The more onerous requirements for thermal insulation in cold stores are met primarily by choosing sufficient thickness of an appropriate insulating material as the core of the panel. However, particular care has also to be taken in the design of the joints which must be air and vapour tight and which must not act as thermal bridges. In most European countries, because of the particular jointing requirements, special panels are used for cold store construction.

Figure 1.5 illustrates a solution to the problem in which the edges of the core material are milled in order to enable a precisely adjusted joint to be made. A preformed mastic is inserted between the edges of adjacent panels during construction. This is a relatively expensive solution but this degree of sophistication is justified in order to obtain the level of thermal and joint efficiency necessary for cold stores.

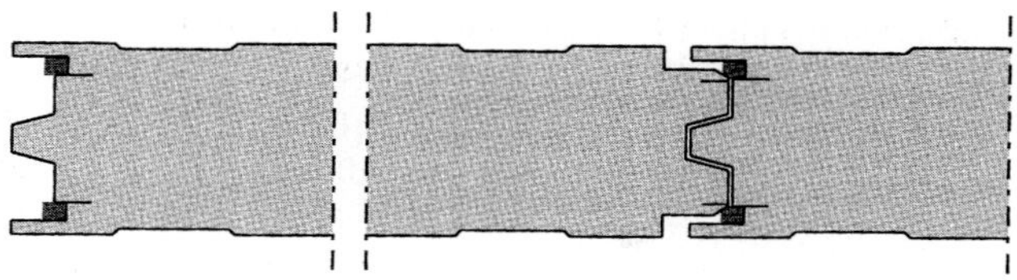

Fig. 1.5 Panel for cold store construction.

1.6 ROOF PANELS

Most types of roof panel have a strongly profiled outer face. This is essentially because, until recently, it was necessary to consider that the loads, particularly the long term loads due to self-weight and snow, were carried entirely by the profiled face. It is only relatively recently that there has arisen the possibility of calculating the stresses and deflections in multi-span panels with profiled faces including the effect of creep in the core material so that only recently-designed panels profit from the composite action as load is transferred from the core to the facings. Provided that the effect of creep is calculated, it is now considered possible to use flat or lightly profiled panels in roof construction, whereas formerly it was considered essential to use profiled panels in order to contain the creep of the core.

The thickness of the outer sheet of roof panels tends to be greater than that of wall panels because roof elements are walked on during erection and maintenance work.

Figure 1.6 shows a typical roof panel. As shown, the longitudinal joints are usually lapped in a similar manner to conventional trapezoidally profiled sheeting. Depending on the application, suitable sealing details must also be incorporated because all construction and manufacturing tolerances must be accommodated here. Sidelaps are usually fastened by either screws or blind rivets through the overlapping sheets at 0.5–1.0 metre centres. The panels are fastened to the supporting structure by means of self-tapping or self-drilling screws which usually pass through the top flange of the ribs. More recently, bottom flange fixing using screws with sealing washers is becoming increasingly common.

These examples of wall and roof panels can only give a limited selection of the many profiles and joint details that are in practical use today.

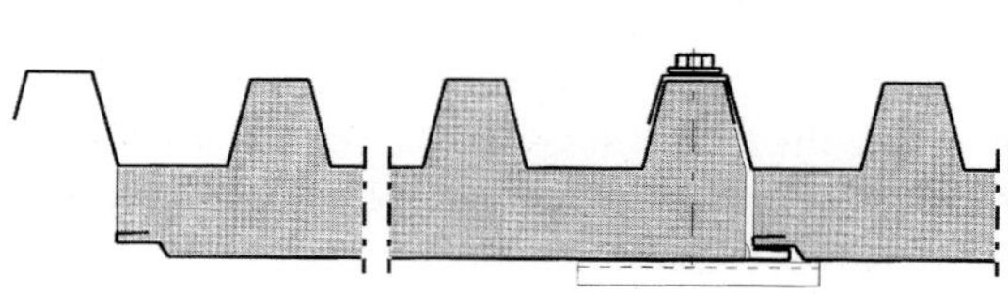

Fig. 1.6 Typical roof panel.

1.7 METHODS OF MANUFACTURE

A variety of production methods are available for the manufacture of sandwich panels. The most economical procedure is influenced by the following factors:

- the geometric design of the panel bearing in mind:
  - the shape of the facing material (flat, slightly or deeply profiled)
  - whether formed edges are required on all four edges or only on the two longitudinal edges
  - accommodation of additional non-metallic edge members
  - the dimensions of the panels
- the basic materials:
  - whether the facing material is available in sheets or coils
  - whether the core material is in the form of pre-formed slabs or foamed *in situ*
- the quantity to be produced and the number of product variations:
  - the number of panels required and when
  - the number of rearrangements of the production line necessary to produce product variations
  - the number of alternative core thicknesses

Furthermore, it must be borne in mind that the technology of sandwich panels and their manufacture is relatively young. Research and development are continuing and new materials, which could require new production techniques, may be anticipated. Research concerned with the improvement of the performance of rigid foams in fire will continue for some time and further development of reinforced or multi-layer cores must be expected.

In principle, the production process for the manufacture of sandwich panels may be considered in two parts, namely, (a) the manufacture of the facings, and (b) the manufacture of the core together with its bond with the facings.

1.7.1 Manufacture of facings

Metallic facings for sandwich panels may be manufactured by the usual methods for cold-formed sections.

If forming of the longitudinal edges only is required, folding, press-braking or roll-forming may be used. Facings formed on four sides may be made by deep drawing or folding. Deep drawing usually requires the use of high quality material with large ductility.

1.7.2 Panels with bonded cores

The core lamina is formed either from pre-formed slabs of insulation that are bonded to the faces using a suitable adhesive or by foaming rigid plastic material directly into the cavity formed by the facings and the edge details. In the former case, the core material usually consists of either polystyrene or mineral fibre. The adhesive is chosen bearing in mind the composition of the materials to be bonded and the production process.

Two different types of adhesive are in use. The first type are solvent based and are applied to both surfaces to be joined by means of sprayers and smoothers. After a short drying time the surfaces are pressed together. Suitable adhesives possess a good initial bonding capacity and the hardening time may be further shortened by applying slight pressure and temperature. These adhesives offer the advantage of easy handling. Their disadvantage lies in the fact that the position of the bonded layers cannot be corrected. Furthermore, precautions have to be taken to collect and neutralise the evaporating solvents if large quantities are used. The second group of adhesives are two-component types based on epoxy resin or polyurethane. The two components are mixed at site. After a predetermined time they suddenly react and harden quickly. The advantage of these adhesives lies in the possibility of correcting the position of the individual layers. The disadvantage is that they must be held in position under pressure for a certain time.

1.7.3 Foamed cores

The recent massive increase in the use of sandwich panels for building applications is due almost entirely to the foamed core. Both the thermal insulation and the bond with the facings originate from the foaming process. The basic material used is polyurethane.

This type of foam is obtained by mixing two liquid components together with certain activators and motive fluids. During mixing, a chemical reaction occurs causing the mixture to foam and finally harden. During the foaming phase, the mixture is very bonding active and adheres strongly to surfaces with which it comes into contact.

The proportions of the various components and additives determine the foam's density, rigidity and other mechanical properties as well as the time required for the various phases of the reaction. There are numerous alternative recipes, some known only to the producers, which have been developed as a result of experience. There are also a number of different processing techniques, often linked to specific machinery. Nowadays, high-pressure foaming units are widely used for large-scale industrial production.

In mould forming, the panel is made in a closed mould whose dimensions are those of the finished panel. The lower facing is laid on the bottom of the mould and the upper facing placed in position supported on spacers. Simple foaming moulds usually have a solid bottom frame and a lid which is clamped to the frame. They must be of robust construction as significant pressures arise during foaming.

Prior to foaming, any required edge profiles are positioned in the mould. The exact quantity of foam is then sprayed into the cavity through a nozzle introduced through the side of the mould, an operation which takes only a few seconds. After foaming, the panel is left in the mould for up to 40 minutes, after which it may be removed and the mould prepared for the next panel.

The advantage of this method is that panels with complicated shapes can be produced. The disadvantage is that the process is relatively slow, although various techniques for batch processing in which a team of operatives work cyclicly on several moulds have been developed.

For mass-production, continuous automatic foaming lines are used. The principles are shown in Fig. 1.7. Two metal strips which form the faces are run-off coils and pass through roll formers where the surface profile and edge details are formed. They are then heated to the required temperature, which is a pre-requisite for an optimal chemical reaction. The two-component foam mixture is then introduced before the strips enter a double conveyor, which is a type of travelling mould that resists the foaming pressure and keeps the faces at the required distance apart. The sides of the panel are likewise formed by lateral formers that are similar to small chain belts. This section of the line may be as much as 30 m long. When the continuous panel emerges, the foam has hardened and it may be cut to the required lengths by a flying saw. The individual panels are then stacked ready for delivery.

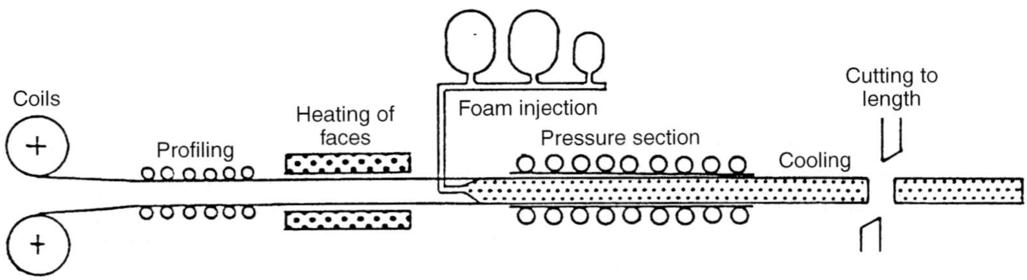

Fig. 1.7 Continuous foaming line.

Such a plant may produce approximately 500 000 m^2 of sandwich panels in a single shift at an average speed of 6 m/min.

1.8 DESIGN REQUIREMENTS FOR STRUCTURAL SANDWICH PANELS

Structural sandwich panels usually form the outer skin of a building and must be designed to resist the usual loads caused by wind, snow, access, etc. In addition, they must be designed to accommodate the additional stresses and deflections caused by thermal bowing when one face is heated relative to the other face by, for instance, strong sunlight. A further design consideration is the tendency of roof panels to creep under long-term loading.

Bearing in mind that the shear strain in the core material has an important influence on the performance of the panel, it follows that the design of sandwich panels includes a number of considerations that are not found in other similar structural components. These considerations have been considered in depth by the European Convention for Constructional Steelwork (ECCS) Committee TC 7 through its working group TWG 7.4 and *European Recommendations for Sandwich Panels* have been written. These have been published in two parts, namely *Part I: Design*, which was published in 1991, and *Part II: Good Practice*, which was published in 1990. Part I was subsequently enhanced by the International Council for Building Research, Studies and Documentation (CIB) Commission W56 in order to give more consideration to mineral wool and other slabstock materials and published as *CIB Publication 148*.

These European Recommendations have been drafted to harmonise with Eurocode 3 *Design of Steel Structures* and, although they have no legal status, may be expected to have a profound influence on European design and practice. The relevant sections of this book will be in accordance with these documents.

At the time of writing, these European Recommendations have been updated by a joint ECCS/CIB Committee and the new version will be published at about the same time as this book. Furthermore, a European Standardisation Committee (CEN/TC128/SC11) is drafting a European product standard for sandwich panels. This latter task is also well advanced and the authors are fully involved in the work and are thus aware of the latest thinking. This book is intended to be compatible with all of these developments and, therefore, reflects the best available design information within Europe.

1.9 STRUCTURAL DESIGN OF SANDWICH PANELS

As an introduction to the structural behaviour of sandwich panels, consider the simply supported panel with flat faces shown in Fig. 1.8(a). The panel is subject to a distributed load, shown downwards of the diagram, from wind or snow.

The shear force and bending moment diagrams arising from this load are shown in Figs 1.8(b) and 1.8(c) respectively. For the purposes of this discussion, it is sufficient to assume that all of the shear force is resisted by a uniform shear stress in the core and that the bending moment is resisted by axial stresses in the faces as indicated in Fig. 1.8(a). Thus the behaviour is rather like that of an I-beam in which the metal faces correspond to the flanges and the core behaves as the web. The adhesive bond between the faces and the core will carry a shear stress equal to the shear stress in the core.

An essential function of the core material and its adhesive bond is to prevent the upper face slipping relative to the lower face. Figure 1.8(d) shows, in an exaggerated form, the slippage caused by using an adhesive with too low a shear strength, while Fig. 1.8(e) illustrates the consequences of using a core material with insufficient shear strength or stiffness. Prevention of this undesirable behaviour requires a core with a sufficiently high shear modulus as well as adequate shear strength and these two quantities feature prominently in the structural design of sandwich panels.

As the upper face of the panel is in compression, another important function of the core material and its adhesive bond is to restrain the upper face so that it does not suffer local buckling in compression, a phenomenon generally termed 'wrinkling'. This behaviour is illustrated in Fig. 1.8(f). The stress at which wrinkling failure takes place is dependent mainly on the stiffness properties of the core and this is another reason why the mechanical properties of the core play a critical role in the structural design. The final mode of failure, which is shown in Fig. 1.8(g), is tensile yielding of the lower face. This is generally the least likely of the alternative modes shown in Fig. 1.8.

For the structural design of a sandwich panel, it is necessary to consider all of these potential failure modes. Furthermore, the situation becomes more complicated when the sandwich panel is continuous over one or more internal supports or when one or both of the faces are fully profiled, as shown in Figs 1.4(b) and 1.6. It follows that the structural design of sandwich panels is far from trivial and the later

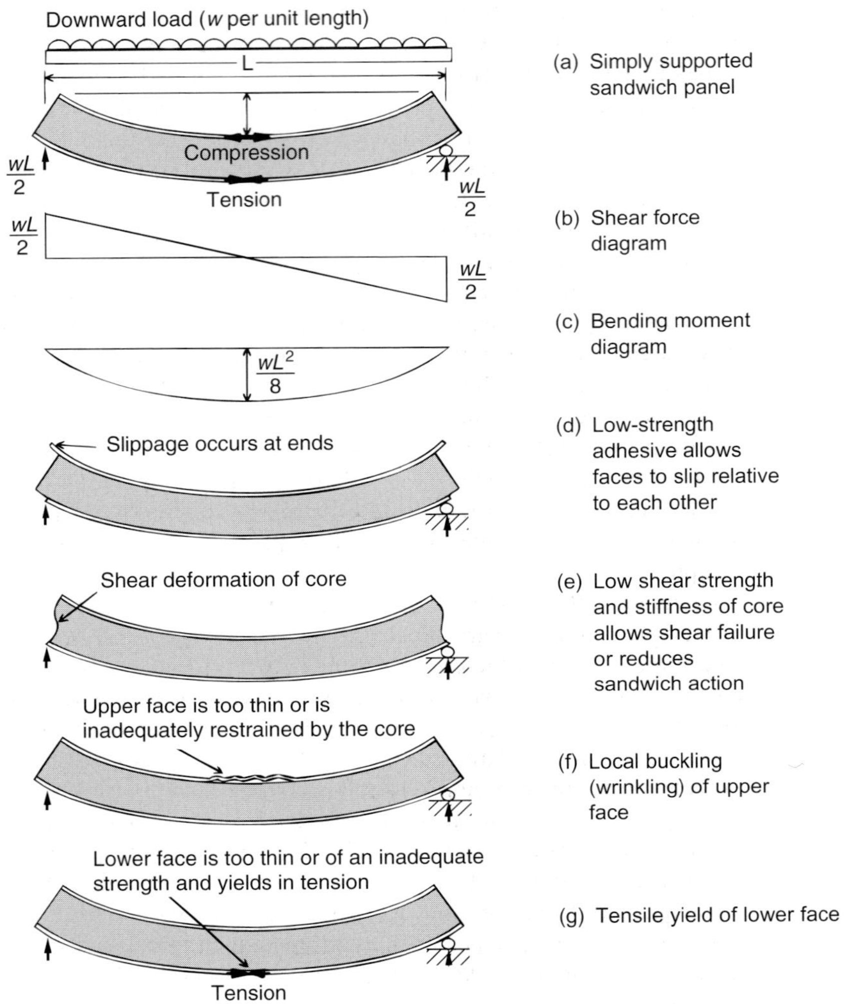

Fig. 1.8 Structural behaviour of a simply supported sandwich panel.

chapters of this book, particularly Chapters 9 and 10, present this in considerable detail.

1.10 LAYOUT OF THIS BOOK

The practical user or specifier of sandwich panels requires a great deal of information ranging over a wide spectrum of subjects. The authors have attempted to present this in a logical and systematic order.

The most difficult decisions concern the choice of materials for the faces (including finishes) and the core. This aspect is considered in Chapter 2 where the characteristics of the main alternative materials are discussed in detail.

Although sandwich panels are often specified because of their extremely efficient resistance to load, it is evident that the various aspects of building physics also play a major role in their performance. Paramount is their thermal efficiency, which is considered in Chapter 3 together with the related subjects of air- and water-tightness. Chapter 4 then considers their acoustic performance. One of the more problematic aspects is the response of sandwich panels in fire situations. This has aroused a great deal of debate in recent years and this continues at the time of writing. Chapter 5 considers this aspect both from a fundamental point of view and also in the light of the governing legislation.

Sandwich panels have now been in regular use as the cladding of buildings for more than 30 years. Their basic durability is therefore well established although, as new materials are frequently introduced, it is necessary to pay continuous attention to their long-term durability. This is the subject of Chapter 6.

Chapter 7 considers the special problems associated with the use of sandwich panels in the construction of cold stores and chill rooms. Particular attention is paid to panels that form the outer walls of large low-temperature storage buildings.

Evidently, for panels that form the visible outer walls of building structures, architectural and aesthetic considerations play a major role. These are discussed in Chapter 8.

Chapters 9, 10 and 11 then go on to describe in great detail the structural design of sandwich panels to carry the loads arising from wind, snow, thermal bow, etc. The authors consider these to be particularly significant chapters because this is a complex aspect of sandwich panel design which engineers tend to find difficult. The authors have, therefore, attempted to give a complete account of the subject, while, at the same time, isolating the essential equations and design aids so that the process can be made as simple as possible.

Chapter 12 *Additional Structural Considerations* considers a number of issues that are important yet do not fall within the logical scope of the preceding chapters. These include fastenings to the supporting structure and joint details, the effect of openings, and the use of sandwich panels as load-bearing walls.

It is implicit in Chapters 9, 10 and 11 that, as far as possible, the structural design of sandwich panels should be undertaken on the basis of well-established calculation procedures. However, the mechanical properties of core materials are more variable than those of many engineering materials and, particularly in the case of foamed *in situ* panels with polyurethane cores, can be quite specific to the manufacturing plant and process used. It follows that certain of the mechanical properties used in the design calculations have to be determined by test and suitable test procedures are described in Chapter 13.

This book concludes, in Chapter 14, with some worked examples of structural design.

Chapter 2
Materials

2.1 INTRODUCTION

Sandwich panels used as the wall cladding and roofs of buildings must compete with traditional products in order to gain new markets. It follows that their design must provide an effective solution for a combination of requirements including mechanical properties, building physics, durability and aesthetic considerations, as well as efficiency of manufacture and economy. Sandwich panels are well able to meet this challenge but the materials must be chosen with a number of potentially conflicting requirements in mind.

One particular advantage of sandwich panels, in contrast to other products and methods, is that they involve the prefabrication of the entire skin of the building. A penalty that has to be paid for this is that efficient production equipment requires standardisation of design with relatively few variations. It follows that almost all producers of sandwich panels in Europe concentrate their efforts on panels with metal facings and rigid plastic foam or mineral wool for the core material. Although other material combinations are possible, and may indeed be advantageous in particular circumstances, they form only a small part of the total production because they require special methods of manufacture.

A typical sandwich panel has a three-layer structure. The rigid faces with a relatively high modulus of elasticity are kept apart by the much lighter core which has a shear stiffness sufficient to carry most of shear force. The core also acts as a highly effective thermal insulation layer. Variations of this pattern also exist. In multi-layer panels, the faces consist of more than one component and the purpose of the extra layers is usually to improve the performance in fire.

Unless they are related to specific standards, the values given in this chapter for the properties of materials should be regarded as indicative and for comparison purposes only. At the time of preparing this chapter, intensive work was in progress to produce new European EN standards in subjects related to it. Unfortunately, at the time of writing, these are still far from complete. The designations and values for materials given here should be checked against the completed standards in due course.

2.2 FACING MATERIALS

Relatively thin, high-strength sheets are generally used as facing materials. These must satisfy the manufacturing requirements with regard to roll-forming and bending, the functional requirements with regard to wind, water and vapour tightness, the structural requirements in their capacity as components of a composite panel and their ability to resist local loads and, furthermore, they must have adequate resistance to corrosion and fire. Not all of these requirements are of equal importance in any given application but it is clear that they are economically satisfied by metal sheets, especially steel or aluminium sheeting. The metal faces are produced from coils and can be easily profiled by roll-forming and incorporated in a continuous manufacturing process using a laminating line.

2.2.1 Steel facings

Thin steel sheets are the most frequently used facing material. In general, only sheets with both metallic and organic (plastic) coatings should be used. Organic coated sheets without a metallic corrosion protection layer should not be used.

Figure 2 1 shows two possible coating variations. The steel itself is usually cold-rolled after pre-coating with a metallic corrosion protection layer, the most common being hot-dipped pure zinc but zinc–aluminium or aluminium–zinc alloys are also applied to steel sheeting. The zinc layers alone do not generally provide sufficient corrosion protection and their aesthetic appearance is poor. At the same time, a reliable bond between the core and the metallic surface of the face is hard to obtain so that additional organic coatings are usually added. These protect the metallic layer from mechanical and chemical wear.

Both sides of the steel sheet are usually coated first with a primer (thickness 5–10 μm) that provides a good bond with the metallic layers (Fig. 2.1(a)). Further organic coatings will adhere easily to this primer layer and are usually applied immediately after the primer.

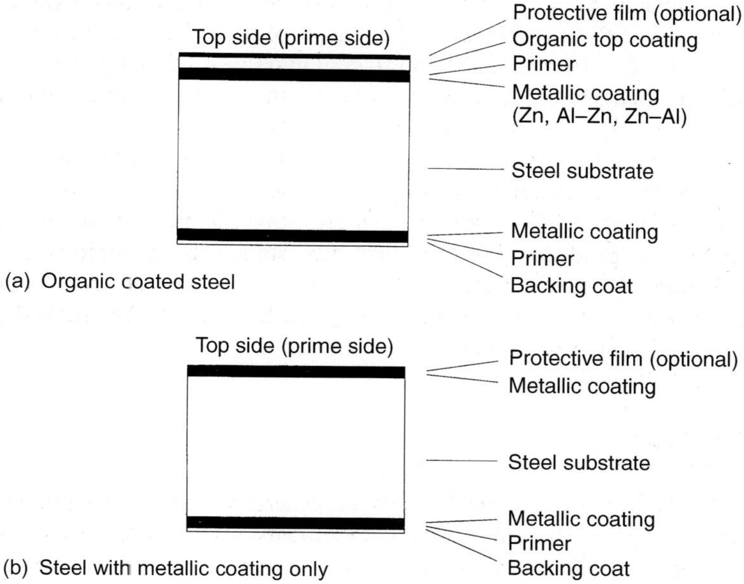

Fig. 2.1 Steel sheets with alternative protective systems.

The organic coatings are then applied. Usually, the thickness of the coating applied to the reverse side backing coat is 7–15 μm and rather more than this on the top surface. There may also be additional primer or other layers.

Figure 2.1(b) shows the alternative case when the metallic corrosion layer alone is considered to be sufficient for the top side. The primer layer and backing coat are still needed on the reverse side in order to ensure a reliable bond between the core and the face.

All of the coatings are produced under controlled conditions in the steel mill or a special painting line and have a good bond to the layers beneath which can survive the cold-forming process undamaged. An aluminium–zinc layer alone gives good corrosion protection. Galvanised steel without an organic coating may also be used for indoor applications.

The galvanised material covered by the European standards EN 10142 (low carbon steel) and EN 10147 (structural steel) is the basic facing material for surfaces exposed outdoors. The standards EN 10214 and EN 10215 deal with other metallic coatings. Sardemann *et al.*[2.1] gives a review of the new types of metallic coatings. The range of steel thickness used in practice is typically from about 0.5 to 1.5 mm. The recommended minimum thicknesses are 0.3 mm for the inner face and 0.5 mm for the outer face. Table 2.1 summarises the basic properties of the most commonly used steel sheets. Note also that the capacities are usually linearly dependent on the actual thickness of the steel substrate. The delivery conditions (tolerances for the thickness) should be consulted in order to determine the proper design thickness (see European standard EN 10143 for the tolerances).

In sandwich construction, the yield stress of the steel is generally of reduced interest because the load-carrying capacity is usually determined by wrinkling of the face in compression or by shear failure of the core rather than by yield of the face material. In flat panels, wrinkling follows rather immediately after elastic buckling of the face so that the buckling stress is dependent only on the elastic properties of materials. Low-carbon steel must be used in the upper faces of roof panels if the seams are made by a concealed fixing technique that requires a nominally zero bending radius and therefore a high degree of ductility. When making such seams on site, preheating may be required when the temperature is below 10°C.

However, in panels with profiled faces, some advantage is gained by the use of a higher yield strength steel because wrinkling of the faces can only take place after significant redistribution of stresses following initial elastic buckling and also because both the support reaction capacity and the shear capacity increase with an increase in yield stress. The yield stress also has some significance in connection with the resistance to local loads and site damage and the steel facing material frequently, therefore, has a yield stress in excess of 320 N/mm^2.

The choice of the most suitable system of corrosion protection for a satisfactory panel life under given environmental conditions is important and has been the subject of much discussion. There is no single corrosion protection system which can resist all types of atmospheric attack. It is therefore necessary to identify the relevant environmental factors at a given site, taking into account such factors as rainfall, local pollution and the deposit of grime on the surface. The expected

Table 2.1 Properties of common steel sheets and metallic coatings

Low-carbon steel sheet EN 10142			Structural steel sheet EN 10147		
Steel grade	Yield strength[a] max. (N/mm²)	Tensile strength[b] max. (N/mm²)	Steel grade	Yield strength min. (N/mm²)	Tensile strength min. (N/mm²)
DX51D	—	500	S220GD	220	300
DX52D	300	420	S250GD	250	330
			S280GD	280	360
			S320GD	320	390
			S350GD	350	420
			S550GD	550	560
[a] A minimum value of 140 N/mm² may be expected [b] A minimum value of 270 N/mm² may be expected					
Available zinc coatings (EN 10142, EN 10147)					
Hot-dipped zinc	Total amount of zinc on both surfaces (g/m²)	Thickness of coating on one surface (μm)	Alloyed zinc	Total amount of zinc on both surfaces (g/m²)	Thickness of coating on one surface (μm)
Z100	100	7	ZF100	100	7
Z140	140	10	ZF140	140	10
Z200	200	14			
Z225	225	16			
Z275	275	19.5			
Z350	350	25			
Other available metallic coatings (EN 10214, EN 10215)					
Hot-dipped zinc–aluminium[c] EN 10214 (95% Zn/5% Al)			Hot-dipped aluminium–zinc[d] En 10215 (55% Al/1.5% Si/rest Zn)		
	Total amount of coating on both surfaces (g/m²)	Thickness of coating on one surface		Total amount of coating on both surfaces (g/m²)	Thickness of coating on one surface (μm)
ZA200	200	15	AZ150	150	20
ZA255	255	20	AZ185	185	25
ZA300	300	23			
[c] tradename *Galfan* [d] tradenames *Aluzink, Galvalume, Aluzinc, Zalutite, Algafort*					

internal conditions in the building, the likelihood of concentrations of chemical fumes and the occurrence of condensation will also influence the choice of the most appropriate coating system.

The choice must be based carefully on the estimated local conditions and the functional requirements and a compromise should be made in order to achieve the best minimum life time. For example, for roofs, the appearance is of minor importance compared with the corrosion protection whereas, for walls the aesthetic appearance has a high priority.

The European Convention for Constructional Steelwork (ECCS) has given some guidance regarding the expected lifetime of galvanised and coated steel in different environmental conditions.[2.2] The recommended corrosion protection systems for trapezoidal steel sheets are equally applicable to the skins of sandwich panels and the data which follows is taken from the ECCS document and panel manufacturers' trade literature.

The expected life of galvanised steel sheets with different weights of galvanising in different environments is given in Table 2.3. The zinc layer generally has a weight between 225 and 350 g/m^2, which corresponds to a thickness between 16 and 25 μm per side. An aluminium–zinc layer usually has a weight of 185 g/m^2 (25 μm). It is clear that plain galvanised sheets are only suitable in a rural atmosphere or if a short life (about 5 years) is required. They may also be used internally in buildings where no condensation is anticipated, i.e. in buildings with a dry internal atmosphere where the temperature is maintained within the range 5–25°C. Conversely, an aluminium–zinc protection layer has good corrosion resistance.

The most common corrosion protection system consists of galvanised steel sheet with an organic coating, as shown in Fig. 2.1(a). The organic coating is primarily responsible for the resistance against corrosion and the zinc layer remains passive until the protection of the organic coat ceases. The zinc is only active at cut ends of the sheet or if the organic coating is damaged. The lifetime of organic coated sheet can be extended without limit if any damage to the organic coating is maintained by cleaning and proper repainting. This is possible for most of the organic coatings used in practice. The critical factor in durability is very often the fasteners. These can usually be replaced, but only at high cost.

The future European standard EN 10169 deals with organic (plastic) coatings. Part 2, EN 10169-2, gives guidance on the selection of the protection system. The most common coatings are polyvinyl chloride plastisol (PVC(P)), polyvinylidene fluoride (PVDF or PVF2) and polyester (SP). Polyvinyl fluoride (PVF(F)) and silicone-modified polyester (SP-SI) are used in small quantities. Acrylic (AY) is seldom used because the service life is very limited. Polyamide-modified polyurethane (PUR-PA) is being introduced on to the market. All of these except PVDF and PUR-PA can be repainted.

Nowadays, there are available a great number of different colours for each type of coating and the performance may be colour-dependent. PVC(P) is rather soft and the coating is relatively thick so that the surface can be textured, handled and formed easily at room temperature. However, the stability of the colours is not good because the resistance against ultra violet (UV) light is limited. This poor UV-resistance is more pronounced with dark colours. Furthermore, the formability is poor at low temperatures. PVDF has good corrosion resistance and very good colour stability with respect to UV attack. It also tolerates rather high temperatures but it is hard and must be handled with care in order to avoid scratches. PUR-PA should have properties that exceed those of PVDF but it is also vulnerable to scratches. PVF(F) has properties which are comparable to those of PVDF coatings. SP has good corrosion resistance and colour stability but it is hard and has restricted formability. SP-SI has similar properties to those of SP but even more restricted formability. Table 2.2 summarises the formabilities of coatings of typical thickness. The expected life time in different environments is given in Table 2.3.

Table 2.2 Typical thicknesses and formabilities of common coatings (*note:* values are only indicative and the final check should be made with the manufacturer)

Coating	Typical thickness of coating (μm)	Value that service temperature range goes down to (°C)	Resistance to cracking on 180° bending, T^a
PVC(P)	100–200	−20	0
PVDF	25–30	−40	3–4
SP	25–30	−40	4–5
PVF(F)	40–50	−40	3–4
SP-SI	25–30	−40	6–10
PUR-PA	20–50	[b]	3–4

[a] In this column, T is the ratio of the minimum bending diameter and metal thickness when the sheet is bent through 180° with the coated side outwards without any visible cracks in the coating (standards ISO 1519, BS 3900/E1, ECCA T7). The steel grade may also limit the bending diameter. Note that the values are for room temperature.
[b] Check with manufacturer.

The resistance to general outside corrosion of the alternative organic coatings can be roughly expressed in a scale from 1 (poor) to 5 (excellent), as indicated in Table 2.4. In this context, the resistance is mainly concerned with the time to first maintenance rather than the total deterioration of the sheeting. The stability of the appearance (colours) is also shown on a similar scale.

Coatings must also resist scratches and wear. Table 2.5 gives some properties of coatings with respect to wear and other durability aspects. The pencil hardness indicates resistance against the penetration of sharp objects. Specular gloss, i.e. the ability to reflect light, is also given.

There are several standard tests for wear which give different results. Table 2.5 is a compilation of information from several sources that are not always consistent. It should therefore be taken as indicative only.

Some quantitative guidance can also be given about the expected time to first maintenance. With the commonly used PVC(P) 200 μm coatings, this is from 10 to 20 years for roofs in central Europe. For walls, one can expect a maintenance-free time from 15 to 30 years in urban, industrial or rural atmospheres. In coastal areas the time is shorter, from 10 to 20 years. The time varies also according to the choice of colour. With proper maintenance, the total lifetime can be more than 40 years.

Galvanised and organic coated sheets with a coating based on polyester with a thickness of about 25 μm can be used in rural and urban atmospheres for a planned medium life (about 10 years) and in an industrial atmosphere for a short life but not in coastal areas. PVDF and PVF(F) systems give a long panel life in rural, urban and industrial areas but not in coastal areas. PVC(P) coating systems are also recommended, particularly for maritime climates, with, however, certain reservations as to their relatively low thermal and colour stability. In sandwich panel applications, AY- and PVDF-coatings on the reverse side are too smooth to ensure a reliable bond to the core.

The behaviour of coatings in fire is considered in Chapter 5 *Fire*.

Table 2.3 Durability of galvanised sheets with or without commonly used organic coatings

	Coastal 1–20 km	Coastal $\leq$ km	Industrial	Urban	Rural
Zinc only					
Z200	—	—	—	—	+
Z275	+	—	—	+	+ +
Z350	+	—	+	+	+ +
Coated galvanised steel					
PVC(P)	+ +	+ +	+ +	+ +	+ +
PVDF	+ + +	+ +	+ + +	+ + +	+ + +
PVF(F)	+ + +	+ +	+ + +	+ + +	+ + +
SP-SI	+	—	+	+ +	+ +
SP	(+)	—	+	+	+ +
	Internal environment				
	Room climate $+5°C \leq T \leq +50°C$			Cold room	
	No condensation	Occasional condensation	Permanent condensation	$-20°C < T < +5°C$	$T < -20°C$
Zinc only					
Z225	+ + +	—	—	—	—
Z275	+ + +	+	—	(+)	—
Z350	+ + +	+	—	(+)	—
Coated galvanised steel					
PVC(P)	+ + +	+ + +	+	+ +	—
PVDF	+ + +	+ + +	—	+ + +	+ +
PVF(F)	+ + +	+ + +	+	+ + +	+ + +
SP-SI	+ + +	+ + +	—	+ +	+ +
SP	+ + +	+ + +	—	+	+ +

The values given are indicative and the advice of the manufacturer should be obtained. The given expected lifetimes are: + + + long lifetime, + + medium lifetime, + short lifetime, — not recommended. For organic coatings, these indicate the maintenance requirements rather than the full protective life of the coating. The table is based on Reference 2.2 where further information on the above topics is given.

2.2.2 Aluminium sheeting

Sandwich panels with facings made of bare aluminium sheet are sometimes used in applications where there are special requirements for corrosion resistance or hygiene; for example, in the production or storing of foodstuffs. The material thickness is usually in the range 0.7–1.2 mm. A thickness of 0.7 mm is often considered to be the minimum in order to avoid local damage during handling but 0.5 mm is sometimes used.

The recommended alloys are AlMn (manganese alloyed in aluminium), AlMg (magnesium alloyed) and AlMgMn (both manganese and magnesium alloyed). A typical alloy may have a tensile strength in the range 200–250 N/mm^2 and a 0.2% yield strength below 200 N/mm^2. Table 2.6 gives some mechanical properties according to the standard EN 485-2.

Table 2.4 Qualitative comparison between the alternative coating materials

Material	Resistance to corrosion	Colour stability
PVC(P)	5	3
PVDF	3–4	5
PVF(F)	3	4–5
SP-SI	3	4
PUR-PA	4	4
SP	3–4	4
AY	3–4	3–4

Table 2.5 Comparison of durability between alternative coating materials

	Pencil hardness[a] ASTM D3363 ISO1518/ECCA T4	Scratch test[b] BS 3900/ECCA E2 (g)
PVC(P) 200 μm	N	> 3500
PVDF 25 μm	F–H	3000
SP-SI 25 μm	H–2H	3000
SP 25 μm	H–2H	2800

[a] Scale from lower to better resistance: N, F, H, 2H
[b] Greater value means better resistance

	Abrasion[c] ASTM D4060 (g)	Sand abrasion[d] ASTM D968 (l)	Impact resistance[d] ASTM D2794 ECCA T5 (J)
PVC(P) 200 μm	10–12	465	18
PVDF 25 μm	16	33	11
SP-SI 25 μm	30	27	1
SP 25 μm	20	19	16

[c] Smaller value means better resistance
[d] Greater value means better resistance

	Specular gloss[e] ASTM D523 ECCA T2 (%)	Salt spray resistance[f] ASTM B117 ECCA T8 (h)	Light- and water-exposure[f] ASTM G53 ECCA T10 (h)
PVC(P) 200 μm	10–20/50[g]	1000	2000
PVDF 25 μm	20–35	1000	3000
SP-SI 25 μm	30–50	750	1500
SP 25 μm	30–50	500	1500

[e] Values are for a smooth surface: larger value means a brighter surface
[f] Larger value means better resistance
[g] Textured/smooth surface

Table 2.6 Typical properties of aluminium alloys according to EN 485-2

Alloy and temper	0.2% yield point minimum (N/mm^2)	Tensile strength minimum (N/mm^2)
EN AW 3004 [AlMn1Mg1] H14	180	220
EN AW 3005 [AlMn1Mg0.5] H16	175	195
EN AW 5005A [AlMg1] H18	165	185
H denotes the temper		

The strength and formability and also the corrosion resistance of the alloys are highly dependent not only on the chemical composition but also on the temper. Harder qualities may have a very limited ductility for cold-forming. The modulus of elasticity of aluminium is approximately only one third that of steel, i.e. 70 000 N/mm^2, and the density is also roughly one third of that of steel, i.e. 2700 kg/m^3. Conversely, the coefficient of thermal expansion is nearly twice that of steel, i.e. 23×10^{-6} per °C.

In the absence of any coating, plain aluminium surfaces oxidise within a short time. However the resulting layer of oxide is resistant to weather and humidity for pH values within the range 5–8. The oxide skin is susceptible to attack from both acid and alkaline solutions with pH values outside this range as well as chlorine and copper ions. Where such aggressive atmospheres are expected, aluminium facings should be protected by a coating. This is particularly necessary in the vicinity of copper foundries or concentrated fumes from oil heating units. The corrosion resistance may benefit from a careful choice of alloy. For instance, Al Mn1 has generally good overall durability and AlMg- alloys have good resistance, even in coastal environments.

As in the case of plain galvanised steel, the problem with smooth aluminium faces is the relatively poor bond to the core. In order to ensure an adequate adhesive bond, aluminium sheets must undergo a surface treatment involving mechanical and chemical descaling, thermal treatment and pickling. A primer layer applied on the reverse side is the best way to overcome this problem. When using other methods, the optimal surface finish with respect to the adhesive bond is best determined by careful testing and special attention must be paid to the long-term durability. The residual stresses caused by mechanical treatment should also be considered.

Aluminium facing materials are, nevertheless, usually coated with the same organic coatings as steel facings (see the European standard EN 1396). This is usually for aesthetic reasons as it is less important from the point of view of corrosion-resistance. In this case the corrosion protection and other properties are the same as if the metal substrate were galvanized steel.

2.2.3 Other metal face materials

Stainless steel facings may be used for buildings where the hygienic demands are high or where attack from an aggressive internal environment has to be resisted. High quality and maintenance-free facades can be obtained with stainless steel or copper sheetings. No corrosion protection is necessary. The thickness range and forming capability of stainless steel facings are similar to those of coated steel facings.

There are a great variety of stainless steel types (see References 2.3 and 2.4) but generally austenic steels are used. They are usually the so-called 18/8 steels that contain about 18% chromium and about 8% nickel, have a 0.2% yield point of about 200 N/mm^2 and are approximately equivalent to the American grade AISI 304. They have good corrosion-resistance in rural and urban environments. In industrial and coastal environments, steel with some molybdenum content (grade AISI 316, so-called 18/10/3 steel) should be chosen. The corrosion resistance is mainly based on chromium that prevents the oxidisation of iron.

Table 2.7 gives some mechanical properties of stainless steels according to the standard EN 10088. The modulus of elasticity is approximately 200 000 N/mm^2, which is about the same as for mild steel. The density is also nearly the same, namely 7700 kg/m^3 and the coefficient of thermal expansion is somewhat higher, 16×10^{-6} per °C. There are a number of methods for specifying stainless steels. The European system uses the notation X *a* CrNi *b c* or X *a* CrNiMo *b c d*, where *a*, *b*, *c* and *d* stand for number which give the content of carbon (multiplied by a hundred), chromium, nickel and molybdenum as a percentage respectively.

Table 2.7 Mechanical properties of cold-rolled stainless steels

Alloy	0.2% yield strength minimum (N/mm^2)	Tensile strength minimum (N/mm^2)
X 4 CrNi 18 10	230	540 . . . 750
X 4 CrNiMo 17 12 2	240	530 . . . 680

Copper sheets are also an alternative facing material. The corrosion-resistance is based on the tight oxide skin that gradually forms on the surface and is good in rural, urban and even coastal environments. The original light colour darkens because of the oxidisation. Full oxidisation takes place in 4 to 6 years in coastal environments, in 8 to 15 years in urban environments and may require 20 to 50 years in rural environments.

Table 2.8 gives the mechanical properties of copper according to the standard ISO 1634. The European standard EN 1172 is under preparation at the time of writing. The values should be regarded as informative only and manufacturers' data should be used for detailed design. The density of copper is 8900 kg/m^3 and the modulus of elasticity is 118 000 N/mm^2 for the type Cu-DHP. The coefficient of thermal expansion is 17×10^{-6} per °C.

Table 2.8 Mechanical properties of copper

Alloy	Temper	0.2% yield point minimum (N/mm^2)	Tensile strength minimum (N/mm^2)
Cu-DHP[a]	M	50	220

[a] This copper is also delivered outside ISO 1634 with tempers that result in a 0.2% yield point between 180 to 350 N/mm^2

From the manufacturing point of view, the bond between plain stainless steel or copper surfaces and the core is similar to the case of plain galvanised steel or aluminium. In order to ensure a satisfactory bond with the core, the reverse sides should generally be coated with a suitable primer. Also other treatments, mechanical or chemical, may be used but their reliability must be verified by testing.

2.2.4 Other facing materials

Timber-based and other types of building boards can also be employed as faces in sandwich construction. The general problem with these boards is that they are produced in fixed lengths that rarely exceed 3.5 m and are therefore not very suitable for use in a continuous foaming line. They may be used with mould foaming and glueing methods but, when producing long panels, special joints must be used. Wood-based chipboard and plywood can be used in environments where no great variations of air humidity exist but, under long-term loading, creep of the face must be considered as well as creep of the core material. Other building boards that may be utilised are gypsum boards, gypsum boards reinforced with fibres, cement-based boards and plastic boards.

Plywood consists of timber plies that have been glued together so that the grains in one ply are perpendicular to the grains in the next ply and, in the outermost plies, the grains are parallel to the direction of stress. The thickness of the plies is approximately 1.5 mm (birch) and 1.5–2.8 mm (spruce and pine). The total thickness of plywood is generally in the range 4–27 mm and the density varies from 550 to 700 kg/m^3. The maximum dimension of plywood boards is usually 3600 mm which means that, in longer panels, it is necessary to use special techniques to join the boards together. The strains due to the variation of humidity have been largely eliminated in the plane of boards by arranging the grains at right angles to each other but they tend to be rather high in the thickness direction. The average values are 0.0015% and 0.45% respectively per 1% change in RH. Table 2.9 gives some indicative values for the mechanical properties of plywood in dry conditions.

Table 2.9 Mechanical properties of plywood in dry conditions

	Birch	Spruce
Bending strength[a] (N/mm^2)	80	35
Compressive strength[a] (N/mm^2)	60	35
Tensile strength[a] (N/mm^2)	70	30
Modulus of elasticity[a] (N/mm^2)	1.5×10^4	1.1×10^4

[a] Note that only plies that have grains parallel to the stress are effective.

Chipboards are made from wooden chips, 0.2–0.4 mm thick, that are glued together using high pressure and temperature. The amount of glue is approximately 10% by weight and some other additives such as cement may also be used. The thickness range is 2–25 mm and the density is generally between 600 and 750 kg/m^3. The strain due to the variation of humidity is about 0.004% per 1%

change in RH in the plane of boards and 0.2–0.3% per 1% change in RH in the thickness direction. Table 2.10 gives some indicative values for the mechanical properties of chipboard in dry conditions.

Table 2.10 Mechanical properties of chipboard in dry conditions

Bending strength[a] (N/mm^2)	8.5–7.0
Compressive strength[a] (N/mm^2)	5.0–4.0
Tensile strength[a] (N/mm^2)	4.0–3.0
Modulus of elasticity[a] (N/mm^2)	1.9×10^3 to 1.2×10^3

[a] Note that values decrease as the thickness increases.

Gypsum boards, with thick cardboard faces and with the core filled with gypsum, have rather low strength but on the other hand they give good fire protection. The total thickness of these boards is in the range 6–15 mm. The board is itself a sandwich panel and the bending strength is mainly based on the tensile strength of the cardboard faces. Gypsum boards are relatively insensitive to small variations of humidity but they should, nevertheless, only be used in dry conditions. Indicative mechanical properties are given in Table 2.11.

Table 2.11 Indicative mechanical properties of building boards

	Gypsum boards with			Cement based boards with paper/pulp fibres
	cardboard faces only	glass fibres	paper/pulp fibres	
Density (kg/m^3)	1100	1100	1200	1100
Tensile strength (N/mm^2)	1.5		3.5	2.5
Tensile modulus (N/mm^2)	2		3	
Compressive strength (N/mm^2)	6		9.5	20
Compressive modulus (N/mm^2)	2		3	

Note that the mechanical values are given for the longitudinal direction of the boards and that different values are obtained in the transverse direction.

Gypsum boards may be reinforced with paper or pulp fibres (amount of the fibres 15–20% by weight) or glass fibres. The mechanical and other properties are similar to gypsum boards with cardboard faces. Indicative mechanical properties are also given in Table 2.11.

Other suitable facing boards are made of cement and other silicates reinforced with pulp fibres. They have better mechanical properties than gypsum based boards and some typical values for these materials are also given in Table 2.11.

An extra facing layer of gypsum board or of chipboard can be added between the thin metal sheeting and the rigid plastic foam core in order to protect the core in the case of fire. When using this technique, the board must be glued to skin before the manufacturing of the panel.

Plastic boards may also be used as face materials and some building construction applications with glass reinforced plastic (GRP) faces exist in practice. However, the main applications of sandwich construction with GRP faces are in boat construction and, to a lesser extent, on offshore oil and gas rigs. Consequently, a vast technology exists which is outside the scope of this book. The reader is, therefore, invited to consult the literature on marine construction if he or she is interested in this topic.

2.3 CORE MATERIALS

Core materials that are suitable for sandwich panels must have appropriate properties with regard to mechanical strength and stiffness, building physics and manufacture. They are generally either:

- rigid plastic foam, or
- inorganic fibre material

The relevant properties are determined by the production process depending on whether:

- prefabricated lightweight slabs are used and bonded to the faces using adhesives, or
- liquid foam components are mixed together and allowed to expand between the metal faces forming their own bond

The relevant mechanical properties are the tensile, compressive and shear strengths and moduli together with, for foamed *in situ* cores, the adhesive bond to the face material. For all high-polymeric materials, these properties are influenced by the ambient temperature and humidity. This applies particularly to the performance under long-term loads where rigid foam exhibits a visco-elastic behaviour. As well as considering the short-term properties at ambient temperature, it is necessary to assess the mechanical properties of the core material subject to both long-term load and temperature variation as a basis for sandwich panel design.

Figure 2.2 gives ranges for the short-term mechanical properties of different core materials as a function of the density for comparison purposes. Note that, for rigid plastic foams, the strengths and rigidities are dependent mainly on the density. The dependence on the temperature and the duration of loading (creep) must also be borne in mind. For mineral wools, the density is not the major factor in determining the mechanical properties because special structural wools may be produced using specific manufacturing plant technology. The values in Fig. 2.2 must be considered to be indicative only. They are based on large numbers of test results available to the authors, many of which are for 'official' purposes, and the large scatter is an indication of the wide range of values that can be obtained with different manufacturing processes. The relevant design values for a particular application should generally be obtained from tests or from manufacturers' tables.

The required core properties with regard to the requirements of building physics are:

- high thermal insulation
- resistance to moisture absorption
- performance in fire
- sound insulation

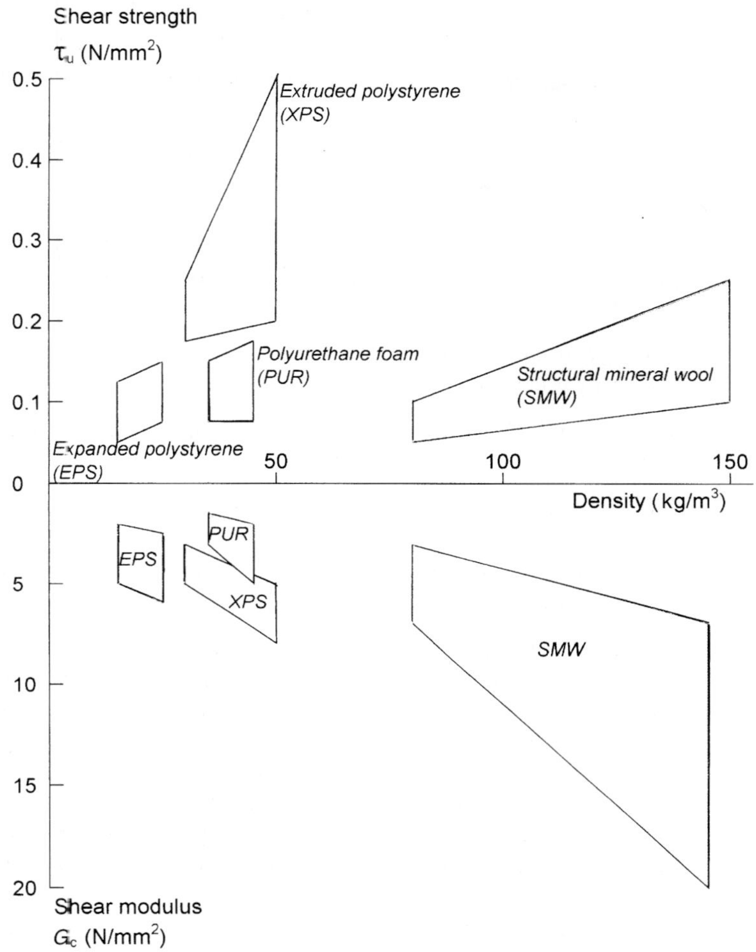

(a) Shear strength and modulus

Fig. 2.2 (See also pp. 28 and 29.) Ranges of the mechanical properties of core materials for comparison. *Note:* The ranges are given only for comparison and the exact values should be obtained from the manufacturers' data or from tests.
The properties of mineral wool materials are more likely to be dependent on the internal structure of the wool than the density.
Figure 2.2(c) takes into account only the strength of the core material itself. With good practice, the bonding strength to the faces should be greater than the material strength so that this table should also apply to the core bonded to the faces.

These requirements conflict with each other and cannot be optimally satisfied by any one material. The choice of core material, therefore, represents a compromise depending on the order of priority of the various requirements.

Most of the new European standards on the testing and characterisation of core materials are under preparation at the time of writing this book.

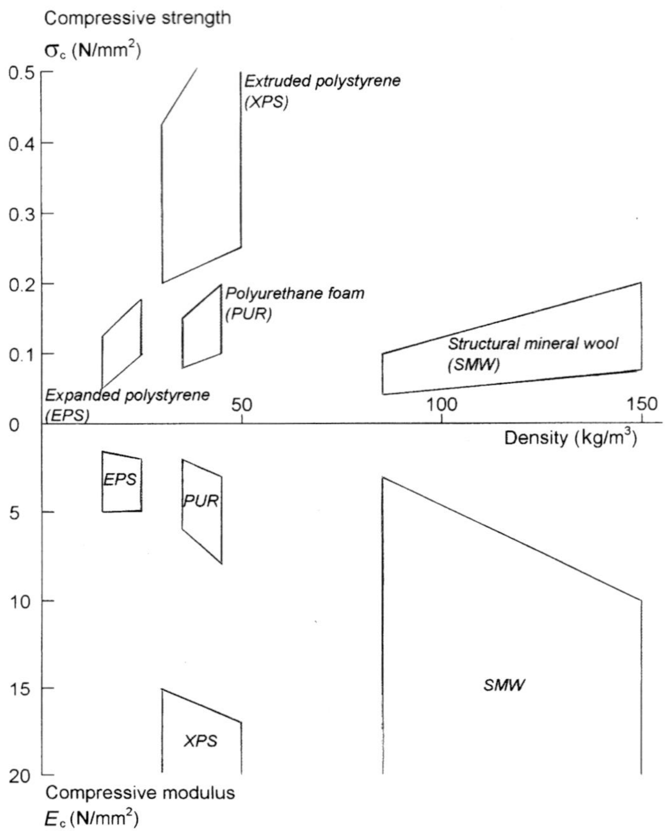

(b) Compression strength and modulus

Fig. 2.2 (Contd).

2.3.1 Rigid plastic foam materials

Nowadays, the materials that give the highest values of thermal insulation are rigid plastic foams. The material properties of the different types of foam vary according to the raw materials used. Moreover it is also possible to vary the material structure within a single group of foams by changing the detailed recipe or manufacturing process. This can result in an open or closed cell structure or in rigid or flexible foam. For sandwich panel construction, a rigid foam material with a predominantly closed cell structure is required. The most frequently used materials are:

- polyurethane/polyisocyanurate (PUR/PIR)
- either expanded or extruded polystyrene (EPS and XPS)
- phenolic resin (PF)

Each of these materials has certain particular properties of its own and these will be considered later. They also have certain properties in common, in particular, a low

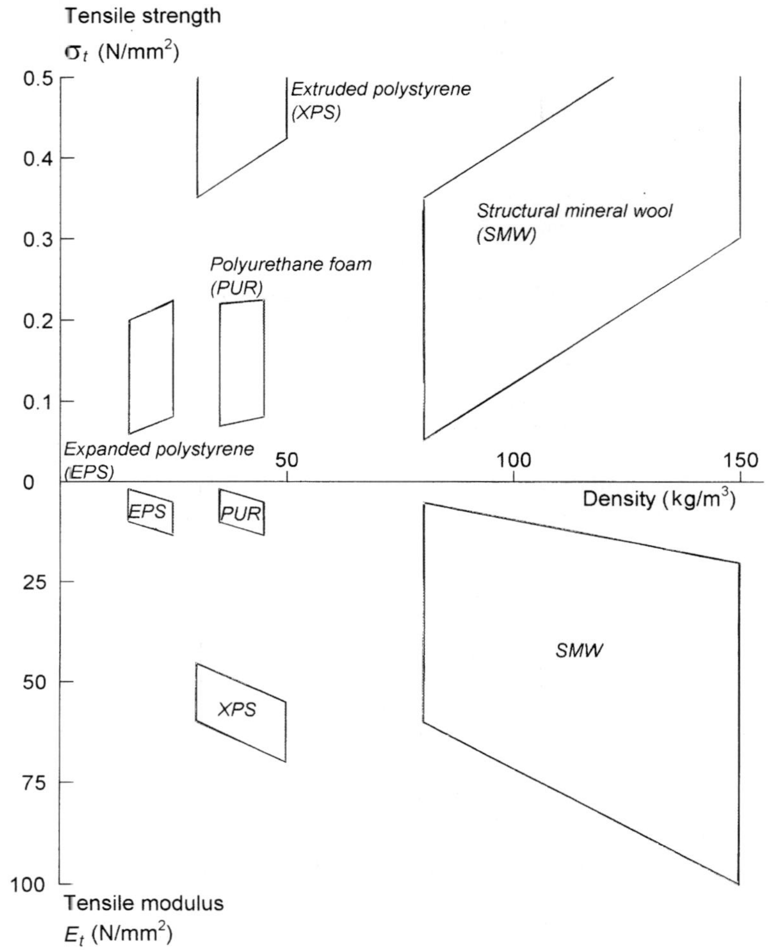

(c) Tensile strength and modulus

Fig. 2.2 (Contd).

density, a cell structure which generally has about 90% closed cells (EPS having an open cell structure is an exception), and a low thermal inertia. Their common properties are discussed in more detail in Section 2.3.2.

2.3.1.1 *Polyurethane/polyisocyanurate (PUR/PIR)*

These are *thermosetting* materials, which means that they can only be formed once, after which the shape cannot be changed because of the extensive chemical cross-linking between the molecules. The two materials are grouped together because there is no clear line of distinction between them. Within the general description, there is scope for a considerable diversity of raw materials, additives and blowing agents, all of which may influence their properties. The main components of rigid polyurethane foam are polyol, isocyanate, a blowing agent and activators that control the reaction. Until recently the blowing agent used was almost invariably fluorotrichloromethane known better as CFC11 or R11 or by the trade name of *Frigen*. CFC-gases (chlorofluorocarbons) are known to be one of the causes of depletion of the stratospheric ozone layer and their use is now banned according to the Montreal Protocol. Nowadays there are a range of blowing agents for use in sandwich panel manufacture including: hydrochlorofluorocarbons (HCFC22 or R22, HCFC142b or R142b, HCFC141b or R141b), hydrofluorocarbons (HFCs such as R134a), various forms of pentane and water that produces carbon dioxide CO_2 when reacting with isocyanate.

After mixing the constituent chemicals, the liquid foams and starts to expand rapidly. In a continuous foaming line, the foam expands mainly in the vertical direction. In vertical mould manufacture, the foam flows like a wave from one bottom end to the other end at the same time rising vertically. The time from the first mixing of the components to the time when the foam core becomes hard is between 3 and 6 minutes depending on the thickness of the core. Because the chemical reaction is exothermic, the core materials of panels thicker than about 100 mm may reach temperatures in excess of 150°C. It is therefore necessary to store thick elements for at least 24 hours in order to complete the hardening and cooling process before shipping them to site.

Polyisocyanurate (PIR) foams differ from pure polyurethane foams only in the ratio in which the primary components, polyol and isocyanate, are mixed. This ratio is approximately 100:150 compared with about 100:100 for PUR. There is, therefore, more isocyanate in PIR than in PUR. The foaming processes and the mechanical and physical properties are otherwise generally similar though certain properties may differ because of the different chemical structure.

Earlier PIR foams showed superior thermal properties but had high friability and were difficult to process. Modification of these earlier formulations with suitable polyols has resulted in the PIR modified polyurethane polymers which are now exclusively used. Although the selection of the polyol is important, so too is the choice of catalyst which influences the mechanical properties as well as the processing.

PIR modified foams are used solely for their superior thermal stability and performance in fire. While pure polyurethane foam gradually decomposes on exposure to temperatures in excess of 250°C, PIR modified foam generally sustains temperatures above 350°C before decomposition commences. It also forms a more stable char which significantly improves the fire resistance.

This improvement in fire behaviour is at the cost of a more expensive process of manufacture because the chemical reaction requires a temperature of approximately 40–45°C, i.e. about twice that required for a polyurethane reaction.

After the foaming process, the structure of the foam consists of mainly closed cells that are separated by thin cell walls. The cells are filled with blowing agent and normally also with some carbon dioxide, CO_2. The CO_2 escapes through the cell walls very rapidly compared with other gases and, after some time, the result is closed cells that contain mainly blowing agents that have excellent insulation properties. After further time has elapsed, some air may diffuse into the foam but this has little influence on the insulation properties. In metal faced sandwich panels the core is protected with diffusion-tight skins and the cell gases can only enter or leave through the panel edges.

With a continuous foaming line, the quality of the foam may vary in the panel thickness direction[2.5]. It is necessary to overdose the mixture slightly so that foam will rise to the upper face and fill the panel under a small pressure. This means that the density of the foam is somewhat greater close to the faces than at the mid-height of the panel where the expansion has been free. Because the line is continuously moving, the cells are usually egg-shaped and they are orientated in the direction of foaming. However, continuously laminated foam cores may have a very plant-specific cell structure and properties.

With the mould foaming technique, the properties of the foam also vary in the plane of the panel[2.6]. In the bottom of the mould, where the mixture was inserted, the foam is heavier because the flow is more restricted. In the panel thickness direction, the lowest density is found at the mid-height. This technique may also give rise to some vorticity during foaming and gaps may be formed in the foam because the flow is not uniform. The local insertion of the liquid and the presence of spacers between the faces may cause additional non-uniformity.

Under stress, the polyurethane rigid foam structure collapses when the cell walls buckle and fracture (in compression or shear) or break (in tension). The average density that is normally used is in the range 35–50 kg/m^3.

2.3.1.2 *Polystyrene (EPS and XPS)*

Polystyrene is a *thermoplastic* material which means that it can be re-formed by heating. It is known worldwide as a good thermal insulating material. However, it is rather less significant in terms of application to sandwich panels because it does not have self-bonding properties with the faces. Consequently, polystyrene is only available in slabs which must be bonded to the facing material using adhesives. This material is therefore usually used where small quantities of relatively simple design are required because of the low cost of the production equipment.

Expanded polystyrene (EPS) core material (also known as bead polystyrene) is made from a polystyrene granulate containing a blowing agent, pentane, which causes the granulate to foam under the influence of temperature. The boards are produced either in a continuous line or in moulds. In both cases, they are heated with hot steam which causes the granules to expand and release pentane which remains encapsulated inside the expanded pearls. The pearls weld together but do not form a totally closed cell-structure. In a continuous line, the slabs are rolled to the required thickness whereas, in the mould technique, slabs are sawn from larger blocks. Because of the manufacturing process, EPS tends to shrink when the pentane escapes from the foam and the pressure differences become equalised.

When comparing the physical and mechanical properties of polystyrene and polyurethane based foams, it is found that about half the weight of polystyrene is generally equivalent to a given weight of polyurethane. Conversely, in terms of the heat transmission value, PUR is nearly twice as good as EPS and this is an important reason why PUR is generally preferred. The density of the polystyrene used in sandwich panel manufacture is usually between about 15 and 20 kg/m^3 and this can be considered to be the recommended density range.

Polystyrene processed by the extrusion method (XPS) yields a more uniform foam, with a structure of small closed cells and plane, compact surfaces[2.7]. During manufacture, the polystyrene grains are melted and then extruded. The blowing agent gas is mixed in and the mixture is then cooled. Previously, the gas that was used was trichlorofluormethane (R11) but today this has been replaced with gases with less ozone depletion potential such as carbon dioxide CO_2 or R22. This material has a density of about 30–50 kg/m^3 and has excellent physical properties. It is very ductile and relatively strong compared with other rigid plastic foams. Particularly striking is the very low figure for water absorption which, due to the closed cell-structure, is almost zero. There is practically no shrinkage of XPS after manufacture.

Both types of polystyrene board must be glued to metallic facings in order to form a sandwich panel and a solvent-free adhesive is recommended because the solvent resistance of polystyrene is poor. Good results have been obtained using two-component adhesives based on polyurethane. The boards must be planed before gluing and EPS boards have to be first cured in order to eliminate the shrinkage.

2.3.1.3 Phenolic resin foam (PF)

With the increasing pressure to improve the fire security of sandwich panels, rigid phenolic foam is a *thermosetting* material which is also worthy of consideration. In comparison with other rigid foam core materials, it has an excellent performance in fire. This includes not only a high resistance to ignition and slow burning characteristics but also a very low rate of smoke emission. Furthermore, the thermal conductivity may be very low, but this depends on the availability of special products.

Phenolic foam is made from liquid phenolic formaldehyde resin which is mixed with a light volatile solvent as blowing agent and a hardening agent. With the application of temperature, the mixture foams and hardens. Phenolic foam is either produced as slabstock and cut to shape for assembly using adhesives or, alternatively, using the latest developments, panels may be manufactured by continuous lamination. The mechanical properties of this relatively new foam are similar to those of PUR and PIR.

PF foams came into prominence in the late 1960s and early 1970s when it was considered that the fire resistance of PUR rigid foams needed improvement. At that time, it was found that PF had some rather problematic properties which constituted a hindrance to its more widespread use in sandwich panels with metal skins. The material was difficult to process, was rather friable and, during manufacture, a considerable quantity of water was produced which had to be eliminated during the curing phase. This prevented lamination with metallic faces which are both water- and vapour-tight. Furthermore, this increased residual water could be particularly acidic and this gave rise to serious corrosion problems when phenolic foam was used with metal facings.

Quite recent developments have considerably improved these properties with the result that PF must now be considered a serious candidate for sandwich panel construction, including by continuous lamination. For sandwich panels with metal faces it is, however, recommended that caution should be exercised when specifying this material until its properties are more widely understood. Acid-cured PF core material should probably only be used when cut from slabstock and then only after a storage time of at least one week before lamination.

There have also been some reported instances of problems when this material has been used for roofs and ceilings which are subject to significant foot traffic. It appears that the relatively friable nature of PF can lead to an early delamination under repeated impacts. This is a further reason why PF should be used with caution until its properties are more fully understood.

2.3.2 Characteristic properties of rigid foams

2.3.2.1 Density and mechanical properties

The density of the foam material is of great importance because the cost of the raw material is more significant than the cost of manufacture so that it is important to obtain the required physical properties with the lowest possible density. However, most of the mechanical properties (e.g. the tensile, compressive and shear strengths and moduli) are usually related to the density. Nevertheless considerable variation exists in the values obtained when considering materials originating from different plant and manufacturing processes. However the density of foams in sandwich panels does not deviate greatly from the optimal value and neither do the mechanical properties.

The strengths of rigid plastic foams vary with the direction in which they are measured. The significant directions in sandwich panels require that compression and tension are measured and given in two orthogonal planes normal and parallel to the faces and that shear is measured in the plane of panel. The typical stress–strain curve of a rigid foam is usually highly non-linear from an early stage in both compression and shear tests. There is no distinct 'yield point' and usually the strength that is given is either the maximum stress or the stress at a strain of 10%. In tensile tests, the failure is usually brittle and the given strengths are ultimate values. It must be remembered that the strain rate used in tests has a great influence on the values obtained because of creep. Usually a rate between 4% and 10% per minute is recommended according to the relevant standards.

An adequate tensile bond between the core and face layers is of great importance because the core and the bond are in tension when the face tends to buckle upwards during wrinkling. In general, the strength of the bond should be higher than that of the core itself. No exact limit can be given, but analyses show that a value 0.10 N/mm^2 is adequate in typical panels and a weaker bond strength than this indicates poor quality of manufacture.

However, these values are given at room temperature and should be regarded as indicative only. At higher temperatures, rigid plastic foams tend to become softer and more viscoelastic and at low temperatures they become more brittle, stiffer and stronger. Representative values under these conditions may be found in the technical literature. The precise properties depend on the test procedure used and on the direction of loading relative to the cell structure of the foam.

Creep is a predominant feature of plastic foams and needs to be considered when panels are subjected to permanent or long-term loading. Rigid plastic foams behave like viscous fluids under long-term loading so that the stress levels under long-term loading must remain below the threshold above which the creep increases considerably. The practical treatment of creep is considered later in Chapters 6 and 11.

2.3.2.2 *Thermal insulation*

The flow of heat through rigid plastic foams is mainly the result of conduction through the cell gases and the cell structure. The cell gases remain trapped in the cells and there is very little radiative heat transfer through foams. The thermal conductivity is, therefore, significantly influenced by the type of gas present in the cells of the foam and most blowing agents are very favourable in this respect. In PUR, the value of the thermal conductivity is 0.020–0.024 W/m °C immediately after manufacture. Owing to the gas tightness of panels with metallic faces, subsequent changes in the cell gas composition are very limited although the aged long term value may increase to 0.024–0.030 W/m°C (see Table 2.12).

The extremely small value of the heat transfer coefficient of PUR should be noted. It should also be observed that this value is not significantly dependent on the density within the usual range of practical applications in sandwich panel construction.

The thermal conductivity of most insulating materials is also temperature dependent. Usually, the relationship between the thermal conductivity and temperature is approximately linear but this is not the case with PUR. In calculating the thickness of insulation required, the operating temperature for the particular application must be considered. For normal applications, the value of λ is defined at +10°C and is denoted λ_{10}.

Finally, the thermal conductivity may also be influenced by humidity because water has a better thermal conductivity than dry air (see Chapter 3 *Thermal Insulation*). It is particularly important to ensure that moisture does not penetrate the foam in cold stores where the temperature is maintained below the freezing point.

Table 2.12 Indicative values of the thermal conductivity of rigid foams

	PUR and PIR	EPS	XPS
(W/m °C)	0.022–0.32[a]	0.035–0.040	0.025–0.028

[a] This value depends on the blowing agent used so that the spread may be wider than indicated as new blowing agents are introduced.

2.3.2.3 *Resistance to aggressive media*

The resistance to attack from aggressive external agencies depends very much on the nature of the basic core material as well as the aggressive medium so that no general guidance can be given here. However, particular attention should be given to the resistance to solvents. Some types of foam are only available in slab form and require the use of adhesives for bonding to the faces of the sandwich panel. Solvent-type adhesives may weaken the rigid foam structure and this should be taken into account when selecting suitable adhesives.

2.3.2.4 *Combustibility and other fire properties*

As a consequence of their organic base, all plastic foam materials are combustible. See Chapter 5 for the detailed description of behaviour of foams in fire. The fire behaviour can be improved by the selection of suitable raw materials, by special foaming processes, by the use of retarding agents or by the inclusion of inorganic filling material. However, the temperatures at which plastic foams start to decompose chemically and ignite can be only little affected by additives. The additives mainly retard the burning process.

Polyurethane (PUR) and polyisocyanurate (PIR) are *thermosetting* materials which do not melt when exposed to fire but rather form a carbonaceous char. PUR starts to decompose at 150–200°C and becomes flammable at about 300°C and releases heavy smoke when burning. At higher temperatures over 600°C, toxic gases like hydrogen cyanide HCN start to form as well as carbon monoxide, CO.

Expanded and extruded polystyrene (EPS and XPS) are *thermoplastic* materials whose behaviour in fire is dominated by their unfavourable tendency to melt at temperatures only a little above 100°C with the result that they melt before they ignite and then tend to form burning droplets. These materials start to decompose at about 300°C and ignite soon after this. When burning they release much smoke and carbon particles. The main releases are carbon dioxide CO_2 and styrene. Fire retardants may be used, as in the case of PUR foams.

Phenolic foam (PF), like PUR and PIR is a *thermosetting* material with particularly favourable char-forming properties and low smoke emission. It is therefore the rigid plastic foam with the best fire properties. It starts to decompose at 350–500°C and ignites at 530–580°C. When burning, a stable char is formed and the products of combustion are mainly hydrocarbons and carbon monoxide, CO.

The fire properties of rigid plastic core materials are one of the more problematical aspects of sandwich panel technology and the fire behaviour of sandwich panels will be considered in more detail in Chapter 5.

2.3.2.5 *Dimensional stability*

All plastic foam materials are subject to a change of dimensions due to a change in temperature[2.8]. This physical phenomenon occurs because, when the foam is heated, the gas pressure in the cells increases and the cells enlarge. The reverse phenomenon occurs when the foam is suddenly cooled down. Then there is the possibility that the foam structure may collapse because of the negative pressure. This is especially important in the case of PUR foams directly after manufacture when the initial hardening and cooling processes are still going on.

This behaviour must be borne in mind when processing and storing plastic foam products. However, extensive experience of continuously foamed sandwich panels with PUR core material has shown dimension changes in core thickness of only +2% as a consequence of long term exposure to temperatures of 80°C and less than −1% under temperatures of −20°C.

2.3.2.6 *Thermal stability*

Foam core materials in sandwich panels exposed to direct sunlight must be stable at temperatures up to at least 80°C because this temperature may be achieved when the surface exposed to sunlight has a dark colour. While polystyrene material may be just adequately stable at this temperature, PUR is generally stable up to considerably more than 100°C (when PIR modified up to over 150°C) and phenolic material up to more than 130°C.

Some of the more general questions regarding the stability of PUR sandwich panel core material under practical ageing conditions have been considered by Just[2.9] who both observed test panels *in situ*, out of doors, over a number of years and also carried out accelerated laboratory tests. He concluded that the ageing behaviour with regard to the mechanical properties is adequately stable in a temperate climate provided that the core is reliably protected from direct weathering.

2.3.3 Inorganic core material

Slabs formed from inorganic fibres may be used as sandwich panel core material when there is a particular requirement with regard to fire security. The most common raw materials are melted minerals, i.e. rock, glass or blast furnace slag (mainly waste from the steel industry) which give rise to materials termed mineral wool, stone wool (often also called rock wool) and glass wool, mineral wool being the most popular.

Stone wool fibre slabs are produced from a silicate melt. The melt is fibrised by dropping it onto rapidly rotating shafts or plates and the centrifugal force throws liquid drops away from the surface. The rapidly cooling drops take the shape of a thin fibre and these fibres are spun and collected on a conveyor. To bind the fibres together, an organic or inorganic binder can be used, phenolic resin being the most common. The binder content normally varies, depending on the desired properties of the product, from 1 to 10% of the wool weight. Oils may also be added to make the wool water-repellent. The slabs are compressed and cured before being cut to size ready for laminating.

Glass wool slabs can be produced in the same way from a melting mixture of quartz sand, soda and limestone or recycled glass. In another production method (the TEL-method) the melt is pressed or drawn through nozzles by means of compressed air. The properties of glass wool are similar to those of stone wool except for a lower melting point and an increased amount of binding agent which is usually between 4 and 15%.

Stone wool fibres have a mean length of 2–4 mm and a mean thickness of 3–7 μm. Glass fibres are longer, with a mean length of 5–10 mm, but have the same range of thicknesses. By adjusting the belt speed and other process parameters, the density and slab thickness may be readily varied.

There are several other production methods and the size of the fibres varies accordingly. Because of the manufacturing process, all wool slabs are highly orthotropic. Longer fibres are aligned along the conveyor and will remain in the same orientation in the fabricated slab. Shorter fibres have a more random orientation which is the reason why wool slabs are stiffer and stronger in their own plane. In the stiff direction, the fibres act as elastically supported small columns whereas, in other directions, they are elastically supported small beams. The wool slabs derive much of their stiffness and strength from the bonding agents.

Of the above-mentioned types, mineral wool, with natural rock as the raw material, has the best resistance to high temperatures and the best resistance to moisture. This material is, therefore, with certain additives, the most suitable wool type for use as a core material in sandwich panels.

In comparison with slabs of rigid foam, mineral wool slabs have a more open cell structure and a high degree of bending elasticity in the longitudinal direction. Moreover, the fibre structure does not have closed pores so that the slabs are considerably more susceptible to water absorption and vapour diffusion. With suitable additives, the water absorption of stone wool can be reduced to be less than that of polystyrene. Inside a sandwich panel with diffusion-tight faces and seams, and with normal wool densities, there is no risk of convective air movements. As a consequence of their internal structure, mineral wool slabs have rather low tensile, shear and compressive strength in a direction normal to their length. However, mineral wool is non-combustible and this provides the stimulus to study how the mechanical properties can be improved. A simple solution has been discovered. The slab is cut into strips (lamella) at right angles to the orientation of the fibres and with a width equal to the required height of the core. Using adhesives, these strips are assembled to form panels with the fibres orientated normal to the faces as shown in Fig. 2.3. Sandwich panels made in this way achieve significant fire resistance times. The density range of the mineral wools usually used in sandwich panels is 70–150 kg/m^3.

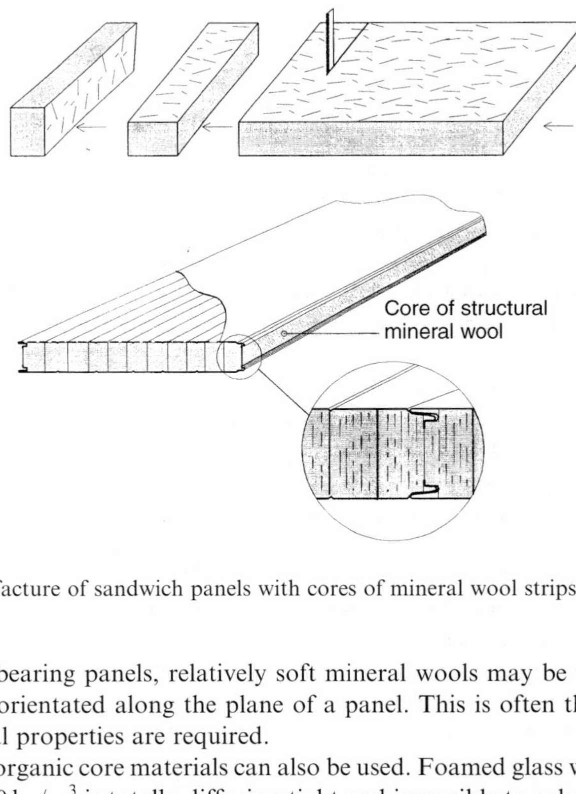

Fig. 2.3 Manufacture of sandwich panels with cores of mineral wool strips.

In non-loadbearing panels, relatively soft mineral wools may be used and the fibres may be orientated along the plane of a panel. This is often the case when good acoustical properties are required.

Some other organic core materials can also be used. Foamed glass with a density in excess of 150 kg/m^3 is totally diffusion tight and insensible to solvents. It has a thermal conductivity λ above 0.09 W/m °C.

In all panels which consist of metallic faces with a core made of slabstock type material, it is important to pay attention to the flatness of core faces. This is emphasised in panels with flat faces where any unevenness in the core is reflected directly in the shape of the faces. If the core is made of several pieces they should be of uniform height for the same reason.

2.3.4 Characteristic properties of mineral wools

2.3.4.1 Density and mechanical properties

Mineral wools are stable materials and both the fibres themselves and also the binding agents retain their properties for a long time. The temperature has little or no effect on the mechanical properties. These wool materials only cease to behave elastically when the fibres and the bonds between them break.

The strength increases with the density but is more dependent on the internal structure of wool than merely on density. The compressive strength normal to the length (and normal to the orientation of the fibres) is typically in the range 0.005–0.08 N/mm^2 for the range of densities 60–150 kg/m^3. The corresponding tensile strength is lower and in the range 0.001–0.01 N/mm^2. The properties in the direction of fibres are much higher. The shear strength varies from 0.03 to 0.20 N/mm^2 depending on the degree of anisotropy of the fibre structure and the corresponding shear modulus varies from 2 to 20 N/mm^2. The tensile strength is between 0.03 and 1.0 N/mm^2 and the corresponding modulus of elasticity between 5 and 40 N/mm^2. The compression strength is in the range 0.10–0.15 N/mm^2 and the corresponding modulus of elasticity is in the range 6–20 N/mm^2. See Fig. 2.2 for a comparison between the alternative core materials.

2.3.4.2 Water absorption and permeability

Under normal conditions of use of mineral wool, the water absorption is small and of the order of 1.5% by weight. In sandwich panels, because of the protection offered by the faces, this is normally reduced to 0.2–0.5%. The water absorption of the wool matrix can be further decreased by using silicone, mineral oil or other additives. The binder content also affects the water absorption. The water absorption of stone wool is less than that of glass wool, even though the binder content is smaller. This is due to a difference in the internal structure of the material.

The fibres and the binding material do not themselves absorb any water but there is a considerable amount (95%) of air-filled space that can be filled with water. The wool materials, therefore, have the potential to absorb considerable amounts of water and also do not resist the internal movement of moisture. This leads to a clear requirement for sound detailing.

2.3.4.3 Thermal insulation

Compared with the closed cell plastic foams, the thermal conduction of air in wool has a greater influence on the heat flow. The thermal conductivity is highly dependant on the temperature. Normally the moisture content is so low that it has no effect on the λ-value but, if the details break down sufficiently to allow an abnormal ingress of water, the adverse effect on the thermal insulation can be significant. On the other hand, the thermal conductivity measured in slabs of material is practically constant ($\lambda = 0.033$–0.034 W/m °C) in the range of densities 60–150 kg/m^3. For the lamellas used in sandwich panels, it is higher and $\lambda = 0.036$–0.044 W/m °C.

Only about 5% of the thermal flow is caused by radiation. Conduction in fibres causes about 20% and the rest is due to the convection and the conduction of the air.

2.3.4.4 Combustibility and other fire properties

Mineral wools with a low organic binder content are practically non-combustible. The difference between glass and stone wool should, however, be noted. The fibres themselves do not burn but rather melt; glass at about 650°C and rock only at about 1000°C. Because the binder content of glass wool is generally higher than 5%, glass wool is not usually classified as non-combustible.

2.4 HONEYCOMB CORES

In its general sense, the term *honeycomb* can be used to describe any array of identical prismatic cells which are nested together to form a repetitive plane structure. The cells can have various shapes but, from the practical point of view, they are usually hexagonal, as shown in Fig. 2.4. Honeycombs can be made from a wide variety of materials such as polymers, metals, ceramics and even cardboard. They have clearly-established high-technology applications and it has already been pointed out in Chapter 1 that sandwich panels with honeycomb cores were an important part of the structure of the Apollo space capsule.

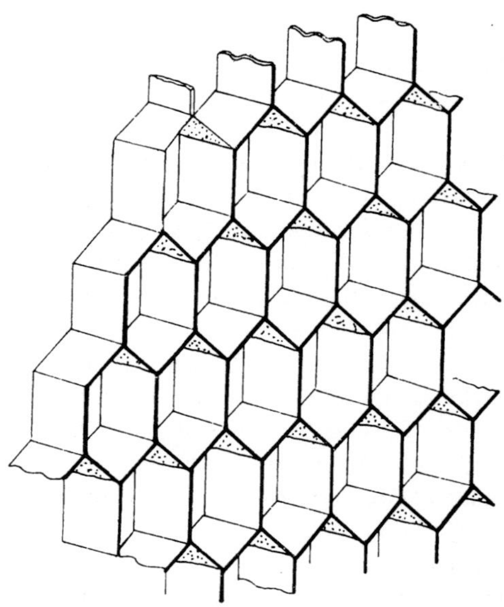

Fig. 2.4 Hexagonal honeycomb core.

In more mundane building applications, sandwich panels with honeycomb cores are rather harder to find, although cardboard honeycombs are widely used in the construction of cheap doors. This suggests that the reasons for neglecting honeycombs in the more general cladding panel applications considered in this book are not necessarily economic. They probably have more to do with meeting the whole range of functional requirements, not least those of thermal and fire performance. However, there is probably some truth in the suggestion that the full potential of honeycomb sandwich panels in building applications has not been fully explored.

An interesting application that has been rather fully researched, though not developed into a practical application, gives an instructive insight into the possibilities. Paper or card honeycombs of the type used in door construction are cheap and commercially available. As produced, they are of little interest for panel construction because they have inferior thermal properties to the rigid plastic foams and poor performance in fire. However, a simple treatment by (repeated) dipping into (or spraying with) a solution of sodium silicate or a mixture of sodium silicate and ball clay changes this situation. The sodium silicate rigidifies the honeycomb and makes it fire resistant. Furthermore, sodium silicate behaves as an intumescent in fire with the result that panels formed of the treated card honeycomb may be stronger and stiffer than their foam-filled equivalents and can have over one hour fire resistance in insulation and integrity. In some of the tests carried out on this construction, the thermal performance was enhanced by filling the cells of the honeycomb with loose perlite or vermiculite. Further details of this development are given in Reference 2.10.

The structural mechanics of honeycombs is well-developed, though relatively complex. Reference 2.11 gives a full account which is outside the scope of this book. Thus the strength and stiffness of regular honeycomb cores can be calculated. They can also be determined by test as with the other core materials in regular use.

2.5 ADHESIVES AND OTHER COMPONENTS

One- and two-component polyurethane adhesives are the most widely used to glue the faces to the core.[2.12] There are a great variety of possible material combinations and adhesives can be tailored to meet specific needs with regard to service temperature or fire resistance. The amount of glue is usually between 200 and 350 g/m^2 per side. Adhesives are generally liquid so that they can be easily incorporated into automated manufacturing processes. They are sprayed on the surfaces and a watertight bond is usually obtained with primed metal facings.

Two-component adhesives are produced in a similar manner to PUR-foamed core material. The base resin (polyol) is mixed with a curing agent that is usually isocyanate (MDI). The base resin may include fire retardants, fillers, etc. and some adhesives may be classified as non-combustible. Two-component adhesives may be used on relatively rigid core materials like XPS and EPS. The adhesive is applied to the faces and the press time (the time needed to keep the faces pressed to the core during curing) can vary from a few minutes to half an hour depending on the recipe. Heating of the faces will also accelerate curing.

One-component adhesives are pre-activated two-component adhesives that continue to cure when exposed to moisture. Water mist needs to be sprayed onto the adhesive in order to cure it. Special fillers can be added to accelerate the cure speed and to improve fire resistance and the bond strength. The press time is usually less then one hour, but may be reduced to below 20 minutes if the faces are preheated. One-component adhesives are suitable for use with soft core materials. They foam during curing and penetrate into the voids of the material.

2.5.1 Mechanical properties

The adhesive layer is of fundamental importance in the behaviour of a sandwich panel. This layer has a complex structure and may have properties that differ considerably from those of the core material which it bonds. Furthermore, the complex structure of coated metallic faces has an effect on the bond between the core and the faces. The bond should not be the weakest link in a panel and it should have mechanical properties that are at least as good as the core material.

In a foamed polyurethane core, the density is greatest close to the faces because rising foam packs more tightly when it encounters a barrier. The higher density results in better mechanical strength but does not ensure a correspondingly good bonding strength. The rise of foam is more easily controlled in a continuous laminating line than in the batch moulding technique where the foam must move along the panel. The consequences of uneven flow of foam or too little foam to fill the panel fully are voids, vorticity and poor adhesion of the faces. A reliable chemical bond requires appropriate conditions that include the correct quantity of foam and the correct temperature of the faces during foaming.

In panels that are manufactured by glueing the faces to the core, the requirements for satisfactory adhesion are very similar. Glue penetrates into the core material and the result is an adhesion zone containing a mixture of glue and core material. It is essential that the correct quantity of glue (and any other materials such as water which may be necessary for the curing process) is used and that it is properly cured according to the manufacturer's instructions.

2.6 FASTENERS AND OTHER COMPONENTS

Screws of carbon steel with possible electrolytic zinc coating are commonly used in internal applications. In outdoor conditions, only screws of stainless steel should be used. In panels with faces of other materials than steel, the choice of fastener material must be based on the usual criteria in order to avoid possible galvanic pairs and other sources of corrosion.

2.7 REFERENCES

1.1 Sardemann, K., *et al.* (1992) Tendenzen bei der Entwicklung kaltgewaltzter und oberflächenveredelter Stahlbleche (The tendencies in the development of cold rolled and coated steel sheets). *Blech Rohre Profile*, **39**, 11 [in German].

2.2 European recommendations for good practice in steel cladding and roofing (1983) European Convention for Constructional Steelwork, ECCS-T7, Publication No. 41.

2.3 Burgan, B.A. (1992) *Concise Guide to the Structural Design of Stainless Steel.* SCI Publication 123, The Steel Construction Institute, Berkshire.

2.4 Baddoo, N. (1993) *Design of Stainless Steel Fixings and Ancillary Components.* SCI Publication 119, The Steel Construction Institute, Berkshire.

2.5 Zöllner, R. (1985) Properties of PU rigid foam. In *Polyurethane Handbook* (ed. G. Oertel). Hanser Publishers, Munich, Vienna, New York.

2.6 Dietrich, W. and Zöllner, R. (1985) Relationship between production methods and properties. *Polyurethane Handbook* (ed. G. Oertel). Hanser Publishers, Munich, Vienna, New York.

2.7 Bukowski, K. and Conway, P. (1990) The application of extruded polystyrene foam as an insulating core material in sandwich constructions. *Design of Sandwich Panels, Symposium*, Salford University.

2.8 Merkel, H. (1988) Dimensionsstabilität von Wärmedämmstoffen aus Polymerschäumen (Dimensional stability of plastic foam thermal insulation materials). *Bauplanung-Bautechnik*, **43** (1) [in German].

2.9 Just, M. (1983) Ergebnisse experimenteller Untersuchungen zum Langzeitverhalten von PUR-Hartschaumstoff-Stutz Kernbauteilen und Schlussfolgerungen für die Anwendung (The results of an experimental investigation of the long-term behaviour of building elements utilising a rigid plastic foam core and conclusions regarding their use), *Ifl-Mitt*, **22** (3) [in German].

2.10 Tajbakhsh, S. (1992) *Developments in sandwich construction*, PhD Thesis, University of Salford.

2.11 Gibson, L.J. and Ashby, M.F. (1988) *Cellular Solids – Structure and Properties*, Pergamon.

2.12 Stroebech, C. (1989) One- and two-component polyurethane adhesives for bonding sandwich elements. In *Sandwich Constructions 1, Proceedings of the 1st International Conference on Sandwich Construction*, 19–21 June 1989, Stockholm.

Chapter 3
Thermal Performance and Water-Tightness

3.1 INTRODUCTION

In residential, industrial and public buildings, the specification of the thermal insulation is becoming an increasingly important part of the building design. In most countries there are standards which specify the lowest allowable degree of insulation. The most stringent demands can be found in the Scandinavian countries, whereas Southern Europe has much lower requirements. As energy costs rise, leading to increasing heating and cooling costs, so the economic motives for better thermal insulation in buildings become more compelling. In this scenario, sandwich panels show to particularly good advantage.

The amount of energy required in order to obtain an acceptable indoor climate in a building depends mainly on:

- transmission losses through floors, walls and ceilings
- ventilation losses as a consequence of ventilation systems and permeable cladding

The transmission losses can be reduced by increasing the level of thermal insulation in the cladding of the building. Ventilation losses can be decreased by an efficient ventilation system and by providing a 'tight' building.

The thermal insulation system must also meet other requirements, such as:

- avoiding condensation
- avoiding too high or too low temperatures
- allowing the rapid heating of rooms

Among the most important properties of sandwich panels are their high thermal insulation and their capacity to enable extremely air-tight structures to be built.

3.2 INSULATION THEORY

In principle, thermal transmission can take place in three different ways:

(1) *Conduction* heat is conducted through solid materials or still liquids.
(2) *Convection* transmission of heat by means of movement in liquids or gases.
(3) *Radiation* transmission of heat into or out of a material by means of electromagnetic waves.

The insulation materials used as the core materials in sandwich panels are usually porous, which means that part of the material volume consists of gas such as air or a gaseous blowing agent. Thermal transmission can therefore take place through conduction, convection and radiation.

The thermal insulation capacity of a material is described by its *thermal conductivity* which is denoted by λ. It is defined as the amount of heat (W h) passing through a 1 m thick layer with a surface of 1 m^2 during 1 hour (h) when the temperature difference across the material is 1 Kelvin (K) as illustrated by Fig. 3.1.

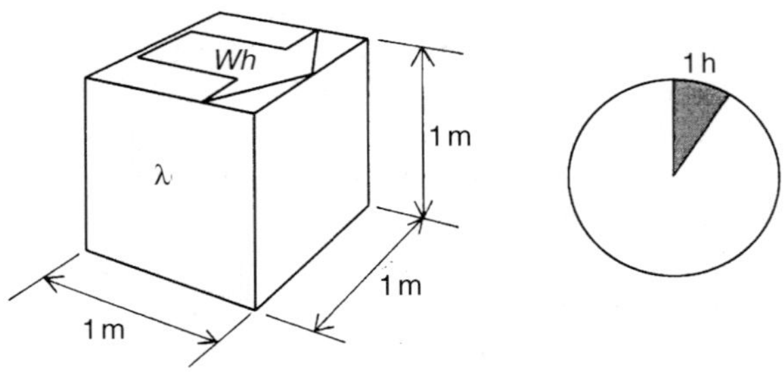

Fig. 3.1 Thermal conductivity λ.

$$\lambda = \mathrm{W} \times \mathrm{h} \times \mathrm{m}/\mathrm{h} \times \mathrm{m}^2 \times \mathrm{K} = \mathrm{W}/\mathrm{m\,K}$$

The lower the λ-value of a material, the better is its insulation capacity. The λ-value is temperature-dependent and therefore should always be given as a numerical value stating the average temperature at which it was measured. The λ-value can be used to calculate the amount of heat transmitted through a material, as illustrated in Fig. 3.2:

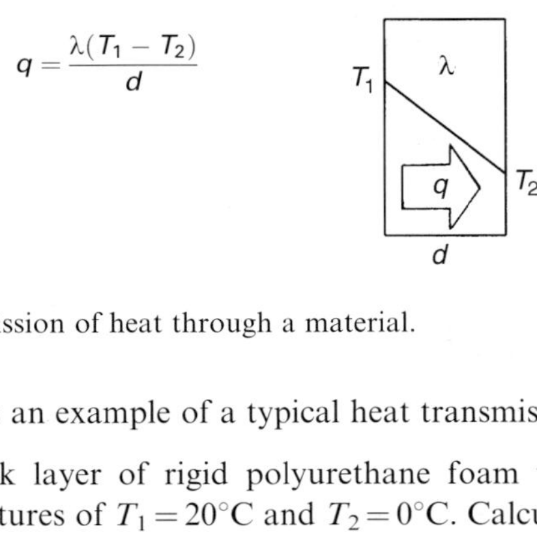

Fig. 3.2 Transmission of heat through a material.

The following is an example of a typical heat transmission calculation:

A 100-mm thick layer of rigid polyurethane foam with $\lambda = 0.026\,\mathrm{W/m\,K}$ has surface temperatures of $T_1 = 20°\mathrm{C}$ and $T_2 = 0°\mathrm{C}$. Calculate the heat flow through the layer.

$$q = 0.026 \times (20 - 0)/0.1 = 5.2\,\mathrm{W/m^2}$$

Compare this with 100 mm of concrete:

$$q = 1.7 \times (20 - 0)/0.1 = 240\,\mathrm{W/m^2}$$

3.2.1 Factors influencing the thermal conductivity of a material

The thermal conductivity λ of typical insulating materials is not a constant during the life of the material but may change as a consequence of a number of factors including the following.

- *Temperature* Usually λ reduces with increasing temperature (see also Chapter 2 *Materials*) as shown typically in Fig. 3.3(a). This has to be considered when using sandwich panels for example in cold rooms or drying ovens. For the materials used in normal construction λ_{10} is used in design for thermal insulation (corresponding to an average temperature of 10°C in a wall with an internal temperature of 20°C and an external temperature of 0°C).
- *Moisture* Increasing moisture content in the insulating material increases the λ-value as shown typically in Fig. 3.3(b).
- *Ageing* Some insulating materials contain a gas (e.g. a blowing agent) with a lower λ-value than that of air. This gas can diffuse out and be replaced by air so that the thermal conductivity may reduce with time. In sandwich panels, the effect of this is minimised because most rigid plastic foams have a high percentage of closed cells and the insulating material is enclosed between gas-tight facings.

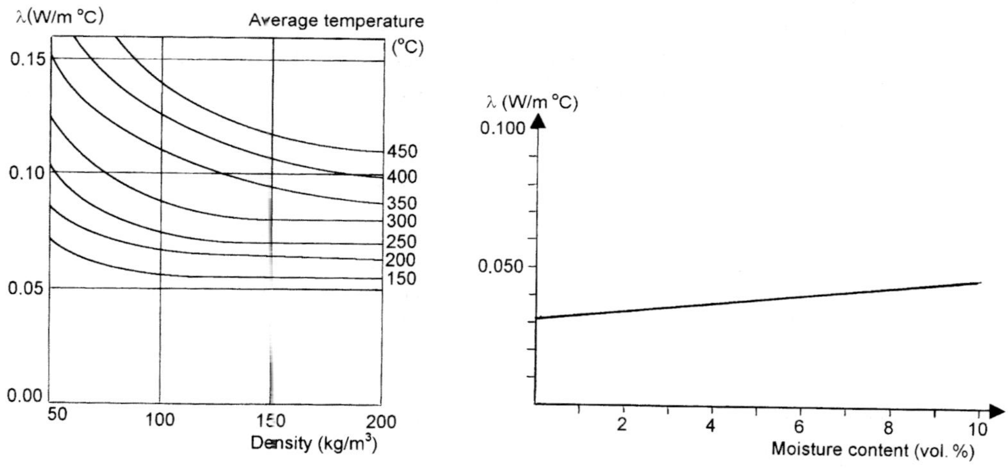

(a) λ for mineral wool as a function of temperature

(b) λ for mineral wool as a function of moisture content

Fig. 3.3 Variation of thermal conductivity with ambient conditions.

When the λ-value is used in practical calculations, it is usually a realistic value which is of interest. This value is based on measured λ_{10}-values with correction terms for ageing, moisture etc. These correction terms are often given in the building standards of the country in question. Thus:

$$\lambda_p = \lambda_{10} + \Delta\lambda_1 + \Delta\lambda_2 + \ldots + \Delta\lambda_n$$

where $\Delta\lambda_1$ etc. are the correction terms.

Table 3.1 gives the thermal conductivity for some typical materials:

Table 3.1 Typical thermal conductivity values

Material	λ_p-value (W/m K)
Polyurethane PUR, 40 kg/m^3	0.024–0.028
Expanded polystyrene EPS, 20 kg/m^3	0.033–0.039
Extruded polystyrene XPS, 25–40 kg/m^3	0.026–0.036
Mineral wool fibres normal to heat flow, 80–150 kg/m^3	0.033–0.039
Mineral wool fibres parallel to heat flow, 80–150 kg/m^3	0.039–0.050
Low-carbon steel sheet	60
Stainless steel	20
Aluminium	160
Concrete	1.7
Air	0.026
Water	0.6
Ice	2.1

3.3 THERMAL RESISTANCE (*R*)

The thermal resistance defines the thermal insulation properties of a product or a structure. The calculation method here is according to the standard ENISO 6946. Thermal resistance is denoted by R and is expressed in units of m^2 K/W, thus:

$$R = d/\lambda_p$$

where d is the thickness of the layer and λ_p is the 'practical' thermal conductivity defined above.

At the transition between air and a wall surface, there is also an additional thermal resistance called the thermal surface resistance. The thermal surface resistances of various building components are usually defined in the building standards of the country. R_{si} is the internal surface resistance and R_{se} the external surface resistance. The resistances given in Table 3.2 are according to the standard ENISO 6946.

Table 3.2 Surface resistances (in m^2 K/W)

Resistance	Direction of heat flow		
	Upwards	Horizontal	Downwards
R_{si}	0.10	0.13	0.17
R_{se}	0.04	0.04	0.04

These values are design values and are mean values for the heating season. For the purposes of defining the thermal properties of components, and in other cases where values independent of the direction of heat flow are required, it is recommended that the values for horizontal heat flow are used. When calculating the interior surface temperatures, the magnitude of R_{si} is of great importance. In all cases where the surface temperatures are critical (e.g. in condensation calculations), R_{si} is required to be accurately estimated.

3.3.1 Total thermal resistance of a building element

The total thermal resistance R_t of a plane building element consisting of thermally homogeneous layers perpendicular to the direction of heat flow, for instance a sandwich panel with a homogeneous core, may be calculated by using the following expression:

$$R_t = R_{si} + R_1 + R_2 + \ldots + R_n + R_{se}$$

where: R_{si} is the internal surface resistance;
R_1, $R_2 \ldots R_n$ are the design thermal resistances of each layer
R_{se} is the external surface resistance

If an element consists of both thermally homogeneous and thermally inhomogeneous layers parallel to the surface, as in a panel with a core of timber frames with insulation between, R_t may be taken as the arithmetic mean of an upper and lower limit of resistance:

$$R_t = (R'_t + R''_t)/2$$

3.4 THERMAL TRANSMITTANCE COEFFICIENT (U-VALUE)

The U-value is the inverse of the overall thermal resistance of a structure and is given in $W/m^2 K$, thus:

$$U = 1/R_t \ W/m^2 K$$

An example follows of the calculation of the U-value for a sandwich panel with a homogeneous PUR core.

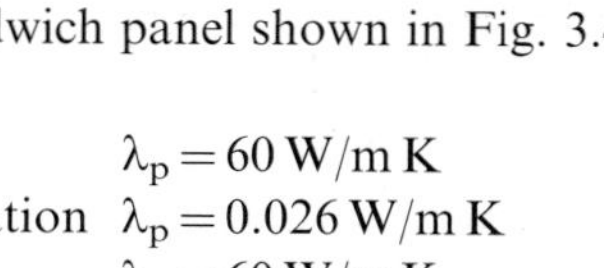

Fig. 3.4 Sandwich panel in example.

The data for the sandwich panel shown in Fig. 3.4 are:

0.7-mm steel sheet	$\lambda_p = 60$ W/m K
60-mm PUR insulation	$\lambda_p = 0.026$ W/m K
0.7-mm steel sheet	$\lambda_p = 60$ W/m K

$$R_t = R_{si} + R_{steel} + R_{core} + R_{steel} + R_{se}$$

$$R_t = 0.013 + 0.0007/60 + 0.06/0.026 + 0.0007/60 + 0.04 = 2.48 \ m^2 K/W$$

$$U = 1/2.48 = 0.40 \ W/m^2 K$$

Table 3.3 shows the U-values for some alternative constructions with the same insulation thickness.

Table 3.3 Typical U-values (W/m^2 K)

Insulation thickness (mm)	70	100	150	200
Panel with PUR core	0.35	0.25	0.17	0.13
Panel with EPS core	0.47	0.34	0.23	0.17
Panel with MW core	0.54	0.39	0.27	0.20
Cassette or profiled steel	0.80–0.60	0.70–0.45	0.60–0.35	0.35–0.30
sheet wall[a]		1.0	0.70	0.55
Lightweight concrete				

[a] with or without thermal bridges

3.4.1 Calculation method for profiled panels

This section describes an approximate method, originating with UEAtc, for calculating the thermal resistance of panels with profiled faces. If more accurate calculations are required, it is necessary to use numerical methods such as the finite-element method.

The method is applicable to panels which have one trapezoidally profiled face with a profile height greater than 5 mm. The other face is assumed to be nominally flat. The symbols used in the calculation are defined in Fig. 3.5.

Panels are divided into five categories depending on the surface ratio, r, which is defined by:

$$r = \frac{b_1 + b_2}{2d}$$

thus:

- Class 1: $r \leq 0.25$
- Class 2: $0.25 \leq r \leq 0.5$
- Class 3: $0.5 \leq r \leq 0.6$
- Class 4: $0.6 \leq r \leq 0.7$
- Class 5: $r > 0.7$ (with a maximum of 0.85)

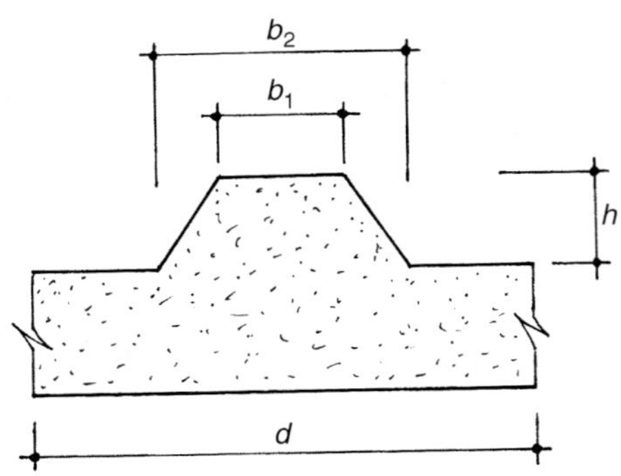

Fig. 3.5 Panel with a profiled face.

Using the terminology defined in Section 3.3.1, the average thermal resistance of the panel, taking into account the influence of the ribs, is then given by:

$$R_t = R_{si} + R_{se} + R_1 + \frac{\Delta h}{\lambda}$$

where R_1 is the thermal resistance for the part between the ribs
Δh is the additional thickness (mm) due to the main ribs according to Table 3.4

Table 3.4 Δh values for panels with one profiled face

Class	Height h of ribs (mm)			
	$10 \leq h \leq 25$	$25 \leq h \leq 50$	$50 \leq h \leq 70$	$h \leq 70^a$
1	1	2	2	2
2	3	5	6	7
3	5	9	12	14
4	7	12	16	19
5	8	15	20	24

[a] Height not exceeding 120 mm. Above this value, more accurate calculation is necessary

The influence of fixing systems, penetrations, etc. in sandwich structures can not be calculated using the methods described above. Such cases require more advanced calculation methods or measurements taken of the actual assembly (e.g. in a guarded hot box or using a measuring method suitable for complete constructions).

3.5 THERMAL CAPACITY

Sandwich panels are lightweight building components, which means that they have a relatively limited thermal storage capacity compared with, for example, concrete structures.

The thermal capacity, C, of a section of a building is obtained by:

$$C = m\,c \text{ W s/K}$$

where m = the weight in kg of the building section
c = thermal capacity per unit weight (W s/kg K)

The approximate thermal capacity/unit weight for some typical building materials is given in Table 3.5.

Table 3.5 Unit weights and thermal capacities of some typical building materials

Material	Unit weight, m (kg/m^3)	Thermal capacity, c (W s/kg K)
Concrete	2300	900
Lightweight concrete	500	1000
Wood	500	2300
Gypsum	900	800
Mineral wool	70–150	1030
PUR	40	1400
EPS	20	1450
XPS	30	1450

As an example of a thermal capacity calculation, a 100-mm concrete wall is compared with a 100-mm thick sandwich panel with a rock wool core and 0.5-mm steel sheets on both sides:

Concrete wall: $2300 \times 0.1 \times 900 = 207 \times 10^3\ \mathrm{W\,s/m^2\,K}$

Sandwich panel: $100 \times 0.1 \times 1030 + 7800 \times 0.001 \times 500 = 14 \times 10^3\ \mathrm{W\,s/m^2\,K}$

In modern buildings it is usually sufficient to use the opportunity for thermal storage provided by the floor and ceiling structures, heavy partitioning walls and internal surfaces to take up the temperature fluctuations.

Contrast this with conventional wisdom which has generally required walls of high thermal capacity. Figure 3.6 shows that, for an interior structure, the whole thermal capacity contributes to the required thermal inertia as it initially has the same temperature throughout. The external wall is less effective as it has an initial temperature gradient.

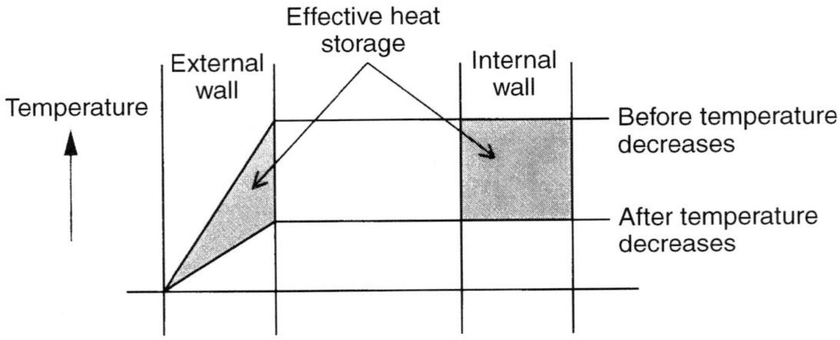

Fig. 3.6 Influence of the initial temperature profile on the effective thermal capacity.

It is important to have structures with a high insulation capacity (low U-value) in order to obtain low energy consumption both for heating and cooling buildings. If, in addition, the structures are lightweight, which means they have a small thermal capacity, little energy is required to heat them up. This means that, in buildings with lightweight sandwich walls, the temperature can be changed quickly and with a small energy consumption if, for example, the temperature is reduced during weekends or nights.

3.6 THERMAL BRIDGES

Thermal bridges usually arise at construction details where a material with a low thermal insulation breaks through a material with a better insulating capacity, for example at the locations of steel studs in an external wall or where fixing screws penetrate through an insulation system.

The impacts of thermal bridges include:

- increased thermal loss
- increased risk of condensation precipitation
- dirt adheres more quickly to cold surfaces than to warm (This can result in external walls which are locally dirty where cold bridges appear.)

In sandwich structures, cold bridging effects are usually minimised as the exterior and the interior surface layers are not in contact with each other. Some examples of the considerations here are as follows.

- *Tongue and groove joints* In the case of thin sandwich panels, there may be a risk of both sheet metal faces coming into contact in the tongue and groove connection because it is often economic to use the same sheet metal width and also the same tongue and groove geometry for all panel thicknesses.
- *Panel fixing systems* Through fixing with small diameter screws has a fairly small influence. It should be observed that stainless screws have a thermal insulating capacity which is approximately three times better than carbon steel. If a fixing system using penetrating steel profiles is used, the influence of the cold bridge may be considerably increased and this should be taken into account both when calculating the U-value and when evaluating the risk of condensation precipitation.
- *Detail solutions* Window and door fixings are often made with U-profiles installed over the edge of the panel in order to connect the faces together. The number of such profiles should be minimised in order to reduce the thermal bridge effect and, if possible, perforated profiles (giving a higher thermal resistance) should be used.

3.7 AIR-TIGHTNESS

An important factor that may strongly influence the energy consumption in a building is the involuntary air leakage through the climate screen. If a building is made air-tight and then provided with a ventilation system, the ventilation can be adjusted to the needs of the building independent of the wind pressure etc. Also thermal recycling of the outlet air can be arranged.

It is not only from the energy point of view that air-tight buildings are important. The automotive, electronics and medical industries also demand tight structures in order to exclude dust and for reasons of hygiene. It is also important to prevent the transport of moisture into the building structure.

The sandwich panel is an intrinsically air-tight structure because impervious facings are glued on to both sides of the core material. This fact, combined with well-designed details, results in very tight buildings. However, special attention should be paid to the following points.

- The joints between sandwich panels should be well sealed with a suitable sealing tape or mastic sealant in the panel joints.
- Details at the foundation and eaves joints should be designed with care.
- Window and door details should also be given careful attention.
- The installation of panels on site should be properly supervised. It is important that the panels have a tongue and groove structure forming a tight joint.
- The panel assembly should be sealed from the warm side, otherwise there is a risk that moist air can diffuse into the structure and accumulate there.

The air-tightness of buildings is often described in terms of the air replacement/hour (l/h) or, alternatively, as a leakage factor ($m^3/m^2/h$). The tightness is measured with a pressure difference of 50 Pa between the internal and external sides of the building. Typical values for sandwich panel buildings lie between 1.5 and 2.0 $m^3/m^2/h$ compared with 2.5–7.0 $m^3/m^2/h$ for on-site built-up steel sheet walls. The large spread of values is due to the fact that the accuracy of manufacture and installation are of great importance in the case of steel sheet walls.

At the usual serviceability limit state, the leakage is approximately 5–10% of that measured at 50 Pa. However, it should be observed that, in the case of tall buildings, considerable pressure differences can be obtained between the internal and external faces just below the roof as a result of the thermal stack (or chimney) effect. If there are leakage points, cold air will tend to leak in at floor level and out at roof level. It follows that it is of great importance that the joints between walls, floors and roofs are carefully made.

In order to calculate the energy loss W_{vent} due to air leakage, the following formula can be used:

$$W_{vent} = 0.33\, n\, V Q \text{ (W h)}$$

where n = air replacement per hour
V = building volume (m^3)
Q = thermal consumption value for the location (°C h)

As a numerical example:

Thermal consumption value for the location:	80×10^3 °C h
Building dimensions:	$30 \times 20 \times 6 = 3600\ m^3$

This gives an air leakage area (walls and roof): 1200 m^2

The values given above for a pressure difference of 50 Pa suggest a typical air leakage of 2 $m^3/m^2/h$ for the sandwich panel building and 6 $m^3/m^2/h$ for the on-site built-up steel sheet building. Air leakage at the usual serviceability limit state is then approximately 10% of the value at 50 Pa.

$$W_{vent} = 0.33 \times \frac{1200}{3600} \times 0.2 \times 3600 \times 80 = 6.3 \text{ MWh/year}$$

for the sandwich panel building

$$W_{vent} = 0.33 \times \frac{1200}{3600} \times 0.6 \times 3600 \times 80 = 19 \text{ MWh/year}$$

for the 'on-site built-up' building, there is a difference of 12.7 MWh/year.

Compare this with the thermal transmission through the walls and roof. Assume that the U-value for walls and roof is $U = 0.33\ W/m^2 K$.

$$W_{trans} = \sum U_j A_j Q$$

where U_j = U-value for region j ($W/m^2 K$)
A_j = surface area of region j (m^2)
Q = thermal consumption value for the location (°C h).
$W_{trans} = 1200 \times 0.33 \times 80 = 31.7$ MWh/year

3.7.1 Test methods for air-tightness

Test methods are available for measuring air leakages in structures either in a laboratory or in completed constructions. A standard, EN 12114, is now available for laboratory measurements. The measurements are made with both positive and negative pressure differences over the test surface. The measuring range is usually for pressures from 50 to 500 Pa. Using this method, the air-tightness of the joint in a sandwich panel can be measured and Fig. 3.7 shows some typical results.

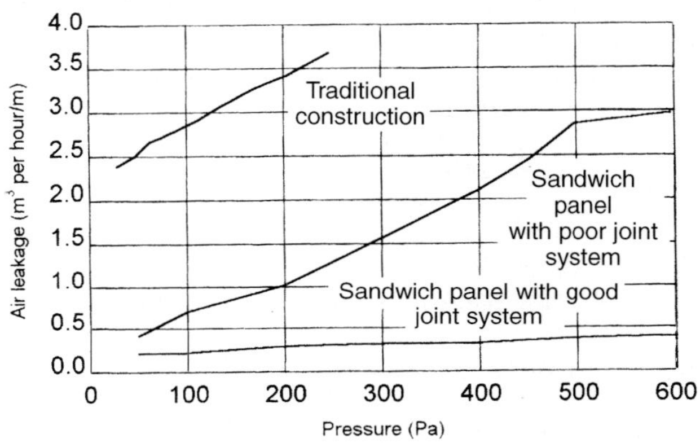

Fig. 3.7 Air leakage in typical panels with joints.

When testing for air-tightness, it is important to include measurements at connection details, for example foundation and column connections. In many cases, it is the detail solutions which are critical for the air-tightness of a sandwich structure.

When taking measurements in finished buildings, the total air leakage is measured excluding the influence of ventilation openings. Here closed windows, doors and other entrances are included as well as the accuracy of installation and the leakages are measured at a pressure difference of ±50 Pa.

3.8 WATER-TIGHTNESS

An important function of the external skin of a building is to provide protection against water penetration from rain, snow or ice. Water within a wall or roof structure reduces its thermal insulation properties and also affects the durability and susceptibility to corrosion.

For sandwich panels, with their extremely tight surface layers in the form of steel sheets which inhibit capillary suction effects, the weak points lie in the joint details and the solutions adopted for making the connections to the supporting structures. In heavy rain, a water film is formed over the outermost surface which may make the joints temporarily air-tight. If the wind then blows against the facade, the pressure difference increases over the joints and water may be forced in. In addition, depending on the joint structure, there may be capillary suction so that water may penetrate past the joint even if the joint height difference exceeds the wind pressure height. Therefore, in sandwich panel construction, the joint details are of great significance for the satisfactory functioning of the wall or roof. Figure 3.8 shows some alternative solutions for joint details.

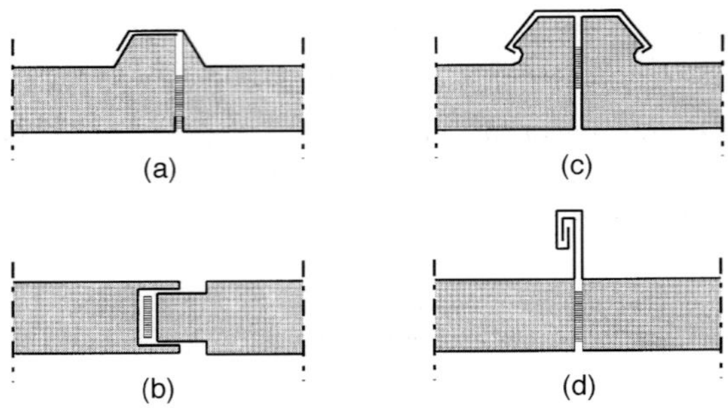

Fig. 3.8 Alternative joint details for sandwich panels. (a) overlapping of external steel sheets; (b) tongue and groove joints; (c) covering using flashings; (d) machine-sealed joints.

These solutions are often completed using sealing material in the form of tapes or sealing compounds in the joints. In normal buildings, the sealing is applied to the inner surface of the panel but, in chill or cold store applications, it should be applied to the external surface as the moisture gradient in such cases often passes from the external to the internal side (see Chapter 7 and also the following Section 3.9 *Moisture*).

In buildings exposed to heavy wind and rain loads (for example in coastal areas) the external joint should also be sealed. Alternatively rain-tightness can be obtained by using an additional weatherproof cladding to form a two-stage 'raincoat'.

3.8.1 Test methods for water-tightness

The water-tightness of building components can be tested in a laboratory and, today, this is widely used for window structures. There are two alternative principles that can be followed. The first method uses a pulsating pressure difference over the test surface as described in prEN 12865-1 *Hygrothermal*

Table 3.6 Comparison of driving rain test methods

Pulsating air pressure: Pr EN 12865-1	Static air pressure: EN 86
• Amount of water: 2 litres/m^2 per min. • Pulsating air pressure difference 0–150, 0–300, ... up to 1200 Pa • Length of a pulse: 15 seconds • Duration at each step: 10 minutes	• Amount of water: 2 litres/m^2 per min. • Pressure difference increases stepwise by 50–100 Pa up to 700 Pa • Duration at each step: 5 minutes

performance of buildings – Determination of the resistance to driving rain under pulsating air pressure: Part 1: External wall systems. The other method in EN 86, which is used in many countries for testing window structures, uses a static pressure difference over the test specimen. These two alternative test scenarios are compared in Table 3.6.

The test method using pulsating pressure simulates a more realistic rain scenario and can therefore be recommended.

Table 3.7 gives the relationship between wind speed and pressure level. Using either methodology, the criteria for water-tightness are the following:

- no water penetrates through the structure to the inside
- water penetrating into the building construction is in small amounts and is estimated to dry out

Table 3.7 Relationship between wind speed and pressure according to the Beaufort scale of wind force

B	Influence	Type	Wind speed (m/s)	Pressure (kN/m^2)
3	Light flags and banners stretch out wind	Gentle breeze	3–5	0.01
5	Small leaf trees begin to move	Fresh breeze	8–11	0.05
7	Big trees move. Tiring to walk against the wind	Moderate gale	14–17	0.15
10	Trees are torn up with their roots. Considerable damages to houses	Storm	24–28	0.42
12	Violent. Disastrous consequences	Hurricane	> 33	0.7

Building components and detail solutions in structures which are expected to be exposed to heavy rain and wind loads should be tested and be shown to be tight up to a pressure of 1200 Pa. In other less demanding applications, the assembly should be tight up to pressures of 600 Pa.

When testing sandwich structures, not only the joint structures should be tested but also the detailed solutions at, for example, the foundations, columns and corners.

3.9 MOISTURE

The definition of moisture is water in its various phases – vapour, liquid or ice. Moisture exists everywhere and is not hazardous as such, except when it appears in the wrong places and in too large quantities. Then it can cause damage in a variety of forms including:

- dripping
- stains
- corrosion
- mould
- breaking up as a consequence of freeze/thaw cycles
- increased heat flow

Parts of a building can be subjected to moisture from various sources. In walls and roofs and ceilings, where sandwich panels often are used, the following sources can be mentioned.

- *Rain, including snow and ice* Rain- and water-tightness are discussed in Section 3.8. It is important to bear in mind that after water, in the form of rain or snow, has penetrated into a structure it can then migrate in the form of both liquid and vapour and thus cause condensation problems.
- *Residual building moisture* Building moisture means excess moisture which remains in construction materials either as a consequence of the manufacturing process or from rain and snow during the construction phase. In sandwich panels, being pre-fabricated products, the moisture content as a consequence of manufacture is not a consideration. The moisture from the construction phase is also minimal as the usual core materials are water-repellent and the only open parts in a panel are the edges. Nevertheless, it is good practice to protect open panel surfaces during the construction phase in order to completely prevent moisture from penetrating into the structure.
- *Air humidity* Air contains vapour and the content is expressed as the vapour content in kg/m^3. In Fig. 3.9, the saturation vapour content is shown, which means the maximum amount of humidity that air can contain at a given temperature. Of interest in technical calculations involving moisture is the relative humidity, RH, which is defined as the ratio between the actual vapour content and the saturation vapour content and is generally expressed as a percentage.

The temperature at which the actual vapour content equals the saturation vapour content is called the dew point. It is important that the dew point does not fall within the construction.

The moisture content of outdoor air varies during the year and with the weather conditions. Indoors the moisture content is determined by the outdoor vapour content, the generation of moisture indoors (people, processes, etc.) and the amount of ventilation. Thus:

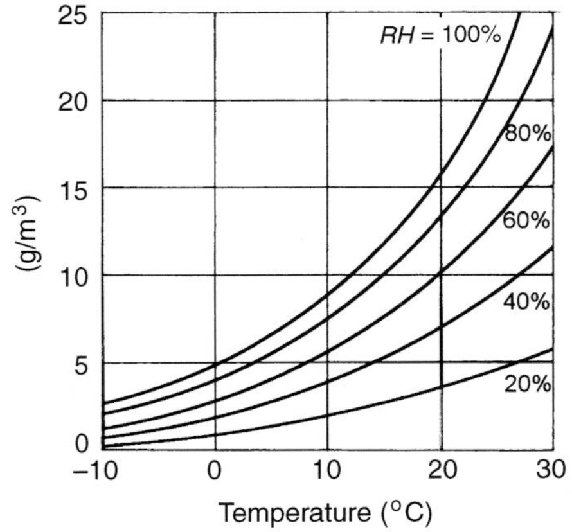

Fig. 3.9 Variation of the vapour content with temperature and humidity.

$$v_i = v_e + GV/n$$

where v_i = indoor vapour content (g/m^3)
v_e = outdoor vapour content (g/m^3)
G = moisture generation indoors (g/h)
n = number of air changes per hour
V = indoor air volume (m^3)

3.9.1 Surface condensation

When the temperature on a surface is lower than the dew point of the environment air, condensation takes place on this surface. The surface temperature at which this takes place can be calculated using:

$$T_s = T_i - R_{si}(T_i - T_e)/R_{tot}$$

where R_{tot} = total thermal resistance of the wall (m^2 K/W)
R_{si} = surface resistance (m^2 K/W)
T_i = indoor air temperature (°C)
T_e = outdoor air temperature (°C)
T_s = surface temperature (°C)

In calculations of condensation, the transition resistance R_{si} has a considerable influence on the surface temperature. At corners, the transition resistance is higher than in the centre of a wall. A general rule, which is usually on the safe side, is to use $R_{si} = 0.5$ m^2 K/W in these calculations.

In many cases, surface condensation is of a local nature, e.g. in places with poor insulation, untight structures, cold bridges or corners or where material has been stored against a wall thus creating additional thermal resistance and therefore lowering the surface temperature. In normal indoor environments and with current insulation thicknesses, there is not usually any problem in buildings where sandwich panels are used. These create a homogeneous insulation without cold bridges. Any critical spots are likely to be at fastenings and poorly-designed details which create local cold bridges. For moist rooms and chill and cold stores, a detailed moisture calculation should be carried out and the construction designed accordingly.

3.9.2 Moisture transfer

Two of the main transfer mechanisms in the vapour phase are *diffusion* and *convection*. Moisture diffusion seeks to level out differences in the vapour content of the air in different locations. The moisture flow takes place from an area with a higher vapour content to another with a lower content. In buildings, in order to prevent this transfer, a vapour barrier can be used. This vapour barrier should always be on the warm side of the structure, which in normal buildings, generally means on the inside of walls or ceilings. In sandwich panels, the facings act as excellent vapour barriers and the diffusion that can take place through the panel joint is of no consequence. Here it is rather a case of moisture transfer in the form of convection.

Moisture convection arises when the total air pressure differences between different parts of a building cause the movement of moist air. Significant here is the airtightness of the structure, the pressure difference over the building part and the air vapour content. A condensation risk may arise where there is interior overpressure in a building. Then, warm moist air can be forced into untight parts of the construction and there condense against colder surfaces. The overpressure causing this may be due to, for instance, the stack (chimney) effect or the adjustment of the ventilation system.

In general, the diffusion process is slow and transfers only small quantities of air, whereas, in the case of convection, the quantities are considerably larger, as shown in Fig. 3.10.

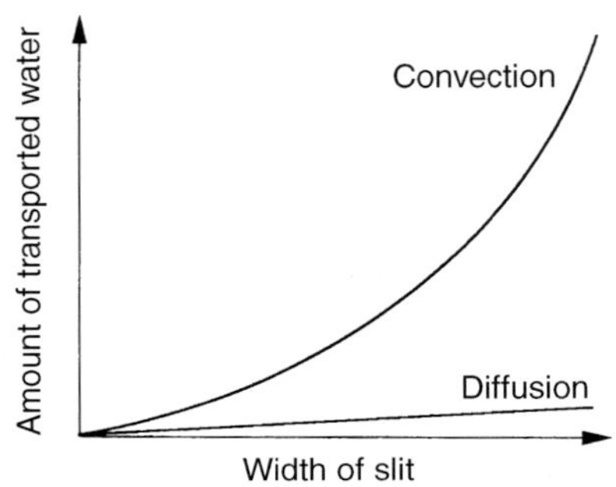

Fig. 3.10 Moisture transfer by convection and diffusion.

Figure 3.10 shows how important air-tight structures are in order to reduce undesirable convection. Section 3.5 demonstrates how sandwich structures give much tighter structures compared with on-site building. This has led to sandwich panels being frequently specified for buildings where these properties are of great importance, such as cold rooms, drying ovens, buildings used by the food industry, etc. The possible leakages in sandwich structures are easy to detect, which makes them easy to correct. There are no hidden places, they are concentrated in panel joints, connections to other building parts and cutouts.

3.10 THERMOGRAPHIC SURVEYS[3.1]

In certain parts of Europe, the use of thermographic surveying has revealed large-scale failures in the thermal performance of conventional built-up construction and has led to significant improvements in insulation standards. The purpose of such surveys is to check that the thermal insulation is in the right place and correctly installed. The early results of such surveys highlighted some frightening deficiencies.

The technique of thermography is simple. It uses a camera which looks and operates like a conventional video camera. It is operated from within the building, generally at night when the temperature difference is at a maximum. In some cases, it may be advantageous to over-heat the building by 10–15°C in order to improve accuracy. The camera records temperature differences to an accuracy of about 0.5°C. In this way, an accurate picture of the variation of U-value in the walls and roof of a building can be quickly obtained.

The results of such surveys are very much what the informed observer might expect. The major losses of heat recorded by the video fall into the two categories of poor installation and poor design. Poor installation may take the form of sections of insulation left out, detailed construction around windows and roof lights poorly carried out, loose pieces of insulation not butted together or allowed to sag in a cavity, etc. Typical design problems are with thermal breaks and details at corners, edge details and flashings. Two particular problems stand out, namely cold bridges and badly made cavities where air whistles up both sides of the insulation, rendering it virtually useless. It should be obvious to the reader that most, if not all of these problems are most easily avoided by specifying sandwich construction. It is also worth pointing out that thermography is not expensive and, in view of the huge energy savings that it can bring about, it deserves to be much better known.

3.11 THE ADVANTAGES OF SANDWICH CONSTRUCTION IN THE CONTEXT OF THERMAL INSULATION

The advantages of using sandwich construction when a high level of thermal insulation is required may be summarised as follows.

- The panels contain no thermal bridges in the form of profiled metal sections or steel frames
- The insulating material is homogeneous and is protected by the surface layers. In traditional steel frame and cassette construction, air channels tend to arise in the insulating layer during installation as a consequence of missing items of insulation, insulation sagging or becoming displaced or being compressed, etc. Thus, with sandwich panel construction, the insulation level is much more reliable.
- Sandwich panels are a prefabricated product that is manufactured under controlled conditions so that no moisture is built into the construction.
- A sandwich panel can be installed in varying climatic conditions and maintains its properties. It is not easy to install a steel frame wall with insulating slabs, vapour barrier and steel sheets in heavy rain and wind at a height of say 25–30 m on a facade. With sandwich panel construction, it is possible for installation to continue under more extreme conditions without any risk to the resulting level of thermal insulation.

3.12 ENERGY SAVING

A reduction in the energy consumption of an individual building can only take place when the climatic screen is well insulated and air-tight. Only then is it possible to take complete advantage of more efficient installations for the production of heat and cold. A high standard of insulation in the climatic casing not only means lower energy consumption, it also lowers the power requirement and thus makes the heating period shorter. It also allows better utilization of free energy (i.e. energy from, for example, machinery or human bodies that is not paid for directly). A high insulation standard is an investment which shows very good profitability during the whole service life of the building without operating or maintenance costs.

Sandwich panels with their good thermal insulation and air-tightness offer excellent possibilities in this respect.

Below, a method is presented for the calculation of an economical insulation standard, the so-called *saving cost* (SC) method. This method calculates the insulation standard that gives the lowest annual costs within the selected calculation criteria. It compares the cost of saving energy with the current energy price. By a stepwise addition of insulation thickness, the marginal cost saving can be calculated. As long as this is lower than the current energy price, the measure is profitable.

The cost saving (SC) can be calculated using the formulae:

$$SC = \Delta I / S \times \alpha$$

where ΔI = increase in investment cost (ECU/m^2)
S = annual energy saving (kWh/m^2), calculated using the formula:

$$S = \Delta U \times Q$$

where ΔU = improvement of the U-value
Q = thermal consumption value for the location in 1000 degree hours/year
α = correction factor calculated using the formula:

$$\alpha = (1 - t^n)/(1 - t)$$

where $t = (1 + q)/(1 + r)$
n = service life in years
q = actual energy price increase (%)
r = desired actual interest payment (%)

As an example of this calculation:

Current energy price	0.06 ECU/kWh
Actual interest, r	4%
Annual energy price increase, q	2%
Service life, n	15 years
Thermal consumption value for the location, Q	80×10^3 °C h/year

These assumptions result in $\alpha = 13.14$.

Sandwich panel thickness (mm)	80	100	120	150	200
U-value	0.49	0.40	0.33	0.28	0.21
Price (ECU/m^2)	30	32	34	36	400
ΔI		2	2	2	4
ΔU		0.09	0.07	0.05	0.07
$S = \Delta U \times Q$ (kWh/m^2)		7.2	5.6	4.0	5.6
SC margin, ECU/kWh		0.021	0.027	0.038	0.054

As long as the SC margin is lower than the current energy price, the investment is profitable. The calculations of the costs for the panels should also include any other additional costs incurred, for example a larger foundation plate etc. The example above shows that the optimum thickness lies at over 200 mm. The average saving cost is determined by calculating the total increase in insulation thickness. For a 200-mm thickness this gives:

$$S = \Delta U \times 80 = (0.49 - 0.21) \times 80 = 22.4\,\mathrm{kW/m^2}$$

$$SC_{av} = \Delta I/S \times \alpha = (10/22.4) \times 13.14 = 0.034\,\mathrm{ECU/kWh}$$

This measure is very profitable as $SC_{av} < 0.06$ ECU/kWh.

3.13 REFERENCE

3.1 Ashley, S. (1991) Therms of endearment. *Building Design*, November 29.

Chapter 4
Acoustics

4.1 INTRODUCTION

In most countries there are regulations for the acoustical environment which include consideration of factors such as:

- noise transmitted from industrial buildings to the environment
- traffic noise entering into buildings
- noise levels inside workshops and buildings
- noise reduction between the rooms in a building

In order to permit the design of wall and roof cladding to satisfy these requirements, acoustic parameters and acoustical calculation methods are needed. In this chapter, some acoustical engineering methods are presented. The acoustical behaviour of sandwich panels is also discussed.

4.2 WHAT IS SOUND?

Sound is something that we hear when particle movement in materials, liquids and gases is registered by the ear as pressure variations. The basic characteristics of sound are frequency and amplitude.

The unit for frequency is the hertz (Hz = cycles per second). Adults have a hearing range from 20 Hz up to 20 000 Hz although our body is also influenced by sound outside this hearing range. Infrasound is in the range below 20 Hz and ultrasound is above 20 000 Hz. Noises are not usually pure tones, but include a range of sound energy spread over a wide band of frequencies. The centre frequencies are internationally standardised and Table 4.1 shows some of the standard frequency bands.

The human ear responds to sound pressure, which is measured in units of Pa (N/m^2). The lowest sound pressure that an average ear can detect is about 2×10^{-5} Pa and the limit for pain is about 20 Pa. Because of this extensive range in pressure, it is impractical to use a linear scale, so sound pressure levels are generally expressed using a logarithmic scale. Thus, noting that the energy in a sound wave is

Table 4.1 International standard frequency bands

Octave band (Hz)		125			250			500	
1/3 octave band (Hz)	100	125	160	200	250	315	400	500	630

proportional to the sound pressure squared,the following equation defines the generally-used unit for sound pressure, the decibel (dB):

$$L_p = 10 \log \frac{p^2}{p_o^2} = 20 \log \frac{p}{p_o} \qquad (4.1)$$

where L_p = sound pressure level (dB)
p = actual sound pressure (Pa)
p_o = reference sound pressure: 2×10^{-5} (Pa)

In practice, about 1–3 dB is the smallest change that can be detected by an average human ear. A difference of 10 dB will be recognised by a human as a doubling or halving of the level of sound. Mathematically, the addition of two identical sources will increase the level by 3 dB and ten identical sources by 10 dB. More detailed rules for the combination of sounds from different sources are given in Reference 4.1.

A human ear has different sensitivity to sound pressure at different frequencies. The minimum level perceptible by the ear at a particular frequency is known as the threshold of hearing. The threshold of hearing varies from person to person and is also a function of the age of the listener. When the sound perceived by the ear becomes louder and louder it reaches a level when it feels like a tickling in the ear. This level is known as the threshold of pain. It is less dependent on frequency than the level of hearing and has a value about 120 dB. These limits, together with approximate limits for speech and music, are shown in Fig. 4.1.

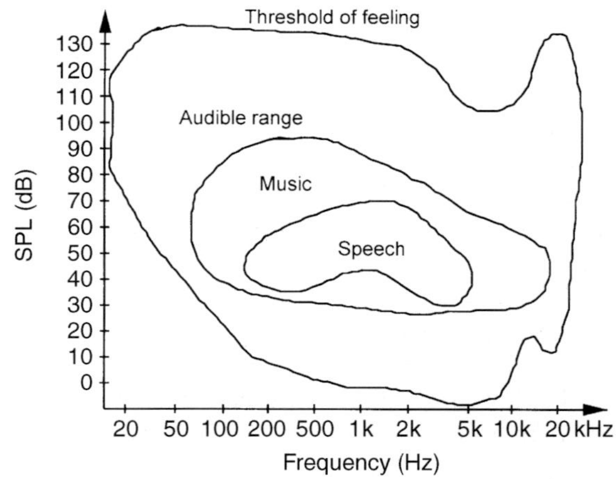

Fig. 4.1 Audible range of frequencies and sound pressure levels (SPL) bounded by the thresholds of hearing and feeling together with approximate thresholds for speech and music.

In order to provide a single-figure value for the sound pressure level that simulates the ear's response, three standardised weighted frequency values are defined A, B and C. The most common is the A-filter (dBA). Figure 4.2 shows some A-filter values for some common sound sources.[4.1]

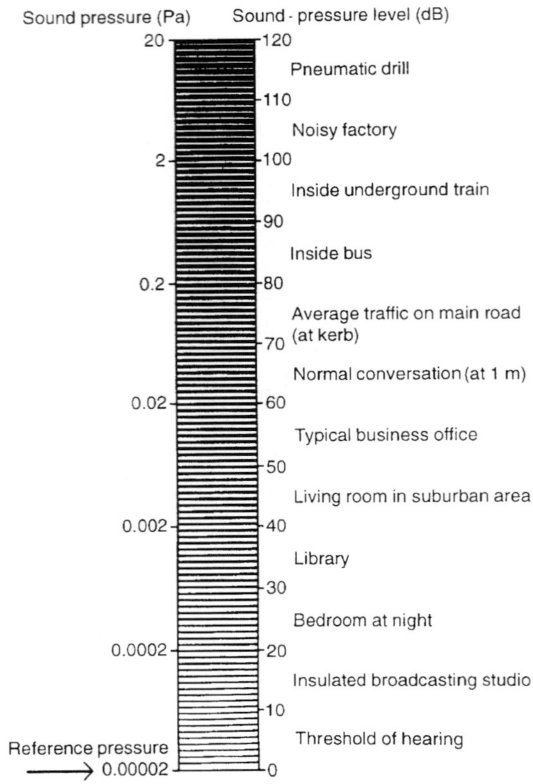

Fig. 4.2 A-filter values for some common sound sources.

A-weighted sound-pressure levels take account of the fact that the human ear notices high frequencies more easily than low frequencies. This characteristic has been incorporated into sound-level meters using a weighting curve so that A-weighted pressure levels may be measured directly.

4.2.1 Decibel arithmetic

We now consider how to calculate with decibels. We first note that the decibel is a logarithmic value which cannot be added or subtracted in the same way as linear values. It is therefore necessary to return to linear units in order to perform the arithmetic and then to go back to logarithmic values.

As an example, we consider how to add the two values:

$$L_{p1} = 40\,\text{dB} \quad L_{p2} = 45\,\text{dB}$$

We first change the units to bels by dividing by 10 and then return to linear values in order to perform the addition thus:

$$10^{4.0} + 10^{4.5} = 10000 + 31622 = 41622$$

Then returning to logarithmic values gives:

$$\log(41622) = 4.62\,\text{bel}$$

So that:

$$L_{p.tot} = 46.2\,\text{dB}$$

Alternatively, Fig. 4.3 may be used to obtain the same result.

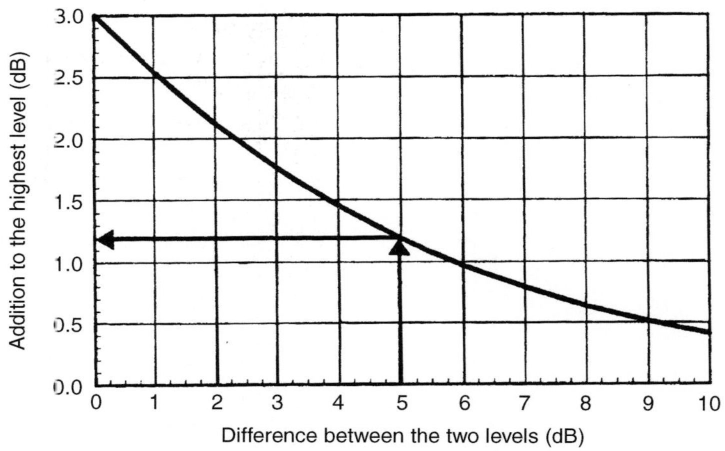

Fig. 4.3 Aid to the addition of two sounds.

4.3 FUNDAMENTAL ACOUSTIC PARAMETERS

Sound may be *absorbed, transmitted* or *reflected.* When a room boundary, e.g. a roof, floor or a wall, is hit by a sound wave, some of the sound energy will be reflected, some is absorbed within the material and some is transmitted through it, as illustrated by Fig. 4.4.

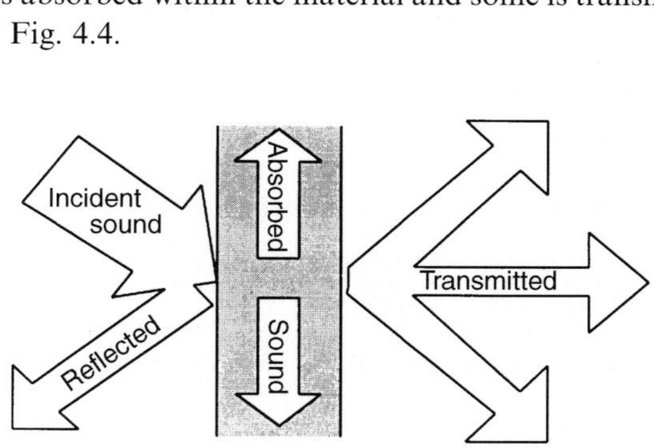

Fig. 4.4 Impact of sound on a room boundary.

The proportion which is reflected, absorbed or transmitted depends on the shape of the material or the construction hit by the sound wave, and the frequency of the sound. Based on this we can define three acoustical parameters.

Absorption coefficient, α = (absorbed sound + transmitted sound)/(incident sound)
Reflection coefficient, ζ = (reflected sound)/(incident sound)
Transmission coefficient, τ = (transmitted sound)/(incident sound)

4.4 SOUND INSULATION

Airborne sound generated in a room or an industrial hall can be transferred to adjacent rooms or to the external environment via different transmission paths, e.g. through partition walls, floors, the building framework, windows, doors and ducts as illustrated by Fig. 4.5. The net reduction of the airborne energy caused by transmission via all of these paths is called the airborne sound insulation.

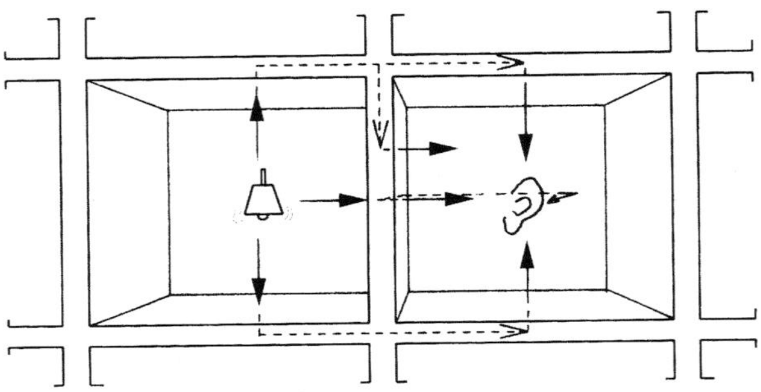

Fig. 4.5 Transmission paths for airborne sound.

The basic measure of the sound insulation provided by a partition is termed the *sound reduction index*, R (dB), or the *sound transmission loss*.

$$R = 10 \log 1/\tau \tag{4.2}$$

where τ = the transmission coefficient.

R refers to values as measured in a laboratory. When the measurements are made in a real building, and include flanking transmission, sound leaks, etc., it is termed R'. The difference between laboratory and field values can be a significant number of dB depending on the construction details and workmanship.

An evaluation of Equation 4.2 gives the following expression for the sound reduction index between two rooms with a diffuse sound field:

$$R = L_S - L_R - 10 \log A/S \tag{4.3}$$

where L_S = sound level in the source room (dB)
L_R = sound level in the receiving room (dB)
S = partition area (m^2)
A = absorption area in the receiving room (m^2)

The absorbtion area of the receiving room is calculated using the Sabine formula which is discussed in Section 4.10.

The transmission coefficient is dependent on the frequency and the angle of incidence of the sound waves. Generally we do not have only a single wave component, instead the sound field in a room is better described as a diffuse sound field. The ideal diffuse sound field is like plane sound waves with equal intensity, which propagate with the same probability in all directions. Normally the sound insulation is measured in a one third octave band interval over a range 100–5000 Hz. The result may be represented graphically or in a tabular form. The standard test procedure is defined in EN ISO 140, where standard methods are given for both laboratory and field measurements.

When sound transmission is measured between rooms in a building, the influence of sound flanking paths will be included. Equation 4.3 can be useful when calculating the estimated level in an adjacent room knowing the sound reduction index for the partition and the level in the source room.

4.4.1 Example 1

An office room measuring 5 m × 7 m × 3 m is adjacent to a factory hall. The dividing wall between the two rooms measures 6m × 3m. The sound reduction index of the wall in the actual frequency band is 40 dB. The general sound level inside the factory hall will be 80 dB. Calculate the estimated sound pressure level inside the office when the reverberation time there is 1 second.

We start by calculating the absorption area inside the office using the Sabine formula (see Equation 4.8 in Section 4.11):

$$A = 0.163\, V/T_R$$

where V = the volume of the room
T_R = the reverberation time.

So that:

$$A = 0.163 \times (5 \times 7 \times 3)/1.0 = 17\,\mathrm{m}^2$$

Equation 4.3 then gives:

$$L_R = L_S - R - 10 \log A/S$$

$$L_R = 85 - 40 - 10 \log 17/18 = 45\,\mathrm{dB}$$

4.5 WEIGHTED SOUND REDUCTION INDEX, R_W

When specifying the acoustic performance of a partition in a more general manner, it can be useful to describe the sound insulation by a single number. The *weighted sound reduction index*, R_w, is a rating method given in EN ISO 717-1. This standard fits a standard reference curve to the measured sound reduction index curve. However, this rating system is not normally applicable to industrial buildings, which is where sandwich panel construction is most frequently used. Here, therefore, consideration must be given to the frequency-dependent sound levels of the noise sources and to the frequency-dependent sound reduction index for the partition as shown in Fig. 4.9(a). The two panels, for which the results are shown in the right-hand diagram, have the same R_w value. In EN ISO 717-1, a rating method is given where the R_w-value is completed by two C-terms which are applied to two models of the noise spectra for various types of noise. These two terms, $R_w + C$ and $R_w + C_{tr}$, also include the frequency area 100–3150 Hz but can be extended to 50–5000 Hz. As industrial and traffic noise often have high sound levels which are also below 100 Hz, it is recommended that the extended frequency area is used.

The summary value, $R_w + C$, gives the reduction value in dBA for a spectrum with a level which is equally high in all third-octave bands. This can be used for:

- living activities (talking, music, radio, TV)
- railway traffic at medium and high speed
- highway road traffic travelling at speeds in excess of 80 km/h
- jet aircraft at a short distance
- factories emitting mainly medium and high frequency noise

The summary value $R_w + C_{tr}$ also gives the reduction value in dBA, but for a spectrum with low-frequency dominance such as:

- urban road traffic
- railway traffic at low speeds
- disco music
- factories emitting mainly low and medium frequency noise

Table 4.2 shows the values of the weighted sound reduction index for some typical structures.

We take for example an industrial building with a noise level of 70 dBA in the production area. The noise has maxima at low to medium frequency. With an external wall structure of sandwich panels with mineral wool core the sound level outside the building is $70 - 29 = 41$ dBA.

Table 4.2 Weighted sound reduction index for some typical structures

Structure	R_w-value	$R_w + C$	$R_w + C_{tr}$
Sandwich panel with 80 mm mineral	31	30	29
Sandwich panel with 80-mm polyurethane core	25	24	21

4.6 OVERALL SOUND REDUCTION INDEX

If a partition consists of different kind of elements, for example a wall with windows and doors which have different sound transmission characteristics, the overall sound reduction index must be calculated. This can be done with the aid of the chart shown in Fig. 4.6 or by using the following formula. This is illustrated by means of an example:

$$R = 10 \log \left(\frac{S_1 + S_2 + \ldots + S_n}{S_1\, 10^{-R_1/10} + \ldots + S_n\, 10^{-R_n/10}} \right)$$

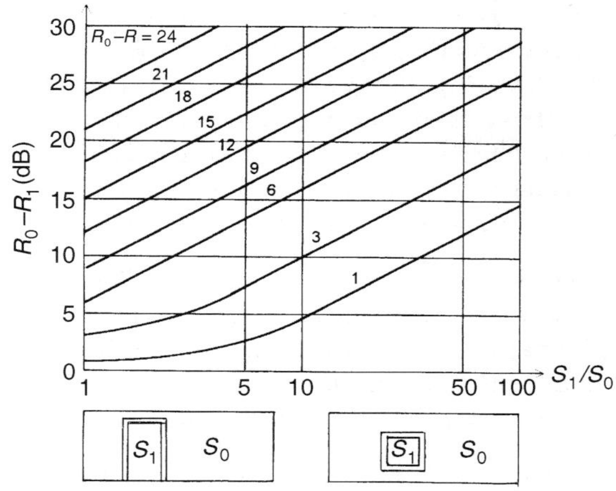

Fig. 4.6 Chart for overall sound reduction index. R = the overall sound reduction index; R_0 = sound reduction index for the wall; R_1 = sound reduction index for the window/door; S_0 = total area for the wall including window/door area; S_1 = area for the window/door.

where R = overall sound reduction index
$S_1, \ldots S_n$ = areas of the different elements
$R_1, \ldots R_n$ = sound reduction index for the different elements

4.6.1 Example 2

If a door measuring 1 m × 2 m with a sound reduction index of 25 dB in the actual frequency band is mounted in the wall of Example 1, Section 4.4.1, what will happen to the sound pressure level inside the office?

Using the chart in Fig. 4.6:

$$S_0/S_1 = (6 \times 3)/(2 \times 1) = 9$$

$$R_0 - R_1 = 40 - 25 = 15\,\text{dB}$$

Where the two values intersect on the chart therefore gives $R_0 - R \approx 6\,\text{dB}$

The overall sound reduction index for the wall with the door is about 34 dB. Without the door, it was 40 dB. The sound pressure level inside the office will therefore be increased by 6.5 dB.

Alternatively, applying the formula to the same example:

$$R = 10\log\left(\frac{16+2}{16 \times 10^{-40/10} + 2 \times 10^{-25/10)}}\right) = 33.5\,\text{dB}$$

4.7 FLANKING TRANSMISSION

Flanking transmission refers to sound that is transmitted through flanking structures to adjacent rooms. When using lightweight sandwich elements, it is necessary to take this into account. Examples of structural solutions where this may occur are where an external wall made of sandwich panels passes an intermediate floor or partition wall which offers a high level of sound insulation or when partition walls with high sound insulation are connected to a lightweight ceiling made of sandwich panels. Figure 4.7b and c shows examples of solutions which reduce the risk of flanking transmission.

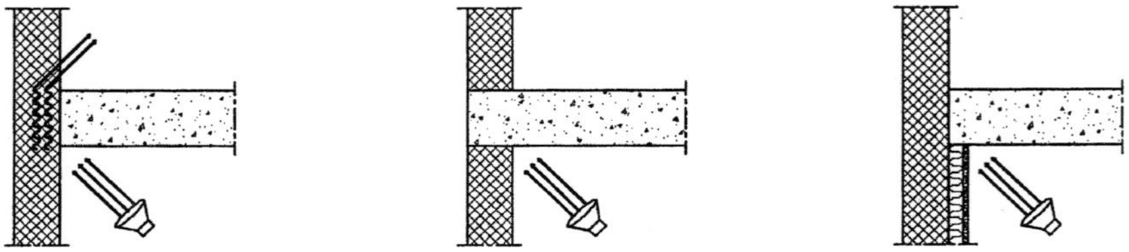

Fig. 4.7 Solutions for reducing the risk of flanking transmission.

4.8 SOUND REDUCTION INDEX FOR HOLES AND SLITS

The sound reduction index for holes and slits is nearly equal to 0 dB. The influence of holes and slits may therefore be important, for instance, at the connections between sandwich panels, at doors and windows without sealing strips, and at any necessary openings in partitions. In the latter case, Fig. 4.6 can again be used to calculate the overall sound reduction. If there is an acoustically absorbing material in the slits, it will give a higher sound reduction index for the slits. It follows that the connection between two sandwich panels with mineral wool cores is not as critical as the connection between panels with rigid plastic foam, which can have rather poor acoustical absorption.

A recognised phenomenon in sandwich structures is the deflection due to the influence of a temperature difference between the faces. As illustrated by Fig. 4.8, this should be taken into account in the detailing, for instance, of connections between the external walls and intermediate floors in order to avoid an adverse impact on sound reduction.

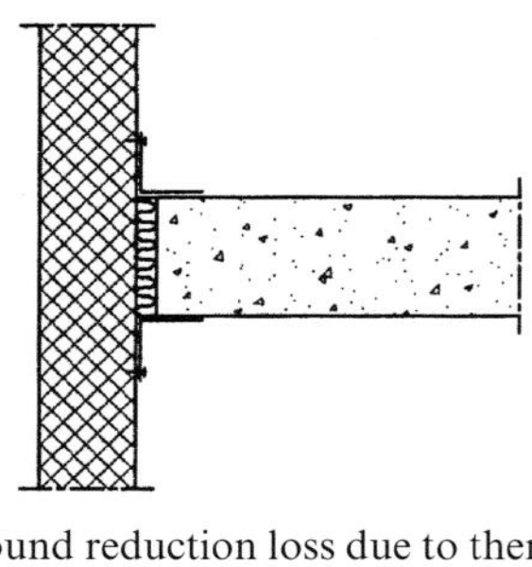

Fig. 4.8 Detailing to avoid sound reduction loss due to thermal bowing of a sandwich wall.

4.9 SOUND REDUCTION INDEX FOR A SANDWICH PANEL

A sandwich panel acts acoustically as something intermediate between a single-skin and a double-skin wall. The sound reduction index curve of a sandwich panel follows that of a single-skin wall (follows the mass law) up to the so-called dilational frequency. At this resonance frequency, where the core acts as a spring and the faces as masses, a large dip in the sound transmission loss curve occurs. This dilational frequency is determined by the density of the faces and the E-modulus of the core and can be estimated using:

$$f_d = \frac{1}{2\pi}\sqrt{\frac{4E_c}{t_c\left(2\rho_F t_F + \frac{\rho_C t_C}{3}\right)}} \qquad (4.4)$$

where E_C = E-modulus for the core in compression (Pa)
ρ_F = density of the skin (kg/m^3)
t_F = thickness of the skin (m)
ρ_C = density of the core (kg/m^3)
t_C = thickness of the core (m)

As an example, consider the polyurethane panel shown in the left-hand part of Fig. 4.9(a), where:

$E_C = 6 \times 10^6$ Pa
$\rho_f = 7200$ kg/m^3
$t_f = 0.5$ mm
$\rho_c = 40$ kg/m^3
$t_c = 80$ mm

The above formula gives the dilational frequency as $f_d = 960$ Hz.

Evidently, this dip in the sound reduction curve can be troublesome if it coincides with a significant sound frequency on one side of the panel. Equation 4.4 may be viewed as a tool to predict the position of the dip and to move it to a more convenient position in the frequency range, if necessary.

As mentioned earlier in this chapter, sandwich panels behave as though they were single skin wall structures within the frequency range below the dilational frequency. The sound reduction for a single-skin wall in the mass-related region follows the mass law:

$$R = 10\log(m) + 20\log(f) - 48\ \text{dB}$$

where R = the sound reduction index at the current frequency (dB)
m = the weight per unit area (kg/m^2)
f = frequency (Hz)

This equation shows that, when doubling the weight or frequency, the sound reduction increases by 6 dB.

The left-hand part of Fig. 4.9(a) shows the sound insulation for two otherwise similar sandwich panels with different types of core material. The diagram also includes the mass law curve for both panels. The higher weight per unit area of the mineral wool panel results in better sound insulation up to the dilational frequency.

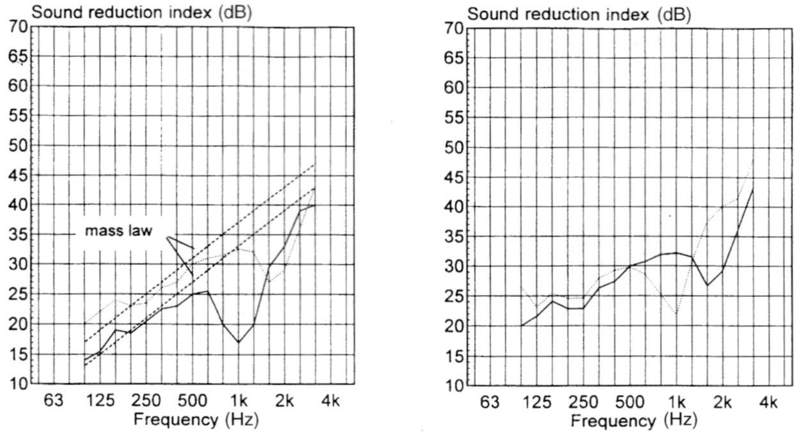

(a) – – – Sandwich panel with mineral wool core thickness 80 mm and weight 18 kg/m^2, R_w = 31 dB

——— Sandwich panel with PUR foam core thickness 80 mm and weight 12 kg/m^2, R_w = 23 dB

——— Sandwich panel with mineral wool core thickness 80 mm, R_w = 31 dB

- - - - - Sandwich panel with mineral wool core thickness 150 mm, R_w = 31 dB

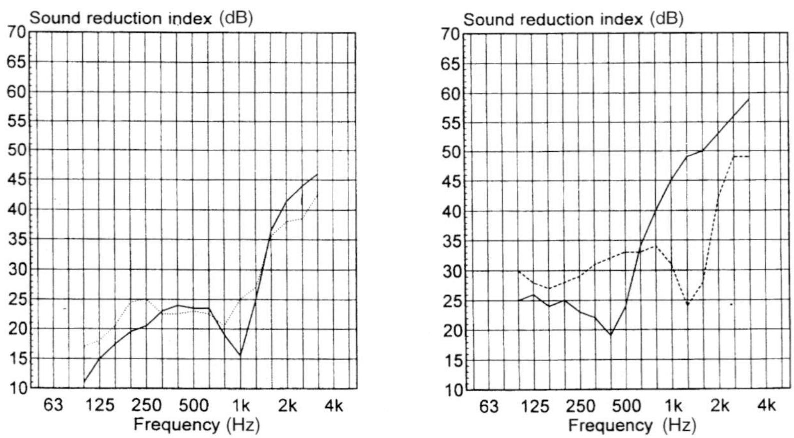

(b) – – – Profiled sandwich panel with a PUR foam core with thichness 66 mm and weight 14 kg/m^2, R_w = 27 dB

——— Sandwich panel with a PUR foam core thickness 80 mm and weight 14 kg/m^2, R_w = 25 dB

——— Sandwich panel with a mineral wool core density 45 kg/m^3, R_w = 33 dB

- - - - - Sandwich panel with a mineral wool core density 150 kg/m^3, R_w = 31 dB

Fig. 4.9 Transmission loss values for some typical sandwich panels.

The right-hand diagram shows the results for two panels with different thicknesses of the core. Here, it can be seen how the dilational frequency reduces with increased panel thickness.

The left-hand part of Fig. 4.9(b) shows the reduction value for a profiled sandwich panel with a polyurethane core and, in the right-hand part, it can be seen how the stiffness of the core material influences the dilational frequency.

Since sandwich panels normally are rather light and stiff, as a consequence of the requirements for the mechanical properties, it is not easy to obtain conventional sandwich panels with high sound insulation. The mass law will give an estimate of the practical upper limit of the sound reduction for a sandwich panel. However, it is possible to build special sandwich structures with high sound insulation. Figure 4.10 shows some results for a double wall structure of sandwich panels. Note that the sound reduction increases by up to 20 dB in some frequency bands when there is an absorbing material in the air gap between the panels. For example, in ventilation and machinery rooms, where often both good sound reduction and sound absorption are required, the panels can be faced with an absorbing material. Figure 4.10 also shows the effect of a mineral wool slab fixed to one side of a sandwich panel.

If the sound insulation is to be improved in a sandwich wall, for example adjacent to an office, a solution such as that shown in Fig. 4.10 can be used. The space between the panel and the gypsum board may be used for electrical installations.

4.10 SOUND IN ROOMS

Sandwich panels are normally used as walls and roofs for factories and workshops where noise levels are often high. If panels with metal skins and no additional sound absorption are used, the acoustic environment in the building may be unsatisfactory as most of the sound will be reflected back into the room. In order to improve the acoustical environment, sound-absorbing materials may be put into the room. These absorbents can be fixed under the roof and on to the walls.

Sandwich panels with metal faces and a core of mineral wool in which one of the faces is perforated provide a good alternative for use as partition walls (where a vapour barrier is not needed) or for machinery enclosures. These panels have both sound absorbing and sound-reduction qualities.

The decrease of the general sound level inside will also help to fulfil requirements for sound levels outside the factory.

4.10.1 Sound absorption

The material constant which indicates the ability of a material to absorb sound is known as the sound absorption coefficient, α. This coefficient indicates the ratio between energy being absorbed and incident energy and therefore varies between 0 (total reflection) and 1 (total absorption). The absorption coefficient is dependent on the frequency.

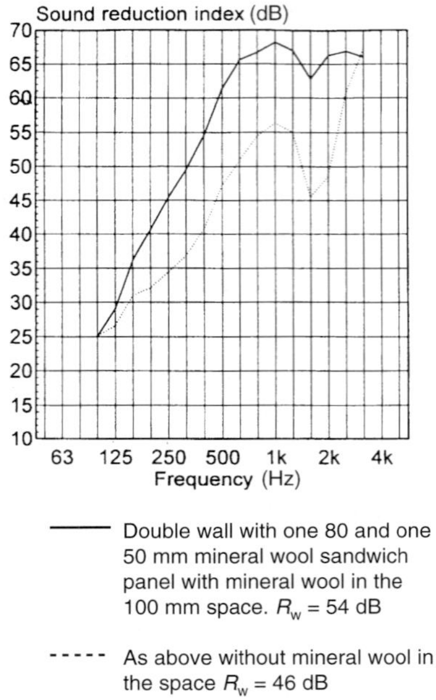

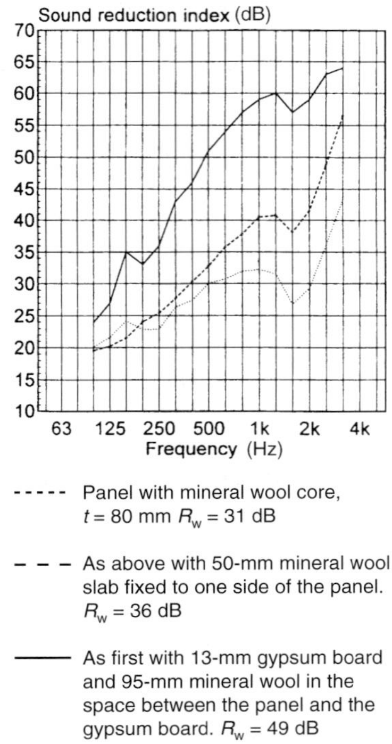

Fig. 4.10 Transmission loss values for some augmented sandwich panel arrangements.

The standard test procedure for measuring the absorption coefficient is defined in ISO 354. The measurements are carried out in a reverberation room.

Knowing the absorption coefficient α, the absorption, A, of a surface can be calculated:

$$A = \alpha S \tag{4.5}$$

where S = area (m).

The total absorption of a space is the sum of the absorptions of the component surfaces:

$$A_{tot} = S_1\alpha_1 + S_2\alpha_2 + \ldots + S_n\alpha_n \tag{4.6}$$

and the average absorption:

$$\alpha_m = \frac{A_{tot}}{S_{tot}} \tag{4.7}$$

Table 4.3 shows the absorption coefficient, α, for a range of materials.

4.10.2 Sound absorption for sandwich panel

Sandwich panels with non-perforated metallic skins offer no significant sound absorption. If the core is of a sound-absorbing material (e.g. mineral wool) one way to obtain good sound absorbing properties is to perforate the skin, as

Table 4.3 Absorption coefficient, α, for a range of materials

Octave band (Hz)	125	250	500	1000	2000	4000
Concrete	0.02	0.02	0.02	0.02	0.03	0.04
Gypsum board on studs	0.2	0.15	0.1	0.08	0.05	0.05
Windows	0.35	0.25	0.18	0.12	0.07	0.04
50-mm mineral wool slab[a]	0.2	0.65	1.0	1.0	1.0	1.0
100-mm mineral wool slab[a]	0.45	0.9	1.0	1.0	1.0	1.0

[a] With a solid backing

illustrated by Fig. 4.11. This type of panel can be used for interior walls where there is not usually any need for a vapour barrier.

4.11 NOISE REDUCTION IN SMALL INDUSTRIAL PREMISES

The reverberation time may be used to describe the sound-absorbing properties of a room. It is determined by the size of the room and its absorption properties. In a small room or hall (volume $\leq 1000\,\mathrm{m}^2$), if the sound field is diffuse and the average absorption is less than 0.3, an empirical formula called the *Sabine formula* is valid:

$$T_R = 0.163(V/A_{tot}) \tag{4.8}$$

where T_R = reverberation time (s)
V = room volume (m^3)

The reverberation time is a measure of how fast the sound in a room will decay and is defined as the time taken for the sound energy produced by a source (e.g. a pistol shot) to decay by 60 dB after the sound source has been switched off. The reverberation time is frequency dependent.

In this type of room we have two different kinds of sound field. Nearest to the sound source is the direct sound field. In this field, the sound transmits directly

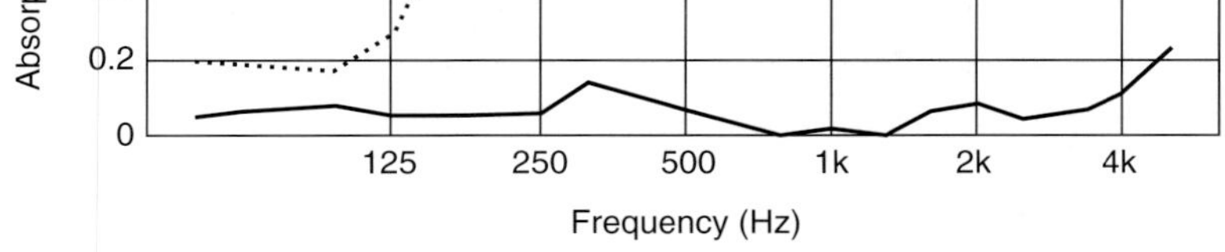

Fig. 4.11 Comparison of a conventional and a sound-absorbing sandwich panel. —— 80-mm sandwich panel with solid metal skins; - - - - - - 50-mm mineral wool sandwich panel in which one face has 23% perforations.

from the source to the receiver without any reflections at any boundary surfaces. For a direct sound field, the sound pressure level is therefore not influenced by any treatment of the surfaces in the room and decreases by 6 dB every time the distance from the source is doubled. Outside this field we have a reverberant sound field where the sound level is constant and depends on the sound absorption in the room.

The radius within which the direct sound field is significant and can be calculated approximately:

$$r \approx 0.2\sqrt{A} \tag{4.9}$$

where A = total sound absorption area in the room (m^2).

The general sound reduction in the reverberant sound field is determined by:

$$L_{dif} = 10\log A_{after}/A_{before} = 10\log T_{R.before}/T_{R.after} \tag{4.10}$$

4.11.1 Example 3

If we now return to Example 1 in Section 4.4.1, and include a sound-absorbing ceiling (e.g. mineral wool) with an absorption coefficient of 0.9 in the actual frequency band, in the construction, the following is obtained:

Using Equation 4.5:

$$A = \alpha S$$

$$A_{ceiling} = 0.9 \times (5 \times 7) = 31.5\,m^2$$

$$A_{total} = A_{before} + A_{ceiling} = 48.5\,m^2$$

The sound level inside the office can now be calculated using Equation 4.10:

$$L_R = 85 - 40 - 10\log 48.5/18 = 41\,dB$$

The sound level inside the office will therefore decrease by 4.5 dB.

4.12 NOISE REDUCTION IN LARGE INDUSTRIAL PREMISES

In industrial halls with a volume exceeding approximately 1000 m^3, the height is normally much less than both the length and the width of the hall. In this case, the height and the furnishing density have a considerable influence on the sound field. In such a hall, the sound field is generally not diffuse and it is therefore not useful to calculate using the Sabine formula. The method presented here is an empirical engineering method.

4.12.1 Reverberation time

The reverberation time in a large hall is influenced by the distance to the roof, the absorption coefficient of the roof and the scattering effect of the sound field from the furniture. If the hall is narrow, the long side of the hall will also have an influence. The following empirical equation can be used to calculate the reverberation time:

$$T_R = 0.15\,h - 1.8\,\alpha + 1.8 + KT \tag{4.11}$$

where h = height of the hall
α = absorption coefficient at 1000 Hz for the roof ($0 \leq \alpha \leq 1$)
KT = room constant (see below)

4.12.2 Room classification

The determination of the room constant KT is carried out in three steps as follows.

(1) *Classify the size of the hall* The width of the room compared with its height has an influence on the room acoustic and sound propagation.
- If width/height ≤ 4, the hall is classified as N (Narrow).
- If width/height > 6, the hall is classified as B (Broad).

It is necessary to interpolate between these values.

(2) *Classify the furniture* In this context, the furniture includes machines, materials and other fixtures. The furnishing density and the height of the furniture compared with the distance to the roof have an influence on the sound propagation. In Fig. 4.12, the furnishing class is found at the crossing point of the relevant lines.

(3) *Determine KT* It is now possible to determine the classification of the room:
- Narrow High (NH) Broad High (BH)
- Narrow Medium (NM) Broad Medium (BM)
- Narrow Low (NL) Broad Low (BL)

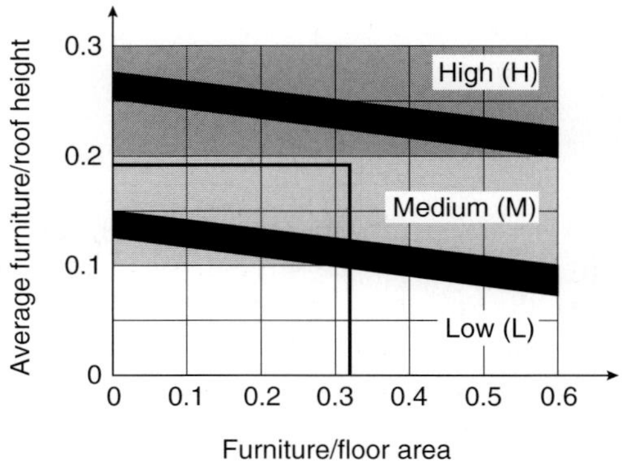

Fig. 4.12 Chart for the determination of furnishing classification.

Table 4.4 can be used to estimate the constant KT. If the input values do not conform to the parameters which are given in the table, it is necessary to interpolate.

Table 4.4 Values of the room classification constant *KT*

	$\alpha = 0.1$			$\alpha = 1.0$		
	h (metres)			*h* (metres)		
	5	10	15	5	10	15
BH	−0.5	−0.3	0	0	0	0
BM	0	0	0	0	0	0
BL	+0.5	+0.3	0	0	0	0
NH	−0.5	−0.3	0	0	+0.5	+1.0
NM	0	0	0	0	0	0
NL	0	+0.5	+1.0	0	+0.3	+0.05

4.12.3 Sound propagation

In a large hall, there is no real direct sound field or reverberant sound field and the sound level decreases continuously with distance. The parameters that influence the decrease are the sound absorption of the roof and the scattering effect of the furniture. In narrow halls, the long side walls also have an influence. An empirical formula to calculate the decrease in sound level is given as:

$$\Delta L = a\alpha + b \tag{4.12}$$

where ΔL = decrease in sound level/distance doubling (dBA)
α = absorption coefficient of the roof
a, b = room constants according to Table 4.5.

Table 4.5 Values of room constants *a* and *b* for sound propagation

	a	*b*
BH	3.0	4.0
BM	2.5	3.75
BL	2.0	3.5
NH	3.0	3.0
NM	2.75	2.75
NL	2.5	2.5

4.12.4 General sound level reduction

At each point in an industrial hall, the sound level is the sum of the contributions from each noise source (e.g. machines) at that point. If the sound absorption is increased, the incremental reduction ΔL will increase and the sound level at each point in the hall will decrease. The reduction of the general sound level is the average value at all points, except for those very close to a sound source.

The general sound reduction depends on the sound absorption properties of the ceiling, the screening effect from machines and equipment and the distance between the noise sources.

$$L_{\text{dif}} = C\Delta - 1 \tag{4.13}$$

where L_{dif} = general sound level reduction (dBA)
$\Delta = L_{after} - L_{before}$ (dBA, see Equation 4.12)
C = room constant, determined as follows:

Distance between noise sources		C
Widely spaced	approximately 10 m	5
Medium	approximately 5 m	4
Close	approximately 2.5 m	3

4.12.5 Example 4

An industrial hall has a roof of profiled steel sheet, walls of sandwich panels and a concrete floor. The noise level in the hall will be controlled by an absorbing ceiling.

(1) *Dimensions of the hall*
Length = 75 m, Width = 30 m, Height = 8 m
Calculated width/height = 30/8 ≈ 3.75
The hall is narrow, N.

(2) *Furniture*
The average height is about 1.5 m and the furniture covers about 30% of the floor area.
Furniture height/height of the hall = 1.5/8 ≈ 0.2
As shown in Fig. 4.12, the furnishing classification is found at the crossing point between 0.3 on the x-axis and 0.2 on the y-axis. This gives the furnishing density as medium, M.

(3) *Type of hall*
The hall is narrow medium, NM.

Calculations before treatment of ceiling
The absorption coefficient $\alpha = 0.1$ for a sheet metal roof.
Table 4.4 gives $KT = 0$.
Equation 4.11 gives the reverberation time:

$$T_R = 0.15 \times 8 - 1.8 \times 0.1 + 1.8 + 0 = 2.8\,\text{s at } 1000\,\text{Hz}$$

For a hall of class NM, the sound level reduction/distance doubling is given by Table 4.3 and Equation 4.12:

$$L = (2.75 \times 0.1) + 2.75 = 3\,\text{dBA}$$

Calculations after treatment
The absorption coefficient for the ceiling with 50-mm mineral wool slab on the underside is $\alpha = 0.9$:

$$T_R = 0.15 \times 8 - 1.8 \times 0.9 + 1.8 + 0 = 1.4\,\text{s at } 1000\,\text{Hz}$$

The sound level reduction/distance doubling is given by Equation 4.12:

$$L = (2.75 \times 0.9) + 2.75 = 5.2\,\text{dBA}$$

General sound level reduction
The distance between the noise sources is approximately 5 metres centre to centre.

$$L_{dif} = 4 \times (5.2 - 3) - 1 \approx 7 \text{ to } 8\,\text{dBA}$$

4.13 REFERENCES

4.1 Building Research Establishment and Construction Industry Research and Information Association (1993) *Sound Control for Homes*, BR238/CIRIA Report 127.

4.2 Friberg, R. (1975) Noise reduction in industrial halls obtained by acoustical treatment of ceilings and walls. *Noise Control and Vibration Reduction*, March.

Chapter 5
Fire

5.1 GENERAL ASPECTS OF FIRE BEHAVIOUR

5.1.1 Introduction

This book has demonstrated that structural sandwich panels are being increasingly used as the external wall and roof cladding of buildings because they are light-weight, energy efficient and can be easily handled and rapidly erected. They are also widely used as ceiling panels and internal partitions and are especially suitable for forming lightweight enclosures with enhanced thermal insulation and hygiene properties such as are required in rooms for food processing. They are used in many types of building ranging from large factories and warehouses to domestic dwellings. Their high thermal insulation makes them especially suited to extreme climates. There are, of course, a number of alternative core materials available for use in sandwich panel construction and their differences are also evident when their characteristics in fire situations are considered.

This chapter concentrates on the fire aspects of sandwich panel construction though it should be noted that these do not always dominate the design. There are generally a number of potentially conflicting design requirements and often a degree of compromise is necessary. Thus, in many applications, it is the thermal insulation requirement which is crucial; in others it may be structural capacity or self-weight. It is when fire requirements are a dominant design factor that this chapter becomes crucial.

The purpose of this chapter is to show how the component parts of a panel (e.g. facings, core and adhesive) together with the engineering design of the panel assembly all influence the performance when exposed to fire and that, with careful design, sandwich panels can meet stringent fire requirements. When properly designed and used, fire-rated sandwich panels may offer the best design solution when they are used in a fire-separating context as a fire wall.

This chapter provides information on the basics of fire, the standard fire tests covering reaction to fire and fire resistance that have been developed within CEN and ISO, examples of test results, the essential design considerations and some engineering solutions.

Nowadays, there is a trend towards basing the assessment of the fire risk in a building on the predicted performance of the building as a whole, rather than on the performance of individual components such as sandwich panels. An important aspect of this assessment is the provision of *active* fire protection measures such as smoke detectors acting together with sprinklers. These can have a profound influence on the outcome of a fire incident. Space precludes any discussion of this important aspect of fire design so that, here, the discussion is restricted to considerations of the sandwich panels alone, together with their supporting members. However, *passive* fire resistance, for example in the form of fire-resistant partitions, may be required by the insurance companies.

Throughout the text, reference is made to stone wool as one of the commonly-used core materials. Stone wool (often called rock wool) together with slag wool and glass wool form a family called mineral wool. Only stone wool with an appropriate high density and fibre orientation leads to high periods of fire resistance when used as the core material in sandwich panels.

Fire design is a large subject, and structural designers are not always familiar with all aspects of it. This chapter therefore includes extensive references which are keyed into the text in the usual way. In addition there are some more general references which are suggested for wider reading on the subject.

5.1.2 Objectives of fire safety design

The decision to use sandwich panels in a building may be influenced by the global fire safety design objectives which include life safety, loss prevention and environmental protection. These functional objectives are, or ought to be, decided jointly by the building designer, the building owner and the building approval authorities. These objectives have been defined in Reference 5.1 in the context of places of public assembly. Here, they may be defined in more detail in the context of sandwich panels as follows.

5.1.2.1 Life safety

The occupants of a building, fire-fighters and members of the public who are in the vicinity of a building can, potentially, be put at risk by fire. The main life-safety objectives have been set down in the *Essential Requirements* of the European Union Construction Products Directive[5.2] *Safety in case of fire*. This requires that the construction works must be designed and built in such a way that, in the event of an outbreak of fire:

- the load-bearing capacity of the construction can be assumed for a specific period of time
- the generation and spread of fire and smoke within the works are limited
- the spread of fire to the neighbouring construction works is limited
- occupants can leave the works or be rescued by other means
- the safety of rescue teams is taken into consideration

Life tenability criteria are given in Section 5.1.4.

In order that the nature and scale of the problem may be appreciated, it should be borne in mind that:

- The majority of fire deaths occur in dwellings so that other occupancies such as places of public assembly, offices and industrial buildings are relatively safe.
- The majority of fire deaths arise as a consequence of inhalation of smoke. There is some mortality from exposure to flames but deaths from collapse of the structure are rare.

5.1.2.2 Environmental protection

A conflagration involving several buildings or the release of quantities of hazardous materials may have an environmental impact that is out of proportion to the size of the original fire. Consideration should therefore be given to the limitation of:

- the effects of fire on adjacent buildings or facilities
- the release of hazardous materials into the environment. (This could involve the release of gaseous products of combustion and fibres into the atmosphere or pollution of the groundwater due to fire-fighting operations.)

5.1.2.3 Loss prevention

The implications of a fire on the continuing viability of a business can be substantial and consideration should be given to the limitation of damage to:

- the structure and fabric of the building
- building contents
- ongoing business viability
- public image

In Western Europe, about 0.2% of the gross national product is accounted for by fire losses and, furthermore, the indirect and consequential losses (e.g. loss of records, interruption of business, widespread smoke damage) can be many times the direct fire losses. Many firms go out of business following quite modest fires.

The requirements of an insurer are, therefore, often more severe than those of the life safety regulator. This is because the insurer is concerned to reduce damage to the building and contents not only during the period in which occupants are escaping (the prime interest of the life safety regulator) but also afterwards, when more fire damage may occur. Thus it is possible that the insurer will want higher periods of fire resistance and require more reliance on the use of materials of low combustibility or fire-rated products than the life safety regulator.

This aspect of fire design is being driven by organisations such as the Loss Prevention Council in the UK and Factory Mutual (USA and worldwide) who both have their own codes of practice and test requirements for sandwich panels.

5.1.3 Phases of a fire

There are three phases in the development of a fire: the growth phase, the fully developed phase and the decay phase, as shown in Fig. 5.1.

The first stage of fire development includes pre-ignition (smouldering) and the stage of fire growth where the heat production is still small. It finishes when the volatile products of combustion ignite. In the growth phase of a fire, properties such as ignitability, combustibility, rate of heat release, flame spread, smoke and toxic gas generation are important and are quantified in small-scale standardised fire tests. Some of these properties may also be quantified in larger-scale tests, for example, the rate of heat release is measured in the ISO 9705[5.3] room test (surface

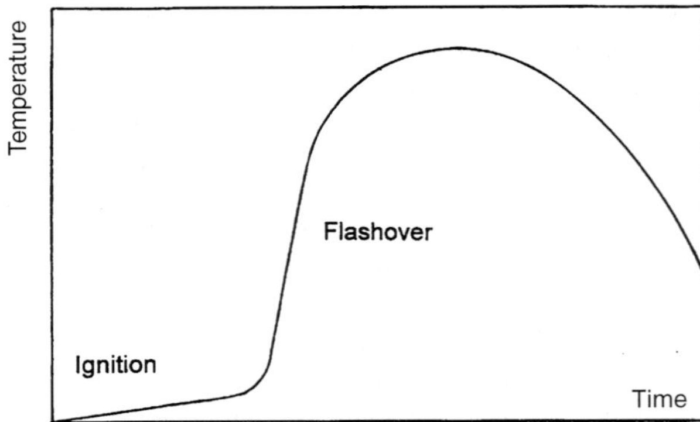

Fire growth	Fully-developed fire	Decay
Reaction to fire	Resistance to fire Non-combustibility	
Ignitability Combustibility Spread of flame Rate of heat release	Fire penetration (containment) Heat release Smoke generation (toxicity, opacity, corrosivity)	

Fig. 5.1 Development of a fire.

products and lining materials) and ISO 13784 (sandwich panels). These properties and tests are described under the heading *Reaction to fire*, in Section 5.2.2. Regardless of the test regime specified, stone wool cored panels have good reaction-to-fire properties but plastic foam-filled sandwich panels generally demonstrate relatively poor reaction-to-fire properties when compared with many other building materials. Nevertheless, with careful design and use, such panels can be safely specified in appropriate cases.

The second phase of fire development – the fully-developed phase – begins with *flashover*, at which point all of the fuel sources in the immediate vicinity suddenly become involved in the fire. This leads to an immediate increase in the combustion gas temperatures to well above 650°C and a sudden increase in the radiation emitted by the fire. Temperatures in the fully-developed phase can exceed 1000^{0}C. It is here that the fire resistance and non-combustibility of the panels are important, properties which are covered under the heading of *Resistance to fire*. It is the flash-over stage and the fully-developed fire which standard time-temperature curves such as the ISO 834[5.4] curve attempt to describe. However, real fires vary greatly and the maximum temperature attained and duration are principally dependent on the supply of oxygen, the amount and porosity of the fuel present, and the thermal insulation of the enclosure boundaries. It follows that standard time–temperature curves are arbitrary and may bear little resemblance to an actual fire, but they are essential in order to enable comparisons of product performance. Natural fully-developed fire scenarios can be calculated and a method which assumes that the actual fire load can be characterised as an equivalent timber fire load is given in ENV 1991-2-2:1994 *Actions on Structures Exposed to Fire*[5.5].

The final decay phase is characterised by a lowering of the combustion gas temperature while the remaining small amount of fire load is consumed. This phase is not normally punishing to the structure.

When attempting to assess the behaviour of sandwich panels in fire, it is helpful to understand the way in which fire develops. When a fire occurs in an enclosed space, hot smoky gases rise to form a layer which, in the absence of venting or fire suppression, at first flows under the entire ceiling or roof and then deepens to fill the whole space. These hot gases radiate heat downwards, raising the temperature below. The fire tends to grow in area and the flames spread to nearby combustibles. The flames increase in height until they reach the ceiling (unless the floor to ceiling height is large) where they are deflected horizontally and, radiating downwards, further accelerating the fire growth. If the ceiling is combustible, it may ignite and add to the volume of flame and speed of fire growth. If the space has insufficient openings to provide a continuing air supply, the burning rate of the fire will diminish as it draws on increasingly vitiated products of combustion, but the gases will then be extremely toxic. Otherwise, at some point in this worsening scenario, a critical temperature will be reached in the combustibles below the plume of smoke and flames and flashover will occur.

In the early stages of a fire the most important effects will be those of smoke and other combustion products. Often smoke will be the first evidence of fire that is detectable by the occupants and it is likely to be the first cause of alarm and possibly panic. When first present, smoke tends (in the absence of any strong air currents) to collect at ceiling level, filling the space from the top downwards. When it extends down to head height it will produce discomfort to the eyes and difficulty in breathing, both of which will interfere with the efforts of occupants to find their way towards the exits. People who are prevented from escaping by dense smoke, or who are delayed from escaping by it, may suffer from the toxic or irritant effects of the combustion products, the asphyxiant effect caused by lack of oxygen, by the temperature and humidity of gases inhaled, or by burns caused by the high level of radiation produced by the hot layer of gases overhead. The most common cause of death in such cases is inhalation of carbon monoxide.

In the scenario described above, the reaction-to-fire properties are important and it is desirable, wherever possible, to reduce the amount of combustibles involved in the fire in its early stage of development and so reduce the amount of smoke generated.

In metal-faced sandwich panels it is the combustible core, not the facings, which adds to the fire load and to the smoke generated. The facings can have a fire and heat protecting function which depends on their insulation and integrity performance.

Although the reaction-to-fire properties of unprotected polymeric foam core materials may not be good, with the exception of foamed polystyrene (which melts and shrinks away from the source of heat), it is possible to obtain useful periods of *fire resistance* by careful selection of materials and detailed design and installation of the whole panel system. The highest levels of fire resistance can be obtained using stone wool cores which have a very low combustibility.

5.1.4 Assessment of the threat to life

5.1.4.1 General

When organic materials such as wood, paper or plastics are burned, hot gases and smoke (termed combustion products) are evolved. Some of the combustion gases produced can prove to be fatal in a short time if they are inhaled in sufficient concentration.[5.6] Coroner examinations of fire victims generally reveal that the most common cause of death is the inhalation of toxic gases, notably carbon monoxide. In addition to its possibly toxic characteristics, the smoke in a burning building obscures vision, makes escape much more difficult[5.7] and hinders fire-fighting. These factors are therefore vital in assessing the dangers of incorporating flammable materials in either the contents or the construction of a building.

5.1.4.2 Toxicity and irritancy of combustion products

Combustion products may provide a life threat at various stages in a fire and at various locations in a building. A person in the room of fire origin may die from inhalation of combustion products when the fire is still in its growth phase. It is also possible for a person well away from the room of fire origin to die because the fire has reached the fully developed phase and combustion products are transported to other parts of the building. In both cases, however, the relatively small amount of combustion products produced by burning of a sandwich panel or other element of construction must be balanced against the amount (and duration) of the combustion products produced by the contents of the building.

When mixtures of toxic products are present, as they are in all fires, the effects of different products are approximately additive. The accumulated dose for each toxic or irritant gas can be expressed in terms of the exposure dose, which is the product of concentration and exposure time.[5.6]

Non-combustible materials yield very low levels of smoke and toxic products. Those materials that have been produced at high temperatures, e.g. stone wool and slagwool, are completely inert in their basic form but the agents used to bind the fibres together and to bond them to the panel faces can produce small amounts of smoke and toxic product.

The yield of toxic products from polyurethane foams depends upon the decomposition conditions.[5.8] Smouldering/non-flaming fire yields mainly isocyanates and carbon monoxide (CO). Well-ventilated flaming fires and fully developed fires produce CO and, at temperatures above 800°C, HCN in relatively low concentrations.

The main toxic component in fire effluent is CO. However, the combustion of combustible components can produce toxic gases which are lethal in small concentrations. Much research has been published in this subject area[5.9, 5.10, 5.11] but correlations with real fire situations have been poor.

5.1.4.3 Temperature/radiation intensity of combustion products

Hot gases can provide a threat to life in two ways. Damage to the respiratory tract may be caused by inhalation. Burns and intense pain may be caused by radiation from a hot gas layer.

The upper tolerance limit for air saturated with water vapour when breathed is

60°C. The design limit for radiant heat is 2.5 kW/m^2 since, above this intensity, skin pain and severe burns can occur within a few seconds. The maximum temperature of a layer of hot gases under a ceiling needed to produce this radiation intensity can be calculated from a knowledge of the configuration factor and the emissivity and temperature of the gases.

5.1.4.4 *Obscuration caused by combustion products*

It is not easy to calculate the obscuration caused by smoke but it is possible to give guidance on the maximum distance people are prepared to move through smoke.[5.12] For small rooms, the minimum visibility acceptable is 5 metres. For large rooms it is necessary for people to see much further in order to orient themselves and find exits and, for these reasons and to avoid toxicity problems, the minimum visibility should not be less than 10 metres.

5.1.4.5 *Unexpected structural behaviour*

The unexpected structural failure of sandwich panels can, in rare circumstances, pose a life threat. External fire on the facade of the building, caused for example by flames and hot gases jetting through a window opening, can cause an outer facing which is not fully secured to become detached and act as a missile to fire fighters below. In a similar way, delamination of a ceiling panel, which can occur if there is no mechanical attachment of the two facings, may allow the lower facing to fall on to firemen below or may block their escape route. With careful design, these structural failure modes can be avoided.

5.1.4.6 *The application of fire safety engineering*

Life safety requirements in building regulations normally assume that all of the occupants of a burning building should be able to escape unharmed. This is certainly so for deterministic design. In probabilistic design it may be acceptable, in the future, to design so that a given very small percentage of the occupants may die when a particularly rare fire scenario is examined. In order to decide if an occupant may die, it is possible to calculate, by fire safety engineering methods, at any point in time after ignition, (a) where people are in the building as they make their escape and (b) the nature of the fire effluent (e.g. temperature, opacity, toxicity/irritancy) and where it has reached in the building. These can then be compared to see if a life threat exists.

A fire safety engineering approach can provide a more fundamental and economic solution than more prescriptive approaches to fire safety. Indeed, it may be the only viable means of achieving a satisfactory standard of fire safety in some large and complex buildings. Documents on fire safety engineering are being, and have been, developed in a number of countries including Australia, Sweden, the United Kingdom and Japan and, at the time of writing (2000), ISO TC 92 Sub Committee 4 has prepared a series of eight Technical Reports (ISO/PRF TR 13387-1 to -8) which are intended to become international standards. They cover the total fire safety design process. For most buildings, however, the prescriptive recommendations presented in existing regulations and codes may be adequate.

If desired, the techniques of fire safety engineering can be used to assess the effect on life safety when using different sandwich panels, although it is accepted that this approach will rarely be appropriate. A fire safety engineering approach might also be used to assess if the life threat would become unacceptable when using combustible cored sandwich panels due to, for example, the fire load represented by the panels and the production of toxic and irritant gases, taking account of the potentially greater threat posed by the building contents. The approach enables the toxic threat of the products of combustion to be assessed in terms of times to reach (a) incapacitation, i.e. loss of consciousness and (b) death.

For some projects, the fire behaviour of sandwich panels with combustible cores may, on first sight, be a determining factor in their use. However, by undertaking a fire hazard assessment it may be possible to show that the sandwich panels can be used with safety. The fire hazard assessment should take account of the interaction of fire, the building(s) and people for a limited number of life threat and/or property loss scenarios.

5.1.5 Fire severity

5.1.5.1 Effect of fire load

It is sometimes claimed that sandwich panels containing foamed plastic cores add substantially to the fire load in a building leading to increased fire severity. It is therefore appropriate to consider briefly the factors affecting fire severity.

It has been shown from small-scale and full-scale compartment fire tests that the fire severity, expressed as an equivalent amount of fire resistance and sometimes referred to as equivalent time of fire exposure, t_f, in minutes, is given by the following empirical relation:

$$t_f = \frac{L}{\sqrt{A_w A_t}}$$

where:

L is the fire load (kg), i.e. the sum of the calorific energies which could be released by the complete combustion of all of the combustible materials in the enclosure including the walls, partitions, floors and ceilings. The fire load (or, alternatively, the fire load density which is the fire load divided by floor area) may be expressed in terms of heat (i.e. MJ) or in terms of a mass of timber having the same calorific energy (kg). The conversion from timber to MJ is such that 1 kg timber yields approximately 19 MJ.

A_w is the ventilation area (m^2), i.e. the area of an opening(s) through which air can enter to feed the fire and from which hot gases flow out. It is usually represented by a window (or windows) in which the glass has fallen out as a result of the fire, or another area of very low fire resistance which rapidly burns through.

A_t is the area of walls and ceiling of enclosure (m^2), excluding the area of the ventilation opening and the floor.

From the above equation, it can be seen that fire severity is directly proportional to fire load (or fire load density). It should be noted that a more detailed calculation method for the equivalent time of fire exposure (which takes account of the thermal properties of the enclosure surfaces) is given in Part 2.2 of Eurocode 1

(ENV 1991-2-2: 1994).[5.5] Information on the important effect of ventilation and the application of the Eurocode equation is given elsewhere.[5.13]

Figure 5.2 illustrates the important effect of ventilation and fire load on the combustion gas temperature–time profiles obtained in a small compartment (roughly 8 m × 4 m × 3 m high) with ventilation in one of the long walls. It can be seen that the fire severity increases with an increase of fire load and a decrease of ventilation. The temperature–time profiles for 30($\frac{1}{2}$) and 15($\frac{1}{4}$) are very similar – where the first number is the fire load density in kg/m^2 and the bracketed fraction is the proportion of the area of one long wall which is open as ventilation. Also shown in the figure is the ISO 834 temperature–time curve which is used as the basis for fire resistance testing.

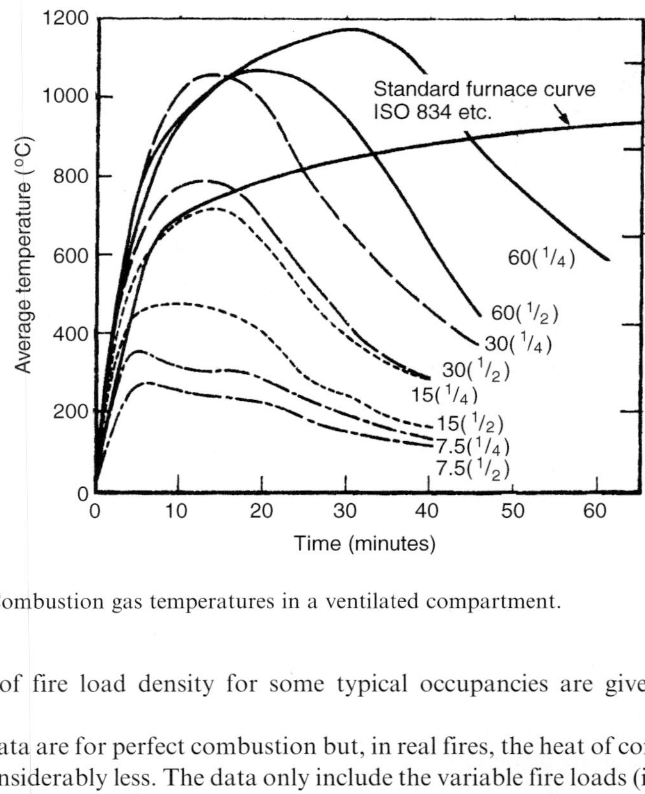

Fig. 5.2 Combustion gas temperatures in a ventilated compartment.

Values of fire load density for some typical occupancies are given in Table 5.1.[5.12]

These data are for perfect combustion but, in real fires, the heat of combustion is usually considerably less. The data only include the variable fire loads (i.e. building contents) and not the fire load in the building construction. The 80% fractile value is the value that is not exceeded in 80% of the rooms or occupancies.

The net calorific value (MJ/kg) of some combustible materials determined according to EN ISO 1716 are given in Table 5.2.

5.1.5.2 *Worked example*

With the above data, it is possible to calculate the fire load provided by sandwich panels and to compare this with the fire load in the building contents.

Assume, as an example, an industrial building the contents of which have an average fire load density of 1000 MJ/m^2. The building is 40 m square and 5 m high

Table 5.1 Fire load densities for a range of building occupancies

Occupancy	Fire load density (MJ/m^2 of floor area)	
	Average	80% fractile
Dwelling	780	870
Hospital	230	350
Hospital storage	2000	3000
Hotel bedroom	310	400
Offices	420	570
Shops	600	900
Manufacturing	300	470
Manufacturing and storage	1180	1800
Libraries	1500	2250
Schools	285	360

Table 5.2 Calorific values of some combustible materials

Material	Calorific value (MJ/kg)
Stone wool	0.8–1.8[a]
Wood	19
Rubber	32
Phenolformaldehyde	29
Polyester	31
Polystyrene	40
Polyurethane foam	26
Polyisocyanurate foam	24
Ureaformaldehyde foam	14
[a] Depending on the amount of binder	

and sandwich panels are used as the external walls and the flat roof. The thickness of the core material is 50 mm.

$$\begin{aligned} \text{area of walls} &= 4(40 \times 5) &&= 800\,\text{m}^2 \\ \text{area of roof} &= 40 \times 40 &&= \underline{1600\,\text{m}^2} \\ \text{total} &= &&= 2400\,\text{m}^2 \end{aligned}$$

$$\text{volume of core material} = 2400 \times 50/1000 = 120\,\text{m}^3$$

fire load of core per unit volume (MJ/m^3) = density of core (kg/m^3) × calorific value (MJ/kg)

Typical densities for the plastic foam core materials used in sandwich panels are given in Table 5.3.

Table 5.3 Typical densities of some foam plastic cores

Core material	Density (kg/m^3)
Polyurethane foam (PUR)	45
Expanded polystyrene (EPS)	20
Extruded polystyrene (XPS)	35

Assume that the core material is polyurethane foam:

fire load of core per unit volume $= 45 \times 26 = 1170\,\text{MJ/m}^3$

Total fire load of core $=$ fire load of core per unit volume $\times$ volume
$= 1170 \times 120 = 140\,400\,\text{MJ}$

fire load of the contents $=$ fire load density $\times$ floor area
$= 1000 \times 40^2$
$= 1\,600\,000\,\text{MJ}$

$$\frac{\text{fire load of panels}}{\text{fire load of contents}} = \frac{140\,400}{1\,600\,000} = 0.08$$

If the panels had cores of extruded polystyrene instead of polyurethane foam of the same thickness, the ratio of fire load of panels to contents would be 0.095.

The panel fire load data can also be expressed as MJ/square metre of floor area. Table 5.4 gives the data in this form for the building used in the worked example (i.e. 4 m × 40 m × 5 m high) and for a building twice as high.

Table 5.4 Fire load of panels

Core material	Fire load of panels (MJ/m^2 floor area) for buildings of dimensions:	
	40 m × 40 m × 5 m	40 m × 40 m × 10 m
50-mm PUR	87.5	175
80-mm stone wool[a]	20	40

[a] Includes adhesive for bonding the core to the facings. 80-mm stone wool may be expected to have approximately the same thermal insulation performance as 50-mm PUR.

5.1.5.3 *Conclusions from the worked example*

The following conclusions may be reached from the above study.

(1) The calculation of the additional fire load represented by the panels is simple to do and should normally be done.

(2) Normal thicknesses of foamed plastic core material (e.g. 50-mm thickness) may add about 10% to the fire load of a building in the high fire load density category (e.g. manufacturing and storage buildings). In buildings in a low fire load density category, such as dwellings, offices and schools, such thicknesses of foamed plastic core material may add significantly more to the fire load of a building, although it should be noted that, unless of a temporary nature, these buildings would rarely use foamed plastic cored panels.

(3) In buildings having sandwich panels of enhanced thermal insulation, such as

in cold stores, chill rooms or in the food processing industry, the ratio of fire load of panels to fire load of contents can rise to 30% which means that the fire load of the panels becomes more important.

(4) The fire load of stone wool cored panels is negligible and consists only of the adhesive used to bond the core to the facings and the small amount of agent bonding the fibres together – see Table 5.2.

(5) Although plastic foam cored panels undoubtedly add to the fire load, other important questions which need to be answered are: At what stage in the fire do the sandwich panels become involved and is this a slow or rapid development? Can fire spread within the core of the panel unseen? Does the core material melt and produce flaming droplets which can initiate secondary fires? Are significant amounts of toxic products of combustion released such as to provide a threat to the life of occupants and fire fighters? Is the additional smoke obscuration acceptable? The importance of these questions is discussed later in this chapter.

(6) In buildings or areas of buildings of high fire load, it is good practice to subdivide enclosures with fire walls of high fire resistance as this reduces the potential fire loss and fire severity. Stone wool cored panels may be used for this purpose.

5.1.6 Introduction to reaction to fire

In Section 5.1.3 *Phases of a fire*, it was made clear that the reaction-to-fire properties of a material affect the growth stage of a fire up to the point of flashover. Reaction-to-fire properties include ignitability, non-combustibility, rate of heat release, flame spread, smoke generation and the toxicity/irritancy of combustion products.

- *Ignitability* Behaviour in an ignitability test is important as it determines the difficulty with which a material can be ignited from a small source of heat or flame.
- *Non-combustibility* A non-combustible material has a major advantage over others because it will be impossible to ignite, will not contribute heat to the fire and will produce very little smoke and toxic/irritant products of combustion.
- *Rate of heat release* Some authorities consider the rate of heat release to be the single most important parameter in assessing the fire performance of a material.[5.14] A material having a low rate of heat release is most desirable but it should be noted that the rate of heat release of a material depends upon the incident radiation intensity so that the higher the incident radiation intensity the greater the rate of heat release. It is important that the contents of a building such as furniture and fire-exposed combustible wall and ceiling lining materials should possess low heat release rates, because it is mainly the rate of heat release that determines the amount of fire effluent produced and the rate at which the effluent is transported throughout the building. Information regarding the heat release in fires is given elsewhere.[5.15]
- *Flame spread* The speed with which flame spreads across the surface of a material should ideally be low. Flame spread is affected by the rate at which

combustible volatiles are given off as a result of the feedback of radiation from the flame front. Flame spread is faster upwards than horizontally and is greatly affected by the intensity of the incident radiation. The rate and distance of flame spread in a thin material, such as a paint coating on a substrate, depends on the thermal properties of the substrate. A substrate which readily conducts heat away from the burning surface layer will slow down the rate of flame spread. Such substrate materials have a low thermal insulation (k), high density (ρ), and high heat capacity (c) as illustrated in Fig. 5.3. It follows that a product such as a plastic coated thin steel sheet will achieve a higher rate of flame spread with a substrate of mineral wool or plastic foam than with a substrate of, for instance, a high density fire protecting material such as a gypsum plaster.

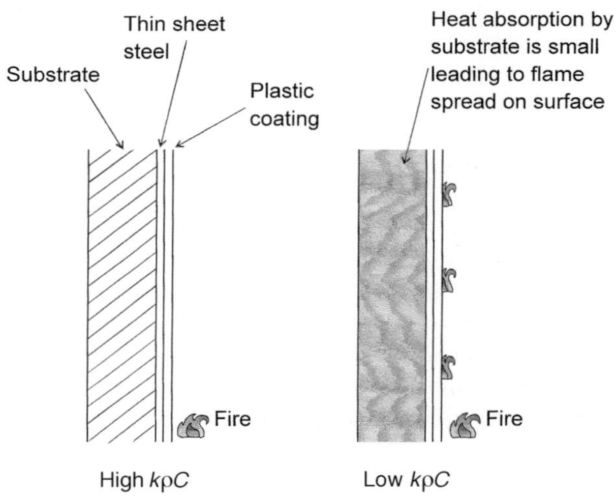

Fig. 5.3 Influence of substrate on flame spread.

Reduced rates of flame spread can be achieved with the use of flame retardant additives or intumescent coatings. Rates of flame spread for all products will be increased as the incident radiation increases.

- *Melting and shrinkage* Some materials, foamed polystyrene for instance, shrink away from the heat source and form molten droplets which flame on contact with the fire. Shrinkage of foamed polystyrene in vertical panels may be acceptable provided that molten burning droplets cannot escape from the bottom of the panels to reach and propagate fire in the construction below the enclosure which is on fire or to spread the fire to the unexposed side of the fire at that floor level. However, the generation of molten burning droplets in horizontal (e.g. ceiling and roof) panels can be problematic: a hot layer of combustion gases under such a panel assembly can cause flaming droplets to fall through joints and cause secondary ignitions away from the fire source leading to rapid fire spread.
- *Corrosivity* The combustion of many of the polymeric materials used in the contents and construction of buildings produce gases such as hydrogen

chloride which corrode metals. These gases may be adsorbed on the surfaces of smoke particles, which are then transported with moisture from the fire to areas remote from the fire where they cause corrosion by hydrochloric acid deposition.[5.16] PVC coatings and some fire retarded polystyrene foams have the potential to cause corrosion in this way. However, the principal source of PVC in buildings is usually electrical cable insulation.

5.1.7 Introduction to fire resistance

In Section 5.1.3 *Phases of a fire*, it was stated that the property of fire resistance becomes important after flashover occurs. Fire resistance is the ability of an element of construction to resist collapse, R, to resist the penetration of flames and hot gases while, at the same time, maintaining structural integrity, E, and to keep the unexposed face sufficiently cool so as not to ignite materials in contact with it, I, all as shown in Fig. 5.4. The determination of the fire resistance involves the exposure of the element to a severe standard fire and the resistance time determined in this way is an important property of fire-separating elements such as walls, roofs and ceilings. Fire-resisting elements provide a barrier to the spread of fire and the amount of fire resistance needed may be as little as 30 minutes giving, in most cases, adequate time for people to escape. These elements are used to prevent fire spread from one compartment to another or from one building to another building for reasons of life safety and reduction of property loss. In other circumstances a high level of fire resistance is needed in order to provide an adequate margin of safety, as in power plant, stores with high fire load, and high multi-storey blocks of flats.

The fire resistance of a building component is measured by testing in a furnace. The designs of furnaces differ throughout the world but the temperature–time curve of the combustion gas is standardised and most countries use the ISO 834 curve. The equation of the (cellulosic) time–temperature curve which is used in the ISO and other related standards is:

$$T - T_o = 345 \log_{10} (8t + 1)$$

where t = time from the start of the test (minutes)
T = furnace temperature at time t (°C)
T_o = initial (ambient) furnace temperature (°C).

A typical hydrocarbon time–temperature curve, which has a much steeper initial rise of temperature is the UK DoE curve which has the equation

$$T = 1100(1 - 0.325e^{-0.1667t} - 0.204e^{-1.4167t} - 0.471e^{-15.833t})$$

where T = increase in furnace temperature at time t (°C).

Sandwich panels may be tested as non-load-bearing walls (internal or external), roofs, or ceilings. The specimen size is typically 3 m square for walls and 4 m long by 3 m wide for roofs and ceilings but larger specimens can be tested in some furnaces.

Most countries have their own fire resistance test specifications which may differ in terms of the criteria for integrity and insulation. Furthermore, the amount of fire

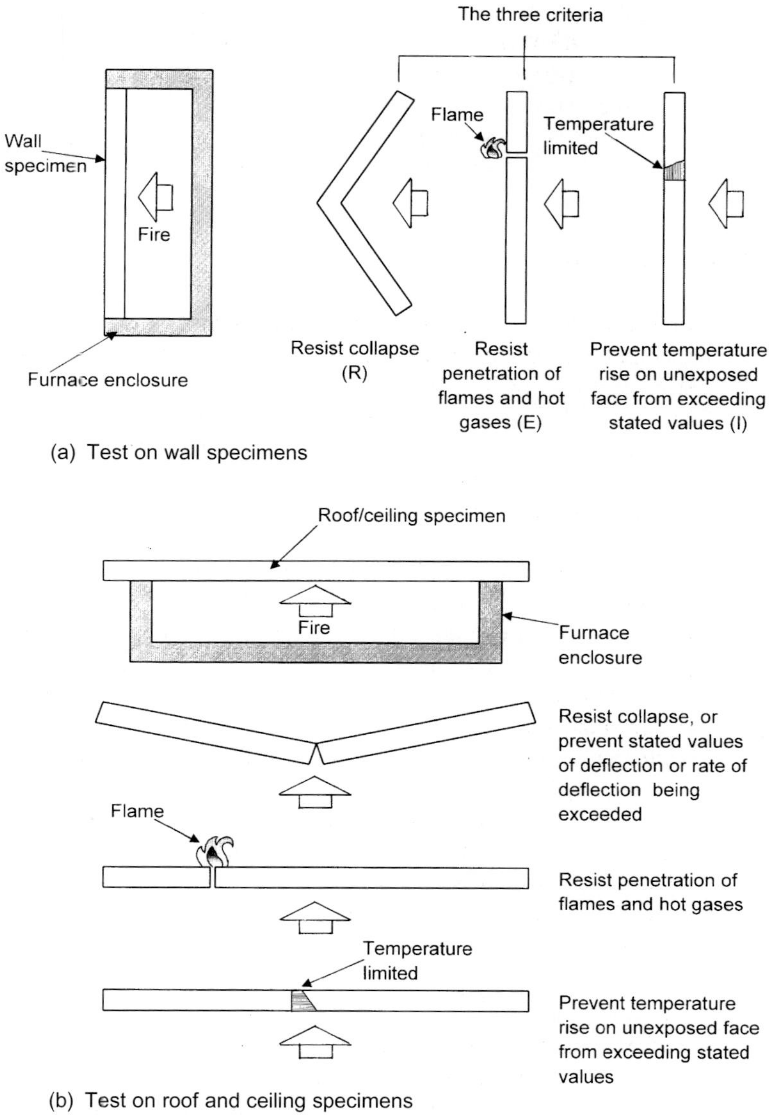

Fig. 5.4 The concept of fire resistance.

resistance specified for life safety and property protection may be different because the objectives are different.

CEN Technical Committee 127 (CEN/TC 127) is preparing a number of fire resistance test standards[5.18] for a range of construction elements in which common criteria for stability (or load-bearing capacity), integrity and insulation criteria will be used at least by EU and EFTA countries.

A system for expressing the results of fire resistance tests has been introduced in the Interpretative Document *Safety in case of fire* which acts as a bridge between the Essential Requirements expressed in the EU Construction Products Directive[5.2] and CEN standards. It is intended that the system will be used by EU and EFTA countries when the relevant CEN fire resistance standards are brought into use. This system uses the symbols:

R for load-bearing capacity
E for integrity
I for insulation

These symbols are suffixed by the recorded performance time (minutes) in one of the following values: 15, 20, 30, 45, 60, 90, 120, 180, 240, 360.

The symbols are used in the following way.

(1) For load-bearing elements:
- REI (time) minimum time during which all criteria (load-bearing capacity, integrity and insulation) are satisfied.
- RE (time) minimum time during which the two criteria, load-bearing capacity and integrity, are satisfied
- R (time) minimum time during which the criteria of load-bearing capacity is satisfied.

(2) For non-load-bearing elements (i.e. the majority of sandwich panels):
- EI (time) minimum time during which the two criteria of integrity and insulation are satisfied
- E (time) minimum time during which the criterion of integrity is satisfied

Thus the following relevant classes may be defined:

REI 15, REI 30, REI 45, ... RE 15, RE 30, ... R 15, R30, ...

so that a building element with a load-bearing capacity of 155 minutes, an integrity of 80 minutes and a thermal insulation of 42 minutes is classified as R 120/RE 60/REI 30, or a building element with a load-bearing capacity of 70 minutes and an integrity of 35 minutes is classified as R 60/RE 30.

The classification can be expanded by:

W when the insulation is controlled on the basis of the radiation emitted rather than by the temperature of the unexposed surface
M when particular mechanical actions are considered (for example impact)
C for doors equipped with a self-closing device
S for elements with a particular limitation of smoke leakage

For unsymmetrical fire-separating elements, the fire resistance classification is based on fire exposure from the side assessed as giving the lowest fire resistance unless the direction of the fire exposure is known.

It must be recognised that a sandwich panel exposed to fire quickly loses its flexural strength and changes into a membrane structure. However, provided that the facings are effectively connected to the supporting structure by, for example, suspending vertical facings from above or by providing a horizontal restraint in the case of a ceiling panel, the resulting air gap may be beneficial and useful periods of fire resistance can be achieved. Fire resistances with respect to insulation of more than 120 minutes are possible for stone wool, but because of the rapid deterioration of the insulation properties, most plastic foam cored panels can only attain small periods of fire resistance, typically up to 15 minutes for a polyurethane foam cored panel without any additional fire protecting layer.

5.1.8 Fire safety legislation

The general absence of legislative control over the contents of buildings reflects the difficulty of enforcing any real control over levels of fire safety. The changing use of buildings introduces different kinds, amounts and distributions of fire load which are difficult to monitor and control. However, where the contents are fixed or where they are especially hazardous, they may be controlled, as in the case of a building used for the manufacture of explosives or the storage of large amounts of flammable liquid.

There is also difficulty in legislating for the control of smoke produced by building materials owing to the difficulty of selecting a suitable test method and the difficulty in stating acceptable levels of smoke production. Within the member states of Europe, only Germany, the Netherlands and the Nordic countries have requirements.

Legislation does, however, control the structure and fabric of a building. With fire safety in mind, the legislation, which may be functional or prescriptive, affects those parts of a building which influence internal fire spread (notably the linings of buildings and fire resisting walls and floors, etc.) and external fire spread which is influenced by the fire resistance of external walls. Each country has its own fire legislation. Differences between countries may be due to:

- different traditional building materials
- different climate
- cultural differences
- different ideas as to what fire scenarios present a life risk
- different experiences of fire, especially national fire disasters involving multiple fatalities, and
- different methods of test for assessing the fire performance of building materials and components

The strategy in Western Europe, which is to create a common market with no technical barriers to trade, recognises that harmonisation of the fire regulations in the member states is so difficult that it is not on the agenda at present. Present work (2000) aims at the harmonisation of design codes for structural materials (e.g. the Structural Eurocodes) and fire test methods, the latter within CEN Technical Committee TC127 (Fire Safety in Buildings) which is engaged in the drafting of a number of parts of standards for the fire resistance and reaction to fire of the full range of building elements and components. CEN Technical Committee TC128 (Roof covering products for discontinuous laying and products for wall cladding) is drafting a product standard for sandwich panels which will reference the work of TC127 as well as other relevant 'horizontal' standards.

In the absence of harmonised fire regulations, the sandwich panel manufacturer or specifier has to identify, for the country of interest, the relevant fire regulations, the relevant fire test methods and the classification system.

5.1.9 Large-scale fire tests and experience of actual fires

The fire performance of a sandwich panel when installed in a building cannot be predicted from the results of standard tests such as the ISO 834 fire resistance test and small or medium scale reaction to fire tests. Apart from the inability to model the real fire scenario, this is because the size of the test specimen cannot accommodate the actual panel size and the range of horizontal and vertical joints between panels which are present in actual building construction. Because the combustible nature of some sandwich panels has sometimes raised questions by the building owner, the life safety regulator or the insurer, it has been the practice for many years for product manufacturers to sponsor ad hoc fire tests to demonstrate that their products are acceptable in terms of, for example, fire resistance, fire spread within the core, smoke generation, heat release and flame spread.

Ad hoc tests of this nature have been performed on panels with polyurethane cores since 1969.[5.19–5.22] Tests on polystyrene-cored panels have only appeared more recently.[5.23–5.25] When assessing the results of such tests, care should be taken to check that the fire is of sufficient intensity to challenge the assembly fully. Sometimes a timber crib is used which exposes the assembly to a severe local fire but is too small to generate the radiation levels necessary to investigate potential flashover. Furthermore, such tests may reveal good performance aspects such as the good resistance to fire spread within a polyurethane foam core behind steel facings while saying very little about smoke generation and toxicity which may also represent a problem.

Both the Loss Prevention Council (LPC) and Factory Mutual (FM) have developed their own large-scale tests and quality systems on behalf of the insurance industry and these may prove to be more stringent than national requirements. The large-scale corner wall test operated by FM is particularly interesting because of its size. A satisfactory performance in this test is an important yardstick for insurance purposes. The corner wall test is particularly appropriate to sandwich panels and represents a naturally developing fire at a critical point in the building. The FM test assembly measures 15.2 m × 11.6 m × 7.6 m high and may be regarded as more in the nature of an industry standard than an ad hoc test. Using this test, FM have recently carried out a systematic evaluation of some polystyrene-cored panels,[5.26] particularly in the context of protection by sprinkler systems, and concluded that there may be particular problems with this material.

If medium-scale fire tests are required, the ISO 9705 room test[5.3] offers a standardised arrangement and procedure which has the potential to generate flashover conditions and in which measurements are made of the rate of heat release, smoke generation and toxicity.[5.27] This test has been accepted by CEN as the 'Reference Scenario' for the Single Burning Item (SBI) reaction to fire test. However, it has been criticised by parts of the sandwich panel industry because the room is too small to be representative of many practical applications and because it does not allow panels to be mounted and secured as in practice. Furthermore, it may not be suitable for thick panels. If, for example, the panel thickness is 250 mm, the volume of the room is reduced by 25% with a consequential severe influence on the performance and classification. Recent experience has shown that, in such room tests, it is possible that some PUR panels may reach flashover.

The German panel manufacturers have reported over thirty examples of recorded fires in buildings clad with of polyurethane filled panels in which the panels appear to have performed well.[5.28–5.30] As with all lightweight sandwich panels, they were severely damaged when exposed to the direct influence of a fully-developed fire. However, panels just a short distance from the fire had maintained their stability and integrity. The sandwich panels had not contributed significantly to the severity of the fire and fire spread through the panels had not occurred. They also report similar experience from a controlled fire test on a complete three-storey building.[5.31]

5.2 FIRE TESTS

5.2.1 Use of standard fire tests

As explained in Section 5.1.4, the time–temperature curve used in the standard fire resistance tests does not relate to any particular fire scenario that may arise during the life of the product being tested. Similar considerations also apply to the various reaction to fire tests which may be used to evaluate a particular product. It follows that the purpose of standard fire tests is not to evaluate how a product or assembly may perform in practice but rather to obtain a comparative evaluation of products under conditions which bear some relationship to reality. The results of such tests may then be used to satisfy the regulations relating to fire performance in the country concerned.

Within the countries of Europe, there are a great number of different reaction-to-fire test methods in current use and harmonising these is proving to be a daunting task. Fire resistance testing is less complicated because there is general agreement regarding the time–temperature curve to be used. The problem here is that it is well known that different fire testing furnaces give different results and the reason for this is less clear.

It follows that, although the legislative position is clear within the individual countries, it is likely to be some considerable time before standard fire testing is harmonised within Europe. Here, it is not the intention to describe the test methods and classification systems currently used in the different countries of Europe, but to focus on the harmonised system that will be used in the future.

5.2.2 Reaction-to-fire tests

Most countries use a number of standard fire tests for life safety regulatory purposes. In Section 5.1.6 it was mentioned that tests which assess the contribution to the rate of fire growth of building materials and composites from ignition to flashover form a family called the ‘reaction-to-fire’ tests. These include tests for ignitability, combustibility, rate of heat release, surface spread of flame and, in some countries, smoke production.

Within Europe there are more than forty different reaction-to-fire tests. Each test measures a different property under different conditions with different criteria. The test specimens may be of different size and orientation, and may be subjected to an external source of thermal radiation of different intensity. Attempts to correlate the results of the different tests for a single material have failed[5.32] and this means that a product manufacturers have to send their products to each country of interest in order to have the product tested and approved. This is an expensive and time-consuming process. The criticism is not that the various tests are unreliable but that they differ and may rate a range of different products in a quite different order. It ought to be possible for product manufacturers to submit their products to a limited number of harmonised reaction-to-fire tests in their own country and for the results to be recognised and accepted by the other member states. This is an important objective and much research effort in EC and EFTA countries has been, and will be, applied to reach this objective. At the time of writing (2000), work by CEN Technical Committee TC127 on the development of harmonised reaction to fire tests is continuing and a new test termed the 'single burning item' (SBI) test has been developed.

The Euroclass system, which is largely dependent on the SBI test, has been published[5.33] and is summarised in Table 5.5. A key classification parameter is '*FIGRA*', the fire growth rate as determined by the SBI test. Some other significant parameters which appear in the classification table are '*PCS*', the gross calorific potential, '*LFS*', the lateral flame spread and '*THR*', the total heat release.

The SBI test uses a medium-scale apparatus to determine the rate of heat release and to provide an indication of the surface spread of flame. The specimen is L-shaped and comprises an 0.5-m wide by 1.5-m high panel joined to a panel 1-m wide by 1.5-m high and is contained in a 3-m square room. A propane gas burner of 30 kW output is placed near the inner corner of the specimen. The combustion products are ducted away and various measurements may be made on these, including carbon monoxide and carbon dioxide production and rate of heat release. Flame spread can also be observed together with characteristics such as melting and dripping.

For the highest performance in reaction to fire (class A1), the commission has issued a list of products which may be given this classification without testing. This list includes, for example, metals and stone wool with a binder content below 1%. Products comprising a combination of more than one class A material may also be granted class A according to the proposed regulations. Other materials and products may be classified as class A1 or A2 on the basis of two tests according to ISO 1182 and ISO 1716.

Small- or medium-scale reaction to fire tests cannot measure all of the important fire characteristics of a material or composite. The ISO 9705 room corner-wall test is one larger-scale test[5.2] which, notwithstanding the problems outlined in Section 5.1.9, has been used in the testing of sandwich panels incorporating combustible cores up to and beyond flashover. It was used for this purpose in the 'Eurific' reaction to fire test programme and also provides a 'reference scenario' for the Euroclass system, and for products that cannot be tested using the 'single burning item' (SBI) method. It should be noted that, unlike many other components, metal faced sandwich panels are produced and tested complete with their surface coatings (e.g. PVC, PVDF).

In the ISO 9705 test, the inside surfaces of the non-combustible test room, which is 3.6-m deep by 2.4-m wide by 2.4-m high, are clad with the test panels. The joints are representative of those used in practice but, however, the panels are not fixed as in normal practice. A propane burner is placed close to the specimens in one

Table 5.5 Table of Euroclass products

Class	Test method(s)	Classification criteria	Additional classification
A1	EN ISO 1182[a] and	$\Delta T \leq 30°\mathrm{C}$ and $\Delta m \leq 50\%$ and $t_f = 0$ (i.e. no sustained flaming)	
	EN ISO 1716	$PCS \leq 2.0\ \mathrm{MJ\ kg^{-1}}$ [a] and $PCS \leq 2.0\ \mathrm{MJ\ kg^{-1}}$ [b c] and $PCS \leq 1.4\ \mathrm{MJ\ m^{-2}}$ [d] and $PCS \leq 2.0\ \mathrm{MJ\ kg^{-1}}$ [e]	
A2	EN ISO 1182[c] or	$\Delta T \leq 50°\mathrm{C}$ and $\Delta m \leq 50\%$ and $t_f \leq 20\ \mathrm{s}$	
	EN ISO 1716 and	$PCS \leq 3.0\ \mathrm{MJ\ kg^{-1}}$ [a] and $PCS \leq 4.0\ \mathrm{MJ\ m^{-2}}$ [b] and $PCS \leq 4.0\ \mathrm{MJ\ m^{-2}}$ [d] and $PCS \leq 3.0\ \mathrm{MJ\ kg^{-1}}$ [e]	
	EN xxxx (SBI)	$FIGRA \leq 120\ \mathrm{W\ s^{-1}}$ and LFS < edge of specimen and $THR_{600\,s} \leq 7.5\ \mathrm{MJ}$	Smoke production[f] and Flaming droplets/particles[g]
B	EN xxxx (SBI) and	$FIGRA \leq 120\ \mathrm{W\ s^{-1}}$ and LFS < edge of specimen and $THR_{600\,s} \leq 7.5\ \mathrm{MJ}$	Smoke production[f] and Flaming droplets/particles[g]
	EN ISO 11925-2[i] Exposure = 30 s	$Fs \leq 150\ \mathrm{mm}$ within 60 s	
C	En xxxx (SBI) and	$FIGRA \leq 250\ \mathrm{W\ s^{-1}}$ and LFS < edge of specimen and $THR_{600\,s} \leq 15\ \mathrm{MJ}$	Smoke production[f] and Flaming droplets/particles[g]
	EN ISO 11925-2[i] Exposure = 30 s	$Fs \leq 150\ \mathrm{mm}$ within 60 s	
D	EN xxxx (SBI) and	$FIGRA \leq 750\ \mathrm{W\ s^{-1}}$	Smoke production[f] and Flaming droplets/particles[g]
	EN ISO 11925-2[i] Exposure = 30 s	$Fs \leq 150\ \mathrm{mm}$ within 60 s	
E	EN ISO 11925-2[i] Exposure = 15 s	$Fs \leq 150\ \mathrm{mm}$ within 20 s	Flaming droplets/particles[h]
F	No performance determined		

[a] For homogeneous products and substantial components of non-homogeneous products

[b] For any external non-substantial component of non-homogeneous products

[c] Alternatively, any external non-substantial component having a $PCS \leq 2.0\ \mathrm{MJ\ m^{-2}}$, provided that the product satisfies the following criteria of EN xxxx (SBI): $FIGRA \leq 20\ \mathrm{W\ s^{-1}}$ and $LFS \leq$ edge of specimen and $THR_{600\,s} \leq 4.0\ \mathrm{MJ}$ and s1 and d0

[d] For any internal non-substantial component of non-homogeneous products

[e] For the product as a whole

[f] s1 = $SMOGRA \leq 30\ \mathrm{m^2\ s^{-2}}$ and $TSP_{600\,s} \leq 50\ \mathrm{m^2}$, s2 = $SMOGRA \leq 180\ \mathrm{m^2\ s^{-2}}$ and $TSP_{600\,s} \leq 200\ \mathrm{m^2}$; s3 = not s1 or s2

[g] d0 = no flaming droplets/particles in ENxxxx (SBI) within 600 s, d1 = no flaming droplets/particles persisting longer than 10 s in EN xxxx (SBI) within 600 s; d2 = not d0 or d1. Ignition of the paper in EN ISO 11925-2 results in a d2 classification

[h] Pass = no ignition of the paper (no classification). Fail = ignition of the paper (d2 classification)

[i] Under conditions of surface flame attack and, if appropriate to the end-use application of the product, edge flame attack

corner. This provides an output of 100 kW for 10 minutes followed by 300 kW for a further 10 minutes. Outside a doorway is a hood and duct which are used to collect and transport all of the combustion gases. Within this duct, measurements are made of the heat release, velocity of gases, production of carbon dioxide and carbon monoxide and optical density. Visual observations include the time for ignition of the ceiling, flame spread along the wall and ceiling surfaces and flames emerging from the opening.

A characteristic of the ISO 9705 test is that flashover can be achieved because of feedback radiation from nearby wall and ceiling surfaces. It can, therefore, test the ability of sandwich panels to withstand a fully-developed fire. In contrast, the corner wall tests of which there are several, notably the Factory Mutual test (see Section 5.1.9) are useful for testing the response of sandwich panels to a fire which does not reach flashover and provides only local punishment to the test specimen.

As background information, it is of interest to note that the relationship between product class and the fire behaviour observed in the ISO reference scenario is given as an informative annex to draft prEN 13501-1: 1999 *Fire classification of construction products and building elements – Part 1: Classification using data from reaction to fire tests* as follows:

Classes A1, A2 and B	no flashover
Class C	no flashover with 100 kW but flashover with the 300 kW ignition source
Class D	flashover after more than 2 minutes with the 100 kW ignition source
Class E	flashover before 2 minutes with the 100 kW ignition source

It is also important to realise that this background information has been used by the Fire Regulators to define the class limits. The above standard makes it quite clear that the reference scenario is not an alternative route to classification.

5.2.3 Fire resistance tests

An introduction to fire resistance has been given in Section 5.1.7.

Fire resistance may be defined as the ability of a material, product, assembly or structure to fulfil, for a stated period of time, the required stability, integrity, thermal insulation, and/or other expected duties specified in a standard fire resistance test. Specimens are mounted in a furnace and exposed to combustion gases which are made to follow a standard temperature–time curve and the so-called ISO 834 curve is used for most building components.

The fact that the thermal severities of fire resistance furnaces differ[5.34] (see Section 5.1.7) and that fire resistance test criteria are not harmonised means that the approval authority (building control authority or fire authority) in one country is reluctant to accept the results of a test made in a laboratory in another country. If, however, the approval authority can get a letter of assessment which states that the product (e.g. sandwich panel) tested in the other country would be expected, if tested, to meet the required level of fire resistance in its own country then the product should be acceptable without the need for a retest in the destination country.

As mentioned in Section 5.1.7, the size of product that can be accommodated in a fire resistance test is usually limited to 3 m × 3 m for walls and 3 m × 4.5 m for roofs and ceilings. In practice the product may be several times larger than the test specimen and it may then be necessary to make an assessment based on the test report in order to predict whether increased size is likely to increase or, more likely, decrease the amount of fire resistance. It is possible that the product specifier will not be aware that the size of product used in his project may affect its fire performance.

It is important that the sandwich panel specifier checks that the loading (if any) and boundary conditions (i.e. support and restraint conditions) recorded in the fire resistance test report are similar to those in the building in which the panels are to be used. If they are not similar an assessment may need to be made. More details are given in Section 5.4.

5.2.3.1 Typical results from fire resistance tests

Some indicative fire resistance classifications for different types of sandwich panel and other types of construction are given in Table 5.6.

5.3 MATERIAL PROPERTIES AT ELEVATED TEMPERATURE

5.3.1 Face materials

5.3.1.1 Surface coatings

Surface coatings may comprise one or more materials. For example a steel sheet facing may be protected by a zinc coating overlaid with an adhering thin layer of another material such as PVDF or polyester. In contrast the surface coating on a plywood facing could be a single material such as a sprayed layer of melamine.

At the present time, the fire behaviour of surface coatings is usually investigated by 'surface spread of flame' tests. In the future, this aspect of fire behaviour will be incorporated in the SBI test. The fire behaviour of a surface coating may be affected by any adhesive layer. A thermoplastic adhesive may weaken and allow the surface coating to delaminate and fall off, but, while this is a possible fire phenomenon, it is rarely observed in practice. It can, however, be a hazard when, for example, many layers of oil-based paint have been applied to a substrate during repeated maintenance operations, but this is unlikely in the case of sandwich panels.

Ordinary mild steel sheet facings are normally protected with a corrosion-resistant zinc layer to which the decorative surface coating is added. The zinc layer, applied by the electrolytic or hot dip processes, evolves irritant fumes when exposed to elevated temperatures but, while this may be a hazard when welding or flamecutting, it is not recognised as a primary hazard when fire occurs. Zinc does, however, have the potential to spread fire to areas below the initiating fire due to falling molten zinc droplets. Zinc aluminium alloy coatings are less hazardous than zinc coatings because less zinc is present.

Polyvinyl chloride (PVC) coatings will produce irritant fumes (hydrogen chloride) when exposed to fire, and fire fighters may need to wear breathing

Table 5.6 Indicative fire resistance classifications for different constructions (This is effectively a table for I. All well-detailed panels should provide an E of 120.)

Type of construction	Fire resistance class EI – min						
	15	30	60	90	120	180	240
Sandwich panels							
Stone wool							
80 mm	▒▒	▒▒	‖‖‖				
100 mm	▒▒	▒▒	‖‖‖	‖‖‖	‖‖‖		
150 mm	▒▒	▒▒	▒▒	‖‖‖	‖‖‖	‖‖‖	
200 mm	▒▒	▒▒	▒▒	▒▒	‖‖‖	‖‖‖	‖‖‖
PUR							
80 mm[a]	▒▒						
Phenolic foam[b]							
75 mm	▒▒	▒▒					
100 mm	▒▒	▒▒	▒▒				
125 mm	▒▒	▒▒	▒▒	▒▒	▒▒		
Concrete[c]							
80 mm	▒▒	▒▒					
100 mm	▒▒	▒▒	▒▒	▒▒			
150 mm	▒▒	▒▒	▒▒	▒▒	▒▒	▒▒	
Brickwork[d]							
80 mm	▒▒	▒▒					
110 mm	▒▒	▒▒	▒▒	▒▒	▒▒		
130 mm	▒▒	▒▒	▒▒	▒▒	▒▒	▒▒	▒▒

▒▒ Normal solution

‖‖‖ Special solution or structural stone wool

[a] May require special detailing at joints
[b] Modified phenolic foam, density at least 130 kg/m^3. May require special detailing at joints
[c] Reinforced dense concrete
[d] Clay bricks, load bearing

apparatus for this reason. Hydrogen chloride can also damage the structure and finishes of buildings.

Polyester based coatings give off toxic fumes (e.g. CO) and dust when involved in fire.

Fluorocarbon based coatings, such as polyvinylidene (PVF) coatings or adhesively bonded polyvinyl-fluoride laminates, can produce dust and toxic fumes including hydrogen fluoride when exposed to fire. However the small coating thickness used means that this hazard is not of primary importance.

Surface coatings are usually organic and are therefore combustible. However the small thicknesses which are usually used means that the contribution of the coating to the fire load of a metal-faced sandwich panel is negligible. A more important aspect of the behaviour of a surface coating in fire is its reaction-to-fire and, in particular, the ease with which flames may spread across the surface, and the obscuration, toxicity and irritancy of the combustion gases. The $k\rho C$ factor for the substrate (where k = thermal conductivity, ρ = density and C = specific heat) has an important bearing on flame spread (see Section 5.1.6), and it is misleading to give data on the flame spread of surface coatings without specifying the nature of the substrate used in the test. However, some general information can be given as follows.

PVF, PVDF and polyester (SP) coatings do not spread flames or release heat and smoke and they are classified as non-combustible in Germany. In contrast, PVC fails to achieve the best classification in flame spread and smoke release tests. PVC starts to decompose at temperatures between 200 and 300°C and ignites at 400°C. It also releases corrosive gases which combine with water droplets in the fire effluent to produce hydrogen chloride which damages the structure, services and finishes in a building.

5.3.1.2 *Steel sheets*

Steel sheet has excellent integrity in fire resistance tests. Very thin facings can, however, erode by oxidation in long (i.e. more than 4 hours) exposures in the standard fire resistance test but such high levels of fire resistance are not normally required for sandwich panels. The coefficient of linear thermal expansion of ordinary grades of steel sheet may be taken as 14×10^{-6} per °C over the temperature range 0–600°C which is the range usually considered in structural applications of steel.

Stainless steel sheet also has excellent integrity in fire conditions, but it has a higher coefficient of linear thermal expansion which depends on the alloying additions. For example an architectural stainless steel having 18% chromium and 8% nickel has a coefficient of linear thermal expansion of 19.5×10^{-6} per °C over the temperature range 0–1200°C.

Steel has a melting point of approximately 1550°C, which is well above the maximum temperature of 1200°C reached by either the ISO 834 cellulosic or the hydrocarbon standard temperature–time curves.

It is not possible to generalise on the flame spread behaviour of coatings applied to steel sheets. The thinner the coating and the greater the $k\rho C$ of the substrate the less will be the flame spread.

Plastic coatings on sheet steel can, depending on the thickness and formulation of the plastic, produce corrosive combustion products. During a fire, PVC produces hydrogen chloride which, as mentioned in Section 5.1.6, can corrode steel even if it is protected by concrete as in reinforced concrete members.

5.3.1.3 *Aluminium and aluminium alloy sheets*

The melting point of pure aluminium is 660°C but this may reduce to 600°C for aluminium alloys. Since this temperature is reached in less than 10 minutes in the ISO 834 fire resistance exposure and in a much shorter time than this in the hydrocarbon exposure, it is clear that aluminum alloy sheets cannot be used for sandwich panels which need to be fire rated. The exception is when the core material has good fire insulation properties together with higher density and inherent structural strength, such as structural stone wool, because the unexposed facing and the adhesive bonding it to the core can continue to perform their function long after the facing exposed to the fire has melted away. In practice, however, when fire resistance is needed, aluminium alloy sheet is only used as the outer facing of an external wall panel.

The melting of aluminium is often observed in fire resistance tests and real fires and can cause fire spread to areas below the melted zone.

5.3.1.4 Other metal sheets

Metal sheets, other than of steel or aluminium alloy, are not generally used in sandwich panel construction but, if they are, the important properties are the melting point of the metal used and the thickness of any non-combustible core insulation which may assist in keeping the unexposed facing cool.

Timber, plywood and chipboard facings
Timber chars when exposed to a fire and the char acts as a protective layer. Furthermore, timber has a low coefficient of thermal expansion which minimises the tendency for charred material to become displaced. It also has a low thermal conductivity so that undamaged timber below the charred layer retains its strength.

Nevertheless, timber is combustible and is likely to have poor or unacceptable flame spread properties for most applications. When exposed in the fire resistance furnace, timber chars at a rate of 0.5 to 0.8 mm/min depending on the species. These charring rates can be used to predict the time of burn-through to the core material. Plywood and chipboard incorporate adhesive which, depending on its type, may weaken in fire and allow the material to delaminate.

Glass reinforced plastic
Glass reinforced plastic (GRP) and other similar materials have attracted the attention of the offshore industry in recent years as potential facing materials for sandwich panel construction. On oil rigs and other similar facilities, weight is perhaps the single most important design criterion and this points to the use of sandwich construction for both internal and external wall panels. However, even stainless steel corrodes offshore so that thin metal is not very practicable and there is considerable interest in the development of non-metallic facing materials and GRP, in its various forms, is a primary candidate. However, GRP burns, albeit slowly, and, following the Piper Alpha disaster, the designers of offshore rigs have to pay particular attention to fire safety. This has resulted in extensive research into the fire performance of GRP and its characteristics are now well established.

Evidently, GRP provides an interesting alternative facing material for sandwich panels, especially where its particular non-metallic characteristics are of interest. However, there are a wide variety of resins (epoxy, polyester, phenolic, etc.) and glass types so that the possibilities are many and varied. Space precludes a full discussion of this interesting topic and the interested reader is referred to the specialist literature (e.g. References 5.35 and 5.36).

Glass reinforced cement
Glass reinforced cement (GRC) sheets or mouldings have excellent fire resistance

properties because the randomly oriented glass fibres prevent the cement from spalling. Because the material is non-combustible, it makes no contribution to heat release and flame spread and releases no combustion products.

Gypsum board and other fire protecting boards
Gypsum plasterboard, either with or without glass fibre reinforcement, and an extensive range of other fire protecting boards are widely used in other applications but are seldom used as the structural faces of sandwich panels. They can, however, serve as a fire insulating layer on either the outside or inside of a sandwich panel which has inadequate fire resistance in itself. Ways of using fire protecting boards are given in Section 5.4.11. Most fire protecting boards contract and bow when heated from one side and their use with a metal faced sandwich panel should be proven to be acceptable in a full scale fire resistance test if the fire integrity of the board is to be assured. These materials usually have a very good performance in reaction-to-fire tests.

5.3.2 Core materials

The discussion of core materials so far in this chapter has concentrated on general considerations and only brief mention has been made of the fire properties of the variety of core materials which may be used in composite panel construction. The differences between the performances of these materials has been discussed by Davies[5.37] and is considered in more detail in this section.

5.3.2.1 Foamed plastic

General
All plastic foam materials are organic and are therefore combustible. Their fire behaviour is strongly influenced by their low thermal inertia. This property permits the surface of the material to respond rapidly to any imposed heat flux so that ignition and maximum rates of burning can be achieved very quickly. Their behaviour in fire can be improved by the selection of suitable raw materials, by special foaming processes, by the use of fire retarding agents or by the inclusion of inorganic filling material. However, additives can have only a relatively small effect on the temperatures at which plastic foams start to decompose chemically and ignite. The additives mainly retard the burning process.

The general characteristics of the alternative rigid plastic foam core materials in current use are as follows:

Polyurethane (PUR) and polyisocyanurate (PIR)
These are *thermosetting* materials which do not melt when exposed to fire but rather form a carbonaceous char. With the correct chemical formulation, this char can be stable and therefore particularly advantageous with regard to fire performance. A high isocyanurate index generally indicates good charring properties and some fire retardant additives are also designed to promote charring. PUR starts to decompose at 150–200°C and becomes flammable at 300°C and releases heavy smoke when burning. At temperatures above 600°C, toxic gases like hydrogen

cyanide start to form as well as carbon monoxide. A considerable range of raw materials, blowing agents and fire retardants are used, all of which may influence the performance in fire. The three-dimensional cross-linked chemical structure of PUR and PIR accounts for why they do not shrink back from flames.

Early fire resistance tests on 60-mm thick sandwich panels with steel faces resulted in fire resistance times of only 4 minutes with a pure polyurethane foam core increasing to 7 minutes with a PIR modified foam core. More recent tests have considerably increased these times but achieving the critical fire resistance of 30 minutes remains problematical.

An interesting step towards achieving a 30 minute fire resistance classification has been made by reinforcing the PIR modified foam with milled glass fibre or foamed glass or clay balls. A resistance time of 20–25 minutes has been achieved in this way and, with currently available foams, this could undoubtedly be increased to well over 30 minutes. However, it is by no means certain that such methods will be successful in economic terms and there are other systems under development which show good fire resistance with equivalent properties at a lower price.

Expanded and extruded polystyrene (EPS and XPS)

These are *thermoplastic* materials whose behaviour in fire is dominated by their unfavourable tendency to melt at temperatures only a little above 100°C with the result that they melt before they ignite and then tend to form burning droplets. These materials start to decompose at 300°C and ignite at 360°C. When burning they release much smoke and carbon particles. The main releases are carbon dioxide and styrene. Fire retardants are widely used, as in PUR foams.

In expanded polystyrene, also known as bead polystyrene, hydrocarbons are incorporated during polymerisation and the polymer beads are formed by steam. Low flammability grades of this material are available. Extruded polystyrene is formed by incorporating chlorinated hydrocarbons in an extrusion process in which the foaming is initiated by reaction heat. Both expanded and extruded polystyrene withdraw rapidly from flames due to melting and shrinking so that, frequently, and particularly in small-scale tests, ignition does not occur. This means that small-scale reaction to fire tests on these materials may give misleading results.

Phenolic foam (PF)

These foams came into some prominence in the late 1960s and early 1970s when it was considered that the fire resistance of panels with PUR rigid foam cores needed improvement. Their great tendency to char was a particular attraction. At that time, however, phenolic foams suffered from a number of disadvantages such as difficulty in processing, production of acid during manufacture, low mechanical strength, friability and relatively poor thermal insulation and integrity properties (because of entrapped moisture, the closed cells tended to 'explode' when exposed to fire). More recent developments have largely overcome these problems with the result that phenolic foam is now perceived as being the rigid plastic foam core material with the best fire properties. One of these developments is the so-called modified phenolic foam (a mineral-filled open cell structure with good fire insulation and integrity) which, used in sheet steel faced sandwich panels, can, according to indicative data, achieve a 2-hour fire resistance rating.

Phenolic foam, like PUR and PIR, is a *thermosetting* material with particularly favourable char-forming properties and low smoke emission. It starts to decompose at temperatures between 350 and 500°C and ignites between 530 and 580°C. When burning, a stable char is formed and the releases are mainly hydrocarbons and carbon monoxide.

Fire retardants

There are a wide variety of materials which may be added to the chemical mixture prior to foaming in order to reduce the flammability of rigid plastic foams. These additives are generally of two types. *Reactive* fire retardants contain halogens, notably bromine, and actively inhibit flaming of the material. *Unreactive retardants* contain compounds of phosphorus which react during thermal decomposition to form phosphoric acid which, in turn, combines with the polymer to form a char layer. Thus, the primary mechanism of the second type of fire retardant is to promote charring rather than burning.

Fire retardants are usually premixed into the constituents of the foam before delivery by the chemical supplier. It is quite normal for both types of fire retardant to be used in a given application and there is probably some synergy as the two act together. They add significantly to the cost of the foam and there is also additional smoke roughly in proportion to the quantity of retardant added. The emphasis is, therefore, to use the minimum necessary in order to obtain the required performance in statutory fire tests.

Without initiating comparative tests on a range of modern foam formulations, it is not possible to make any definitive statements regarding the current usage of these products and their significance in terms of fire performance. It is likely that the detailed situation is in a process of change as new blowing agents are introduced with consequential changes in foam chemistry. However, the chemical industry claims that the new blowing agents are not causing them any problems with regard to fire retardants.

Small-scale fire test results

The excellent thermal insulation properties of the core materials used in composite panels, means that only a small amount of heat can be conducted away from the (thin) metal face in contact with the fire. The temperature of the foam core directly behind this face is therefore almost the same as the temperature of the fire combustion gases. Regardless of the core material, polyurethane adhesives are widely used and it follows that the adhesion between the core and this face fails early in the fire, generally at between 150°C and 200°C. Rigid foamed plastic cores then continue to deteriorate rapidly and the influence of the fire may soon reach the unexposed face resulting in relatively short times to insulation failure.

The ignition temperature of a material is the temperature to which it must be raised in air for ignition to take place. If no flame is used, the temperature is known as the auto-ignition temperature. If a flame is present, the temperature is known as the flash-ignition temperature of the material. Some indicative values of these temperatures for core materials are given in Table 5.7.[5.38, 5.39]

Another useful measurement is the minimum oxygen concentration which will allow a material to burn. The higher the oxygen index, the more 'fire resistant' the

Table 5.7 Performance of core materials at elevated temperatures

Material	Temperature for loss of dimensional stability (°C)	Flash-ignition temperature (°C)	Auto-ignition temperature (°C)	Oxygen index
EPS/XPS	≈90–100	245–345	490	18–28
PUR	≈200	285–310	415–500	17–24
PIR	≈200	415	510	30
PF	> 250	490	450	46

material and materials with an oxygen index greater than 26 may be regarded as self-extinguishing. Some values of the oxygen index are also included in Table 5.7.

In ignitability tests, depending on the nature of the test, polystyrene may be unclassifiable or may appear to give excellent but completely misleading results because the material melts away from the flame before it ignites. This characteristic is, however, of no benefit in real fires.

5.3.2.2 Mineral wool

The generic term 'mineral wool' includes glass wool, stone wool (often called rock wool) and slag wool.

- *Glass wool* has a melting point of approximately 600–650°C and, because the binder content is generally more than 5%, glass wool is usually classified as combustible. It therefore has limited application in fire rated panels, and it is not available in a sufficiently high density to possess the required shear modulus to act as a core in load-bearing panel construction.
- *Slag wool* is produced from molten slag which is a by-product of iron and steel production. It can remain stable up to about 900°C. It is often acidic in nature and can cause corrosion problems. It has variable mechanical properties and is generally unsuitable for use as a core material in structural sandwich panels.
- *Stone wool*, more usually called *rock wool*, is widely used in the construction industry as a thermal insulant. It comprises fibres which are inorganic and non-combustible and a binder which holds the fibres together. In board form, the fibres are usually aligned in a preferred direction. However, in order for stone wool slabs to have sufficient tensile and shear strength for use in structural sandwich panels, the density (to which the strength is closely related) must not be less than about 50 kg/m^3 and, importantly, the fibres must be oriented perpendicular to the surface of the panel. The high melting point (approximately 1400°C), high strength, high thermal insulation and limited combustible nature of stone wool make it an ideal core material for structural sandwich panels capable of achieving up to 4 hour fire resistance, while contributing a negligible amount to the fire load.

Mineral wools with a low organic binder content are practically non-combustible. The organic binder and the additive start to decompose at temperatures above 250°C but the fibres themselves do not burn but rather melt. The small amount of combustible material means that only small quantities of gases are released and the thermal insulation properties remain until the fibres melt. For a 100-mm thick sandwich panel with stone wool core, values of fire resistance (according to the standard ISO 834) from 30 to 120 minutes have been measured, the variation being due to the type of stone wool and the joint details.

Nevertheless, it should be noted that the adhesives used to bond the mineral wool core to the metal faces are generally urethane based so that debonding of the exposed face takes place at an early stage, in much the same way as in a panel with a PUR core. It is vital that fire resisting mineral wool panels are detailed in such a way that they retain their integrity under these circumstances.

5.3.2.3 Fire resistant boards

Fire resisting boards are not often used on their own as the core material in structural sandwich panels because these generally require good thermal insulation properties. Such boards are, however, sometimes used in conjunction with a core material of foamed plastic (e.g. polystyrene and polyurethane) in order to improve the fire resistance properties as shown in Fig. 5.11. By virtue of the added mass, they also improve the sound insulation. To be suitable in this context, boards should possess good insulation and integrity in the fire condition and should not exhibit large amounts of thermal bowing. Boards which have proven fire protecting properties when used as wall and ceiling linings are generally suitable for use in this sandwich panel context.

5.3.3 Adhesives

The temperature at which an adhesive used to bond the facings to the core softens should, ideally, be as high as possible so as to preserve the flexural strength of the sandwich panel and prevent delamination for as long as possible when exposed to fire. Most thermoplastic and thermosetting adhesives soften at a temperature of roughly 150°C. Adhesives are available that have high temperature resistance but they are generally not suitable for the production methods generally used in panel manufacture because of problems of curing and reliability in normal service. Such adhesives may be used for adhering fire protecting boards together and to other substrates but, if they are water-based, they may not be suitable for use in external panels where condensation or other conditions allowing the ingress of water or water vapour may precipitate failure.

Panels employing slabs of stone wool, XPS and EPS generally use a polyurethane based adhesive that has a softening point of approximately 150°C. If the normal amount (200–300 g/m^2) of polyurethane adhesive is exceeded the reaction-to-fire properties can be adversely affected.

5.3.4 Mechanical fasteners

The material used for mechanical fasteners in fire rated panels should have a melting point above 1000°C. This excludes the use of aluminium alloy which can melt at 600°C and in practice it means that steel is the main choice. For reasons of corrosion resistance and appearance the steel is usually a chromium stainless steel. Fasteners can be in the form of screws, bolts or rivets.

It should be remembered that fasteners sized to suit every-day performance may, because of their reduction in strength at elevated temperature, become overstressed in the fire condition and this should be kept in mind especially for panels that are far greater in size than that fire tested.

In order to achieve optimum performance from the fasteners, it is necessary to check that the materials being held together by the fastenings are also of a size suited to transmit the forces involved. For example, angles or straps of sheet steel may need to be larger than those used in the fire resistance tested specimen when tall panels are to be suspended from the top.

Premature failure of mechanical fastenings can lead to failure of integrity at an early stage in a fire, which is especially detrimental if the core is of a combustible nature.

5.4 DESIGN CONSIDERATIONS

5.4.1 General

In this section a number of factors affecting the choice of sandwich panels are given and, where possible, preferences are stated. It should be understood that there is often considerable interaction between the context of the use of a panel, its required fire performance and its acceptability. A plastic foam cored panel may be acceptable in one context and unacceptable in another.

Many of the following factors need not be considered if the panel assembly has minimal combustible content since the reaction-to-fire properties will automatically be acceptable, as for example with steel-faced mineral wool cored panels. Hence with steel-faced non-combustible cored panels it is necessary to consider only the likely performance in the post-flashover fully developed fire situation where fire resistance properties of integrity and insulation are important.

The information presented here is intended to offer practical guidelines for the designers of buildings.

5.4.2 Design for fire safety

As discussed in Section 5.1.2, the objectives for the fire-safety design of a building may be the safety of the occupants, the safety of fire fighters inside and outside the building, reduction of property loss caused by smoke and heat damage, reduction of interruption to business and reduction of damage to the environment by pollution of the atmosphere and groundwater caused by contaminated fire fighting water.

The design of sandwich panels can influence how easily these objectives are reached although contamination of the environment by burning sandwich panels rarely needs to be considered. Much depends, for example, on the following.

- The stage in the fire development when the sandwich panels may become involved in the fire; the later the better in terms of safety of life of occupants and fire fighters, and safety of property.
- The fire load of the panels compared with the fire load of the contents of the

building and the fabric and structure of the rest of the building – see Section 5.1.5.
- How easily the occupants can be overcome by the sandwich panel combustion products at the development stage of the fire when these can become significant. Sandwich panels with plastic foam cores should not generally be used in rooms where people may sleep and where there is no early warning such as an automatic smoke detection and alarm system. The exception might be where the panels can be modified to provide a delayed contribution to the production of fire effluent in the room by, for instance, cladding the fire risk side of the panel with a fire protecting board material, or by ensuring that the fire effluent escapes harmlessly to the outside air.
- The amount of compartmentation using fire resisting partitions. The smaller the compartments, the smaller is the area which may be damaged by heat and smoke.
- The level and toxicity/irritancy of the sandwich panel combustion gases. This is especially important if the combustion gases are not able to escape to the outside so that they are transported throughout the building.
- The volume and obscuration of smoke produced by burning sandwich panels: this is a consideration for panels having plastic foam cores, especially polystyrene.
- The corrosivity of combustion gases coming from the sandwich panel. Formulations of core material which, upon burning, produce hydrogen chloride should be avoided where it is important to minimise corrosion of metals especially steel bars used for reinforcing concrete structures. Flame retarded foamed polystyrene containing halogens should be avoided if possible, but this may be relatively unimportant if the building contains a lot of PVC in the contents or in the building services (e.g. PVC insulated electrical cables) since their potential for producing hydrogen chloride may then dwarf that from the polystyrene foam.

5.4.3 Fire regulations

Designers should find out at an early stage what regulations, fire tests and fire test criteria need to be satisfied for the elements of construction for which they intend to use sandwich panels. The regulatory requirements vary from country to country but some typical examples of requirements are given in the following.

External walls in single storey industrial buildings may need to possess no fire resistance if spaced well away from other buildings or the site boundary. External walls in multi-storey buildings and industrial buildings near other buildings may need a small amount of fire resistance when exposed to fire from the inside. External walls close to other buildings may need fire resistance from the inside in order to prevent large amounts of thermal radiation which could cause fire spread to other buildings. Roofs may need to resist flame spread and/or resist penetration from the outside. Internal walls or partitions acting as fire walls may need to possess a large amount of fire resistance. A ceiling may need to possess fire resistance in itself or to contribute to the fire resistance of a floor or roof which it is protecting. Internal and external surfaces of some construction elements may need to possess only some reaction-to-fire properties such as low flame spread but no fire resistance.

A relatively new consideration is that an external wall should be safe with respect to an arson attack from outside. It should not be possible to cause a fire in a building by lighting a fire against an outside wall.

5.4.4 Insurance requirements

The designer should identify the requirements at an early stage. To reduce the potential for large fire losses the insurer may ask for levels of fire performance or periods of fire resistance which are greater than those required under life safety building regulations. Insurers may require evidence that sandwich panels which contain combustible material, have been successfully tested in a room corner test (Loss Prevention Council[5.40] in the UK or Factory Mutual[5.41] for the USA and worldwide), or they may require the provision of other kinds of test data which are relevant to the losses caused by heat and smoke damage before they will insure the building or insure in a low premium class.

5.4.5 Structural support for panel assemblies

There is little to be gained from the use of a panel system which has a given fire resistance unless the structure to which the panels are fixed can provide support for the panel assembly for at least the same period.

Ideally, the panels to be used in a building should have the same or better support conditions than those present in the fire resistance test. However this may not be practicable and it is then necessary to make an engineering assessment in order to predict whether the panels to be used in the building will have comparable or adequate fire resistance. The prediction of integrity performance is often difficult but the prediction of the insulation performance is easier since it is not height- or width-dependent, though it will be thickness-dependent.

Where panels are to be used in the building in the vertical orientation, and where they have a height considerably in excess of the 3-metre high panels used in the fire resistance test, it may be necessary to adopt the suspension concept of support in the building. This concept recognises that a sandwich panel may be able to support its own weight in every-day use but may be unable to do this in a fire because of the loss of strength due to delamination of the fire-exposed facings, whereupon collapse can occur as shown in Fig. 5.5. In order to avoid this, the panel facings should be 'suspended' from the top so that, in a fire, the two facings (exposed and unexposed sides) will hang down like curtains as shown in Fig. 5.6. Although Figs 5.5 and 5.6 indicate an internal wall the same principles apply to an external wall.

The suspension fasteners for tall panels may act locally at the edges of the panels, as shown in section in Fig. 5.7(a), or continuously over the full width of the panel as shown in Fig. 5.7(b). Point fastenings can sometimes be attached to the inward-turned panel face edge, where they are concealed and partly fire protected by the panel facing, using, for example, short steel cleats. The cleats may use a detail which allows free upward movement of the panel head caused by expansion of the fire-exposed facing but prevents the facing from dropping down later in the fire.

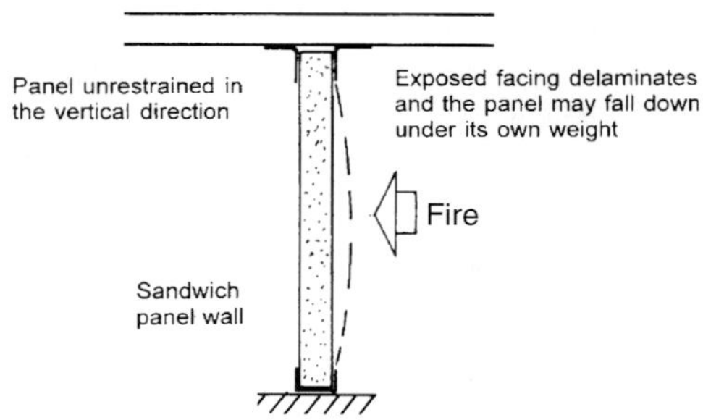

Fig. 5.5 Internal sandwich panel wall supported at the bottom.

Line fasteners may be attached to the outside of the faces, for example using a steel angle. In addition to the tensile force in the facing caused by the self-weight of the panel there may also be membrane forces caused by internal air pressure.

An important feature in the suspension concept is the use of fire stopping which fills the gap between the top of the panel and the underside of the support member above. It is important that the fire stopping, usually of stone wool, is of an appropriate density such that the wool packing is 'elastic' and therefore able to expand after being compressed as the panel head moves up and down in normal and fire conditions. If the fire stopping is unable to do this, it is likely that an integrity failure will occur at the top of the panel.

The suspension concept of panel support would also be appropriate for panels in a building which are the same height as those used in the fire resistance test but where additional fire resistance is required above that obtained in the test. In this case, if the failure mode in the test was integrity, it would also be necessary to check that the insulation criterion is also satisfied for the longer period of fire exposure.

5.4.6 Loading and restraint of the panel assembly

Most sandwich panels, when used as walls, ceilings or roofs, are non-load-bearing so that they are capable of supporting only their self-weight. In some situations, internal walls used as fire separating elements have to withstand the action of fire and internal pressures generated by wind through openings in the external cladding.

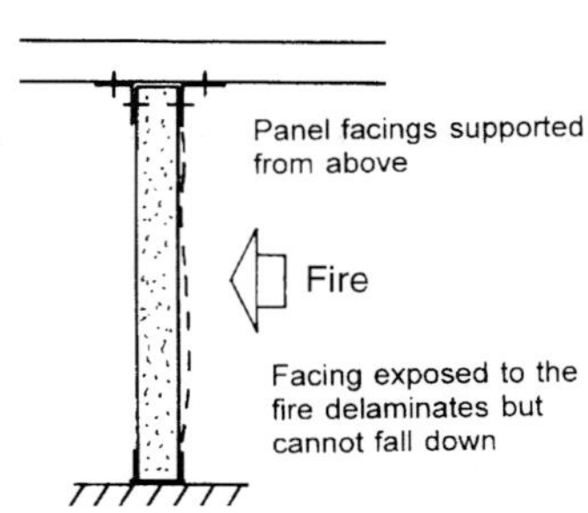

Fig. 5.6 Internal sandwich panel wall suspended from the top.

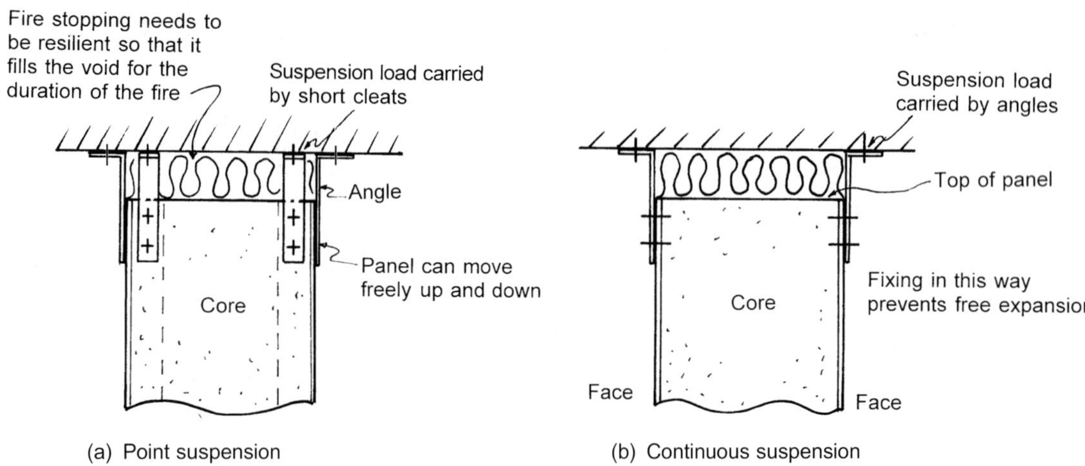

Fig. 5.7 Suspension methods for tall panels – shown in section.

Figure 5.8 compares the forces acting on a wall panel when tested in the furnace with a height of 3 metres and when used in a project when its height is 9 metres. It can be seen that, in the project, the dead load acting on the tall panel is a factor of three greater and the pressures acting on the panel face can be greater or smaller depending upon wind direction. Wind pressure which is additive to the fire pressure is the worst condition because the increased pressure occurs at the top of the panel where failure usually occurs.

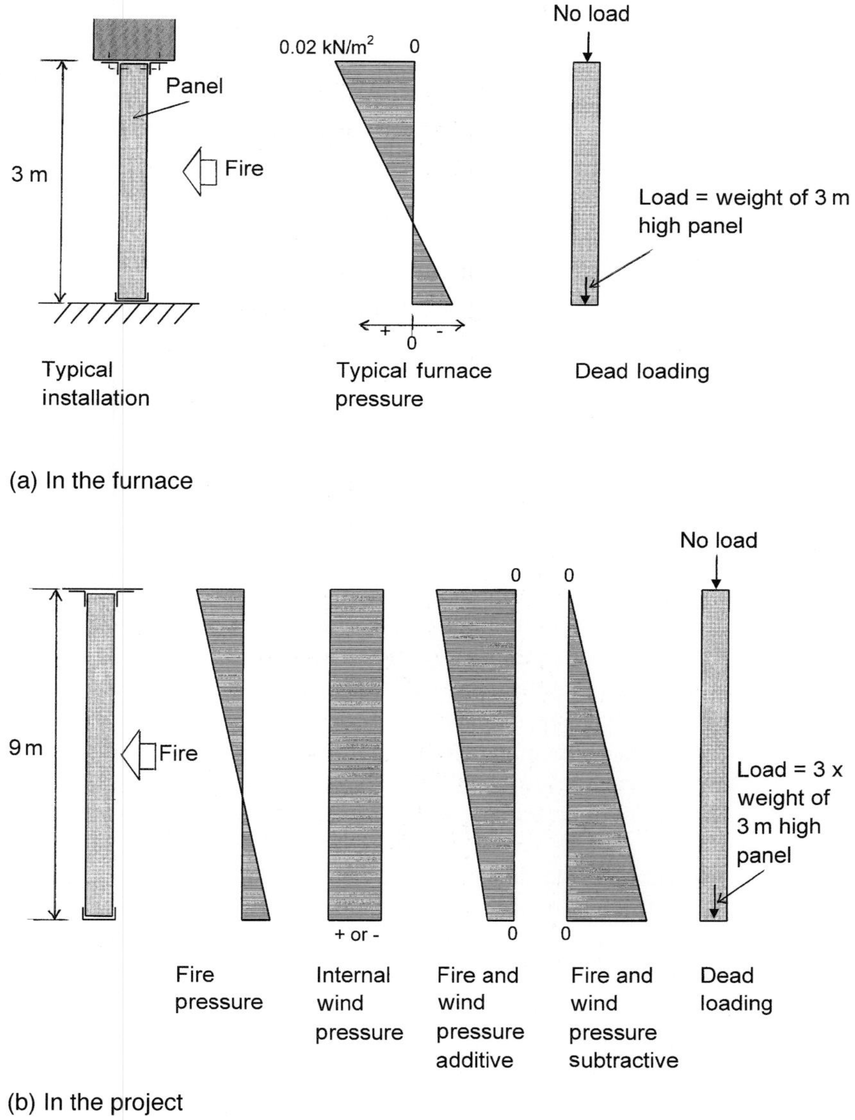

Fig. 5.8 Pressures and loads acting in the furnace and in the project.

The restraint conditions used in the fire resistance test, for example restraint against in-plane thermal expansion at the boundary of the assembly, should be compared with those present in the building and, if they differ greatly, an assessment should be made of the effect. Restraint forces are not shown in Fig. 5.8.

5.4.7 Spread of fire within the core

Fire may spread within some combustible core materials beyond the area originally affected by the fire and this can lead to increased property loss as well as making it difficult for fire fighters to extinguish the fire. Thermoplastic core materials such as polystyrene behave particularly badly in this respect whereas phenolic foam, and current formulations of polyurethane (PUR) and polyisocyanurate (PIR) tend only to char under the influence of heat and thus continue to restrict the passage of the fire away from its source. Several publications[5.28–5.31] provide evidence that PUR has given satisfactory performance in this respect in actual fires although this may not always be the case if profiled faces contain cavities that can feed the fire with air.

5.4.8 Fire stopping

Fire stopping, namely the filling of any gaps between sandwich panels and adjoining or penetrating elements where passage of the fire needs to be prevented, must be dealt with carefully. Penetrations by building service pipes, ducts, fire doors and fire shutters, etc. need to be fire stopped around their perimeter in such a way that the fire stopping achieves the integrity and insulation performance of the sandwich panels and that this is retained as the fire progresses.

An example of fire stopping for a service pipe penetration is shown in Fig. 5.9(a). A hole is cut in the panel on site. The pipe is passed through the hole and the gap between the pipe and the panel is filled with a non-combustible material such as stone wool and sealed with a fire protecting mastic (e.g. an intumescent). Two split halves of a collar of sheet steel or strong fire protecting board are finally added and mechanically fastened in place. The collar is designed to retain the fire stopping in place should the service pipe move axially due to normal temperature variations and fire temperatures and also to help in the fire condition to prevent a collapsing service pipe from tearing the panel facings or causing loss of integrity in some other way.

In the example shown in Fig. 5.9(a), if the service pipe is of steel, it will need external insulation for a short length on each side of the panel, typically about 1 metre, in order to reduce the heat conduction along the pipe from the fire exposed side. For thermoplastic pipes, such as PVC, which melt under fire attack, the collar of stone wool fire stopping should be replaced by an intumescent collar which expands to seal the gap completely as the PVC pipe collapses inwards.

Figure 5.9(b) shows an alternative detail where the penetrated sandwich panel has a formal fire resisting function. Here, a non-combustible sleeve, of a suitable internal diameter to fit snugly round the pipe, passes through the opening together with the pipe and extends for at least 1 metre on each side of the panel. The panel

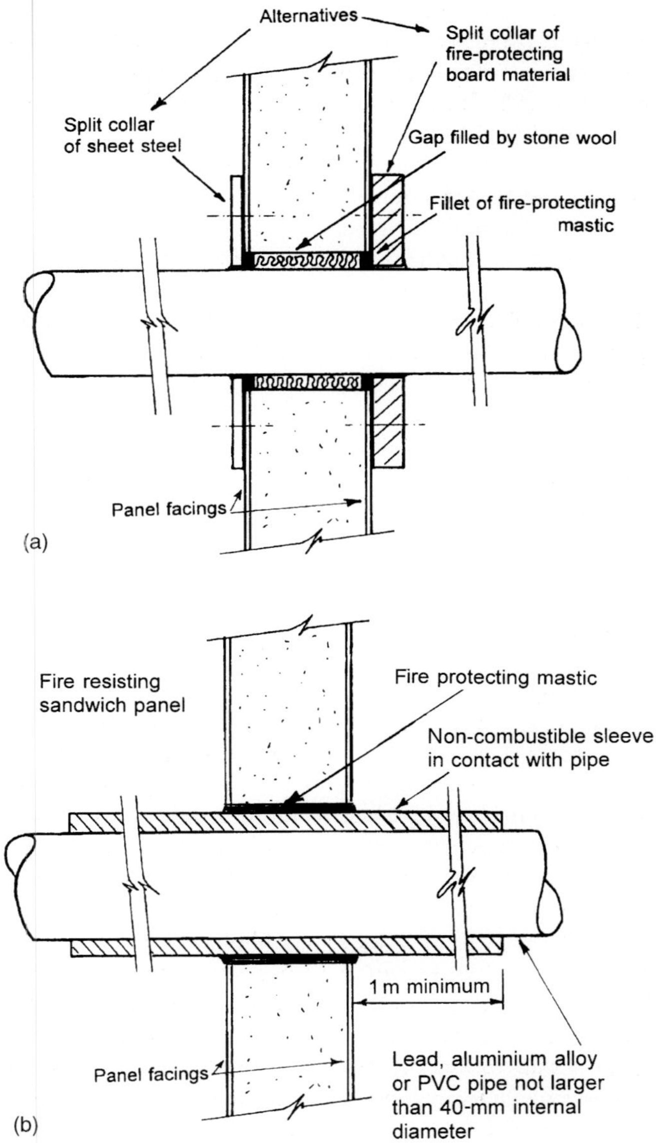

Fig. 5.9 Fire stopping at a pipe penetration.

core is sealed from the fire by a fire protecting mastic. This detail is believed to be in accordance with guidance given by the UK regulatory authorities.

5.4.9 Joints in the panel assembly

In the panel assembly subjected to a fire resistance test, the panels are in one plane. In the building, they may have changes of direction to provide external and internal corners in walls or a change in level in the case of ceilings. Care is needed in the engineering design of un-tested joints in order to ensure that integrity and insulation are preserved. Care is also needed during site assembly.

Most fire-rated proprietary steel-faced sandwich panels rely on interlocking panel edge joints as illustrated in Fig. 5.10. The more substantial the interlock, the less likely it will be that the fire exposed and unexposed face joints will become unlocked due to distortion or separation of the panel edges in fire. In some projects, where the sandwich panel assembly is fulfilling a very important fire-separating function, as in the separation of very high value products, it may be necessary to fasten the panel edges together mechanically over their full length in order to avoid loss of integrity. Edge fastening may also be necessary if the panels used in the project are considerably longer than those tested.

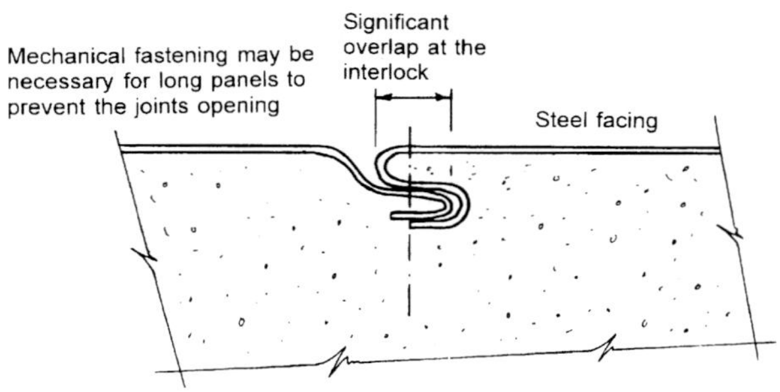

Fig. 5.10 Section through the joint at a panel edge.

The materials used for panel jointing should not, when exposed to fire, shrink or form flaming molten droplets. Shrinkage could cause loss of insulation. Flaming droplets could, if they were to find their way past the unexposed face, cause a secondary fire.

5.4.10 Sealants

The volume of sealant used in a sandwich panel assembly is small and the possession of good reaction-to-fire properties is therefore not important. However sealants should not produce flaming droplets in situations where secondary fires can be initiated.

Fire resisting joints between panels or between panels and other parts of the structure can be, and often are, formed by packing the joint with stone wool material (in the form of wool or low density strips) such that a recess is left to be filled in with a surface layer of intumescent mastic. The width of the joint should be designed so that the stone wool/intumescent system can accept the thermal movements associated with normal use and accommodate those movements generated by fire exposure.

There will be many applications in which sandwich panels do not need to possess fire resistance and ordinary sealants can then be safely used.

5.4.11 Fire protection

Fire protecting layers of material, e.g. boards, can be used to improve the fire resistance of a sandwich panel in two ways. A panel already installed in a building can have a fire protecting material applied to the face on the fire risk side. Panels having cores which can only contribute a small amount to the insulation performance in a fire, e.g. panels with PUR/PIR cores, can have their fire resistance improved by incorporating a fire protecting board within the panel during manufacture. These two methods are now discussed.

5.4.11.1 Increasing the fire resistance of installed panels

The fire resistance of an existing sandwich panel construction can be significantly increased by providing an additional fire protection layer as shown in Fig. 5.11(a). This can be a layer of board material (stone wool, exfoliated vermiculite, glass fibre, reinforced gypsum, etc.) attached to the fire-exposed side or, alternatively, a sprayed plaster layer or an intumescent paint spray system may be used. The thicker the insulating layer and the lower its thermal conductivity, the more effective it will be. However, the thermal conductivity is not a constant but is temperature dependent and this must be taken into account when evaluating fire protection materials.

Some materials, such as gypsum board, further delay the increase of temperature by the evaporation of trapped moisture. Gypsum contains about 20% of crystalline water and, during the time that it takes to evaporate this water, the temperature is maintained at about 100°C. Depending on the thickness of gypsum boards, this effect can increase the fire resistance by between 10 and 30 minutes.

The fire protecting layer needs to be fastened so that its integrity is retained during the fire exposure. Most boards with a cementitious base contract upon heating and, unless the fasteners are sufficiently close together, areas of board may fall away. Advice on fixing methods should be obtained from the fire protection manufacturer.

An alternative method of increasing the fire resistance of a sandwich panel is to coat it with an intumescent paint or spray that expands to create a protective char when it is exposed to fire. There appears to be very little experience of the use of intumescent paints in conjunction with metal-faced, foam-filled sandwich panels. Their most likely use is in upgrading the fire resistance of installed panels brought about by a change in use of the building. The main problem in using intumescent systems is the existence of the decorative coating on the fire exposed metal face which is often PVC. This, and other coatings, may not be compatible with the intumescent system and, before using this approach, confirmation of compatibility between the intumescent and the panel surface coating should be obtained from the intumescent manufacturer. Some intumescents are water-based and may degrade if exposed to rain or used in conditions of high humidity. It is also sensible to seek confirmation from the intumescent manufacturer that the fire insulation properties of the intumescent coating, together with the fire insulation properties of the core (if any), are sufficient to satisfy the fire insulation criterion: an indicative fire resistance test can be carried out on a 1-m square specimen in order to check the thermal insulation performance. However, in some applications, such as the fire protection of structural steelwork, intumescents can be very beneficial.

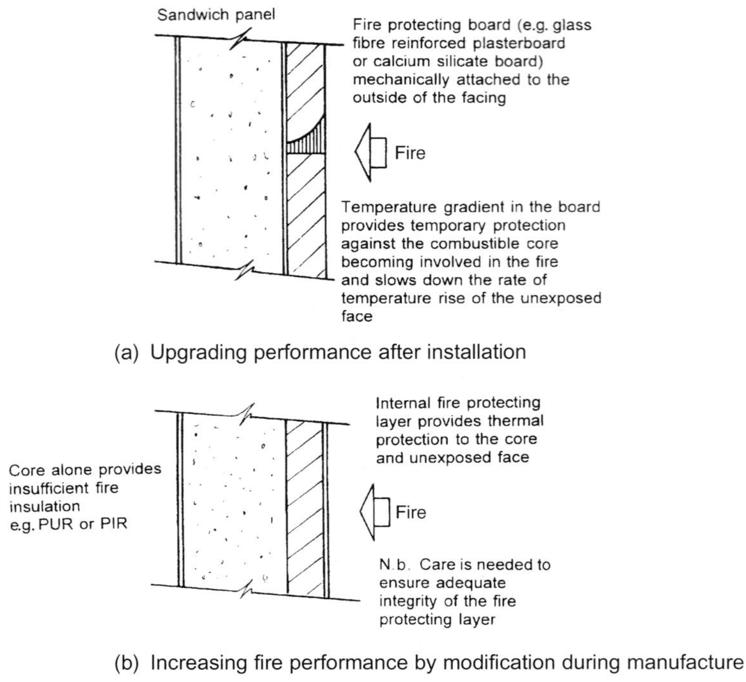

Fig. 5.11 The use of fire protecting boards in sandwich construction.

5.4.11.2 *Increasing the fire resistance during panel manufacture*

Plastic foam cores having poor fire insulation properties can have their fire resistance increased by using a fire protecting layer inside the panel, preferably in contact with the fire-exposed facing as shown in Figure 5.11(b). This means that thermal degradation of the plastic foam is delayed so that the insulation value of the fire protecting layer combined with the plastic foam can reach useful levels. This approach can be used with PUR or PIR cored panels which, without the help of a fire protecting layer, generally only achieve a 15–20 minute insulation in the standard fire resistance test for a core thickness of approximately 80 mm. Care is needed to ensure that the support for the fire protecting layer is sufficient to avoid failure of integrity within the fire protecting layer caused by sections of the layer falling down inside the panel. If this happened, the panel could fail the insulation criterion.

5.4.12 Retention of panel facings

Some metal faced sandwich panels may have facings that are not directly fixed to each other in order to avoid the cold bridge effect. It is important that such facings do not become detached in a fire and cause a hazard to occupants and fire fighters inside or outside the building. Detachment of a facing will also expose the core material to fire which, if it is combustible, will add to the fire load and increase the production of smoke and toxic/irritant gases.

Detachment of facings may occur with both walls and ceilings and some potentially hazardous scenarios are described below.

5.4.12.1 *Fire above a ceiling*

A scenario which causes a life risk is illustrated in Fig. 5.12. It is assumed that the ceiling is assembled from metal-faced panels employing a core material having a high level of thermal insulation. It is assumed that the panel facings are not attached to each other along their sides and ends, thus avoiding the cold bridge problem, and that the adhesive used to bond the facings to the core is not high-temperature resisting. A fire has started and the fire brigade has arrived and has begun to search the building for occupants and to try to find the seat of the fire. The fire has penetrated into the roof void (or may have started in the roof void) and, unknown to the firemen, is spreading unseen above their heads. The ceiling sandwich panels delaminate due to weakening of the adhesive which bonds the fire-exposed (upper) facing to the core. Since the lower facing is not mechanically attached to the upper facing, it and the core material suddenly fall down, providing a missile hazard to firemen underneath or a tangle of sheets and burning core material which now obstruct their escape route.

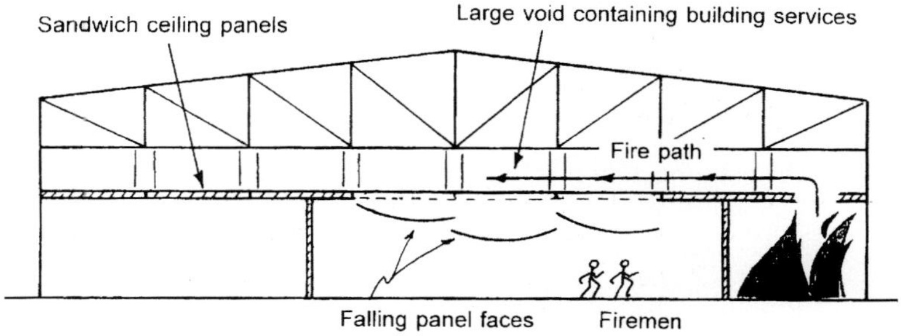

Fig. 5.12 The hazard of unseen fire in a ceiling void.

This scenario is seriously aggravated if the ceiling core material is thermoplastic and melts at low temperature. Not only does the fire spread more rapidly within the core material as air is drawn in, but subsequent collapse of the ceiling deposits a pool of molten burning plastic on the floor below. This eventuality is believed to have led to the death of two firemen in the UK Sun Valley food processing factory fire.

Evidently, this and the following scenario can be avoided by proper detailing of the connections between the ceiling panels and the supporting structure.

5.4.12.2 *Fire below a ceiling*

A layer of hot gases below a ceiling, as shown in Fig. 5.13, may cause delamination due to weakening of the adhesive and collapse of the lower facing. It is likely that the normal occupants of the building would have left the area before these conditions developed, but firemen may be expected to work in these conditions and could be at risk from collapse of the lower facing.

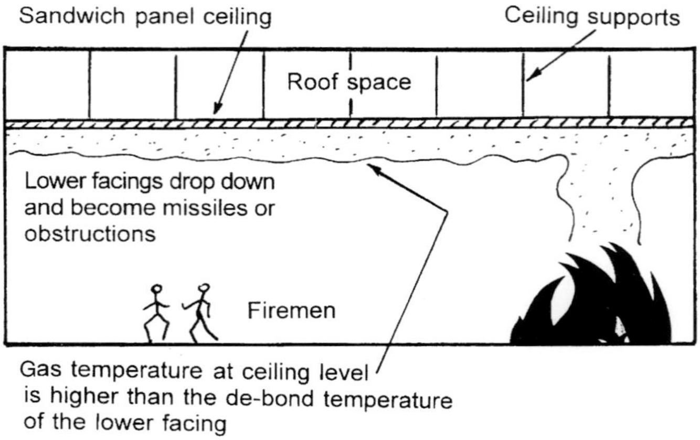

Fig. 5.13 Hazard of fire below a ceiling.

The scenario described above emphasises the need to examine the fire resistance test report in order to establish how the panel assembly was tested. The designer may have assumed that the assembly of sandwich panels has been fire resistance tested with the panel ends resting on the top of the furnace as shown in Fig. 5.14(a). The lower facing is, to an extent, trapped at its ends and this, together with the fact that the ends of the panel are cool where they extend beyond the line of support and are thus not directly exposed to fire, means that the lower facing remains in place for longer than if it was fully exposed to fire and not supported at its ends. In the building, however, the panels are likely to be suspended from the upper face by secondary steelwork (suspension rods), as indicated in Fig. 5.14(b), so that the lower face is not supported from below. If fire occurs from below, the lower facing can drop when the adhesive bond between the lower facing and core (which usually has a low softening temperature) fails as shown in Fig. 5.14(b). This exposes the core which, if combustible, adds to the fire load and releases smoke and toxic/irritant combustion products. If, alternatively, fire occurs above the ceiling panel assembly as, for instance, when fire starts in the void between the ceiling and the roof as in Fig. 5.14(c), the bond between the upper facing and the core fails, allowing the core and lower facing to drop down which may block escape routes and make fire fighting dangerous. This was the scenario in Fig. 5.12. The problem of falling facings can be overcome by joining the adjacent ends of the lower facing together with mechanical fastenings to form a continuous membrane and this solution is described in more detail in Section 5.5.4.

5.4.12.3 Calculation of the catenary force in a sandwich panel

A panel can be prevented from collapsing due to the loss of flexural strength when a fire occurs by providing a horizontal outward force at the ends of the panel. Making the reasonable assumption that a panel or its facings has negligible bending resistance, it will sag to form a catenary, like a cable. The horizontal reaction force, H, can be calculated for a catenary of parabolic form, as shown in Fig. 5.15, using the equation below:

$$H = wL^2/8d$$

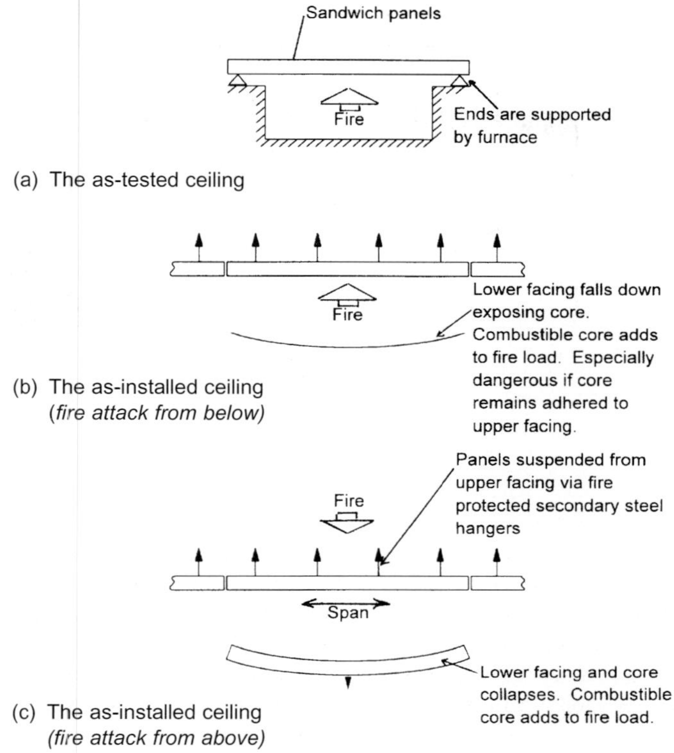

Fig. 5.14 Influence of the method of ceiling support on the fire performance.

where w = uniformly distributed load per unit length (the dead load in this case)
L = span of the catenary (span of the panel)
d = mid-span deflection

Before this equation can be used, it is first necessary to make an estimate of the mid-span deflection, d. This can be done if the temperature of the facings is known or can be estimated. Consider the panel face to be an initially flat member with immovable ends. It is heated such that the temperature rise is T and, since the ends are held in position, it bows into a circular arc. The mid-span deflection is then given by:

$$d = L\sqrt{0.375\,\alpha\,T}$$

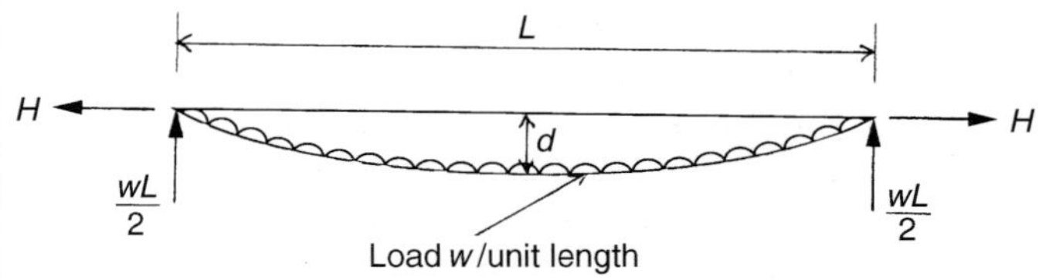

Fig. 5.15 Catenary force diagram.

where L = length of the member
α = coefficient of linear thermal expansion
T = temperature rise

Substituting this equation into the previous equation gives an estimate of the horizontal reaction force.

Assume that the panel is 4.5-m long × 1.0-m wide with steel facings 0.6-mm thick and that the temperature of the exposed facing is 800°C (i.e. a temperature rise of 780°C). The coefficient of expansion is 0.000 014 per °C and the density of steel is 7850 kg/m^3. So:

$$d = 4.5\sqrt{0.375 \times 0.000014 \times 780} = 0.288\,\mathrm{m}$$

$$\begin{aligned} &\text{volume of one facing} &&= 4.5 \times 1.0 \times 0.0006\,\mathrm{m}^3 \\ &\text{weight of one facing} &&= 4.5 \times 1.0 \times 0.0006 \times 7850\,\mathrm{kg} = 21.2\,\mathrm{kg} \\ &\text{weight/unit length} &&= 21.2/4.5 = 4.71\,\mathrm{kg/m} = 46.2\,\mathrm{N/m} \end{aligned}$$

$$H = \frac{46.2 \times 4.5^2}{8 \times 0.288} = 406\,\mathrm{N} = 0.406\,\mathrm{kN}$$

This calculation can be refined to take account of the additional extension of the panel due to the tensile force, H, bearing in mind the reduction in the elastic modulus at high temperature. Here, the force H is quite small so that this extension can be neglected but this may not always be the case.

Fire tests on sandwich panels having a trapezoidal facing have shown that the catenary force, H, is given approximately by $H = wL$ but it is not clear if this empirical relation is valid for a flat facing. Substituting the data used in the above worked example would give an empirical value of $H = 46.2 \times 4.5 = 208$ N.

5.4.12.4 Fire on one side of a wall panel

The detachment of an external wall panel facing may be a hazard in the scenarios illustrated in Figs 5.16 and 5.17. Fire attack on the inside face, as in Fig. 5.16, weakens the bond between inner facing and the core so that the external facing (on its own or with the core) is torn off by wind and acts as a missile and also permits significant levels of radiation to be transmitted to neighbouring buildings. Fire

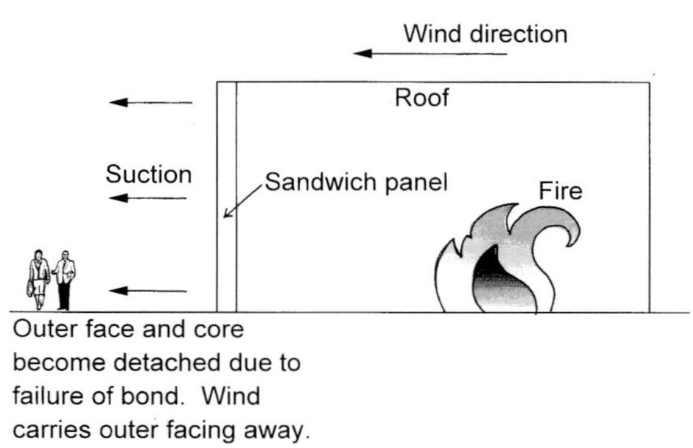

Fig. 5.16 Hazard of a fire on the inside face of an external wall.

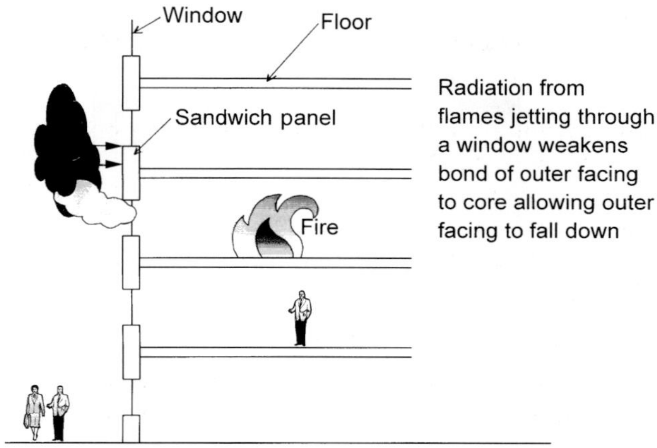

Fig. 5.17 Hazard of a fire on the outside face of an external wall.

attack on the outer face, as in Fig. 5.17, again weakens the bond and allows the facing to act as a missile to people below.

There is a particular problem associated with tall fire walls used in single-storey buildings where the sandwich panels are used in the vertical orientation. Consider the following situation. A sandwich panel has been tested as a 3-metre high wall in the furnace. It is fixed to the test frame with light gauge steel angles as shown in Fig. 5.18(a). The panel system is to be used as a 12-metre high fire separating internal wall in a single-storey factory.

There are two possible problems, both arising from the unknown integrity performance in the large-scale assembly. First, it is clear that once the bond between the fire exposed face and core material fails, which occurs when the facing reaches a temperature of approximately 150°C, the assembly no longer possesses any flexural strength and the fire-exposed facing is unable to transmit the force represented by its self-weight to the base channel with the result that the fastenings at the top of the panel are now required to carry the full load. The mechanical fastenings at the top may not have been designed for this condition and failure may occur due to partial or full collapse of the assembly as shown in Fig. 5.18(b).

The second problem arises from pressure within the enclosure, bounded on one side by the sandwich panel fire separating wall, caused by differential air pressure as illustrated in Fig. 5.19. This could have a significant detrimental effect on the integrity performance. The effect of under-pressure on one side caused by wind and over-pressure caused by fire on the other side would increase the tensile restraint needed at the top of the panel assembly and could cause an integrity failure at this point. Wind generated over-pressure or under-pressure may similarly affect the integrity of vertical panel joints. Careful design can overcome these problems as shown in Section 5.4.9.

Despite the potential problems identified in the above examples, well-detailed sandwich panels with steel faces and lightweight cores can achieve significant fire resistance times in terms of stability, integrity and insulation. However, panels with plastic cores have difficulty in satisfying the insulation criterion and some manufacturers (incorrectly) discount the insulation requirement as being less

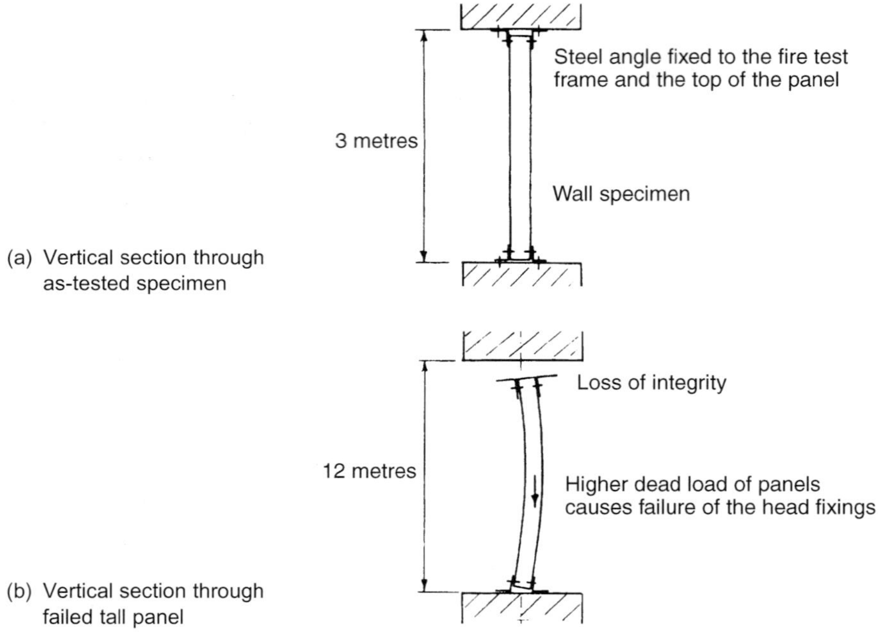

Fig. 5.18 Example of loss of integrity due to panel height.

important and merely quote the resistance times of their products in terms of stability and integrity. However, modern polyurethanes and phenolic foam formulations are capable of achieving useful insulation times. Mineral wool cored panels are capable of satisfying all three criteria for large periods of fire resistance.

5.4.13 Fire load of panels

In most cases, the fire load of the combustible contents of sandwich panels is small compared with the fire load of the building contents. This may not be so for thick panels and it is possible to calculate this comparison and the method is given in Section 5.1.5.1 *Effect of fire load.* In order to decide whether the panel fire load is acceptable, it may be necessary to make a qualitative fire hazard assessment.

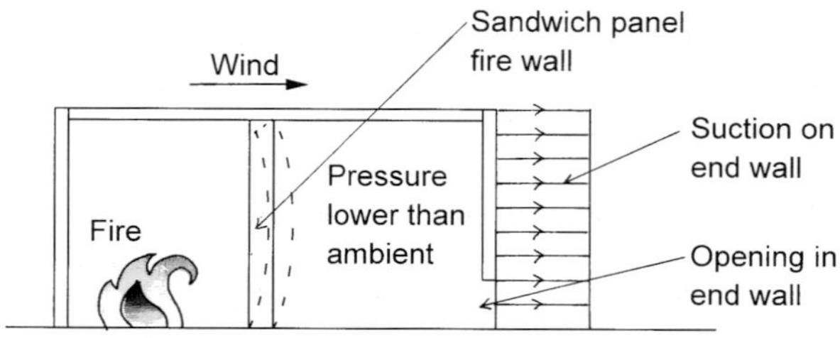

Fig. 5.19 Effect of wind pressure on integrity and stability.

5.4.14 Choice of panel facings

Combustible facings such as plywood or chipboard should be avoided even if the panel is not fire rated. Sheet steel is the preferred facing as, unlike aluminium alloy which melts at around 600°C, it is able to withstand the effects of fire for all practical applications. The surface finish is usually a thin plastic coating whose combustible content can be ignored, but a check needs to be made that the coatings possess the required low level of flame spread and other reaction-to-fire test requirements.

5.4.15 Choice of core material

This is a most important decision. Ideally the core should be of low combustibility. In practice, stone wool comes nearest to this ideal because the binders and glues in the core material make only a limited contribution to the fire and produce low levels of toxic gases or irritant fumes. A stone wool core does not produce flaming droplets and can retain its insulation function in a fully-developed fire for a significant length of time. However, this may not be the optimum solution when other functional requirements such as thermal insulation, self-weight, cost, etc. are taken into consideration.

Other core materials, such as plastic foams, should be chosen very carefully because of their relatively poor reaction-to-fire properties and, ideally, plastic foam cored panels should not be used as the internal partitions or ceilings of buildings when the toxic, irritant and corrosive products of combustion can be released at an early stage in a fire, especially when this would present a risk to life.

Of the major plastic foams used in sandwich panel construction, those having the 'best' fire performance include phenolic, retarded polyurethane and polyisocyanurate foams. Polystyrene foam (either XPS or EPS) has the worst performance because of its tendency to melt away from the fire exposed face and produce flaming molten droplets.

If a fire forms in the space below a ceiling formed by sandwich panels having a core of polystyrene foam, the polystyrene above the fire will first melt onto the lower face of the panel and in so doing it will allow the ready access of air into the space between the metal faces. The molten material will then catch fire and the temperature in the void will rise rapidly. The molten region will quickly grow beyond the area in direct contact with the fire drawing in more air. Soon, and possibly before significant flaming droplets are visible below, the whole ceiling area will be alight. The ceiling will then collapse depositing a mass of burning liquid onto the space below with catastrophic consequences.

It is not possible to postulate a similar mechanism of failure for any of the other materials considered in this book.

A similar, though possibly less spectacular, failure mechanism may occur in walls consisting of polystyrene sandwich panels. When a fire occurs adjacent to the wall, the polystyrene first melts and forms a puddle at the bottom of the void. Air is drawn in and the molten material catches fire. This raises the temperature in the void allowing the mechanism of melting followed by ignition to spread laterally along the wall.

Evidently, joints between adjacent panels that incorporate fire stops could inhibit the fire spread mechanisms described above. Unfortunately, these are not usual in panels with polystyrene core material and there does not appear to be any evidence regarding their effectiveness.

Phenolic, retarded polyurethane and polyisocyanurate foams have the benefit that they form a char and fire is unlikely to spread within a panel beyond the fire-affected zone provided that there are no significant air gaps in the panel construction. PUR/PIR on its own is not usually able to provide more than 15–20 minutes fire resistance because of failure of the insulation criterion. Modified phenolic foam (i.e. mineral-filled open cell phenolic foam) can provide up to 2-hours' fire resistance. These foams do, however, give off toxic and irritant products of combustion and their use in external wall or roof panels of life-risk buildings may be acceptable if the core is protected by a sheet steel facing having good integrity at the panel joints and the combustion products can escape to the outside air rather than be trapped in the building where they can be transported throughout the building so as to cause a life risk.

5.4.16 Choice of adhesive

Most adhesives used in commercial sandwich panel manufacture are one- or two-component polyurethanes which start to soften at about 150°C and can retain a little strength for short-term exposures at temperatures greater than 150°C. The temperature resistance of the adhesive above the initial softening temperature is unimportant in view of the rapid increase in the combustion gas temperatures. This means that the fire-exposed metal facing of a panel using slabs of core material delaminates early in a fire resistance test. The smaller the amount of adhesive used, the smaller is the amount of heat released by its combustion, though this is usually of minor importance.

5.4.17 Required behaviour for horizontal panels (both ceiling and roof panels)

Ceilings should be mechanically fixed to the supporting structure at both ends so that, in the event of loss of core strength in a fire, membrane action may be developed in the bottom sheet thus preventing the bottom face from falling off. Some engineering solutions are given in Section 5.5.4.

Thermoplastic materials that melt at relatively low temperatures (e.g. both expanded and extruded polystyrene) should not be used as core materials for ceiling construction because of their tendency to form pools of molten flaming liquid which may be retained for a time on the lower face of the ceiling panel before being deposited onto the floor below with possibly serious consequences.

Sandwich panels that are designed to carry heavier loads than those which are normal on ceilings generally have to rely on the performance of the unexposed face if they are to retain any significant load-bearing function in the case of fire. The sandwich core then functions primarily as a fire insulation layer which should be capable of keeping the temperature rise of the unexposed face below 180°C for the required period of fire resistance. In this case, it is usual to make the unexposed face from a trapezoidally profiled steel sheet which can be designed to carry the design load under fire conditions with a low safety level appropriate for fire design as shown in Fig. 5.20.

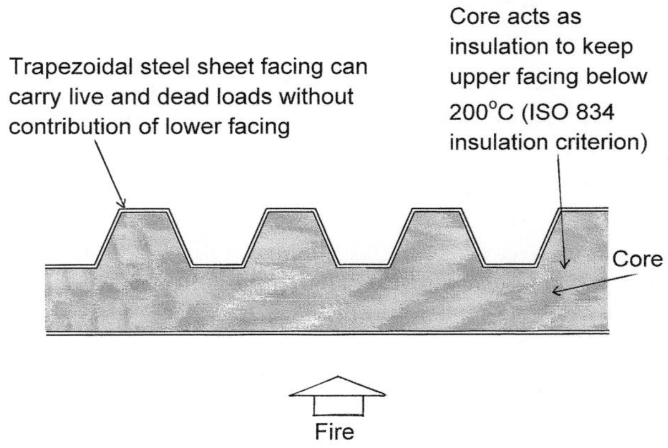

Fig. 5.20 Ceiling capable of carrying live load during a fire.

5.4.18 Required behaviour for vertical panels (internal and external walls)

For the formal fire resistance testing of vertical walls, a standard panel size of 3 m × 3 m is used. This can be a practical disadvantage when walls are to be designed which are significantly higher than 3 metres because the test does not reveal the real behaviour due to the greater vertical span and dead load. Such cases have to be designed with care, especially if the wall has to carry loads in addition to its own weight such as the vertical load from a roof or ceiling or a lateral pressure load. For this purpose, a top fixing system to the primary structure is advantageous.

If polystyrene is used in wall construction, it is essential that the vertical joints between adjacent panels should be fire stopped.

5.5 ENGINEERING SOLUTIONS

In this section a number of fixing methods are illustrated for panels that are required to possess more than half an hour fire resistance. The panels have sheet steel facings and a stone wool core of low combustibility. Although the fixing methods shown are adaptations of the fixing systems for a particular proprietary fire resisting panel system, it is suggested that the methods shown have a wider application and illustrate the principles as well as specific details.

The fundamental engineering principles are that:

- the structure supporting the panel assembly shall be fire resisting for the required period
- both panel facings shall be mechanically fastened to the panel support structure in a substantial way so that they cannot become detached when early loss of the flexural strength of the panel occurs

- the non-combustible stone wool core material is held in place by the adhesive bond between the core and the unexposed facing
- joints shall be fire stopped with stone wool of appropriate density which is retained in place and retains its integrity as the structure deforms in fire.

The fixing methods shown are for:

- external walls
- internal walls
- roofs
- ceilings

5.5.1 External walls

It is possible that an external wall on or near a site boundary or near another building may be required to possess fire resistance when exposed to fire from the inside. This means that the structure supporting the panel assembly and the panels themselves must be capable of retaining their insulating properties and remaining roughly vertical in order to act as a heat shield to prevent massive radiation to neighbouring buildings. Figure 5.21 shows the scenario.

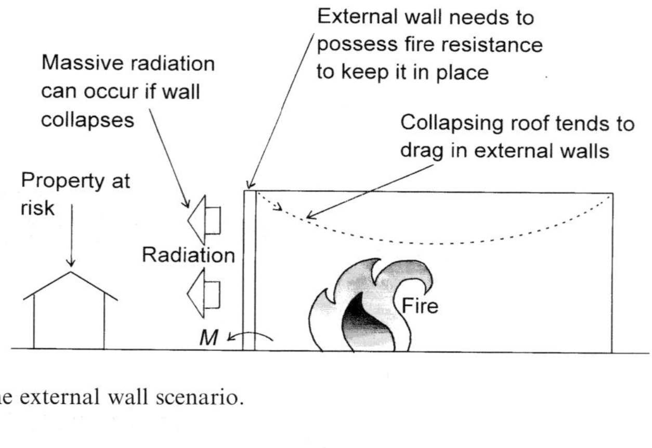

Fig. 5.21 The external wall scenario.

In the case of a steel portal frame building, it is possible to avoid inward collapse of the external wall structure by fire protecting the external columns up to the level of the underside of the haunch and designing the column section size and foundation to the resist overturning moment developed as the roof collapses and forms a partial catenary structure,[5.42] as shown in Fig. 5.22.

Figure 5.23 shows a vertical section through the wall incorporating vertical panels, which has fire resistance from the inside.

Figure 5.24 shows a horizontal section through the wall incorporating horizontal panels, which has fire resistance from the inside.

Figure 5.25 shows a horizontal section through an external wall having fire resistance from the outside. It should be noted that the stone wool packing inside the flashing is only necessary in order to satisfy the ISO 834 insulation criterion

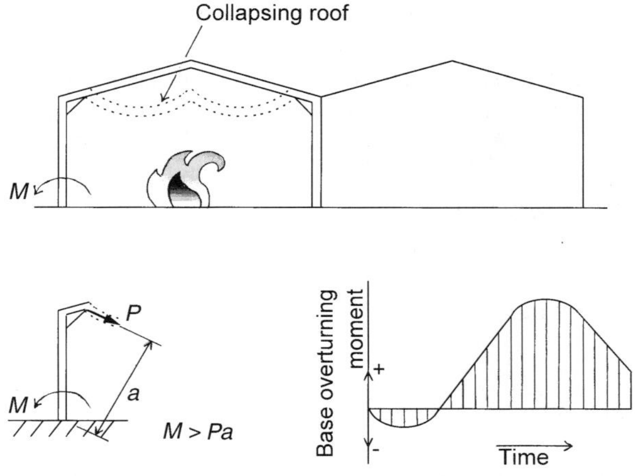

Fig. 5.22 Development of the overturning moment as a portal frame collapses.

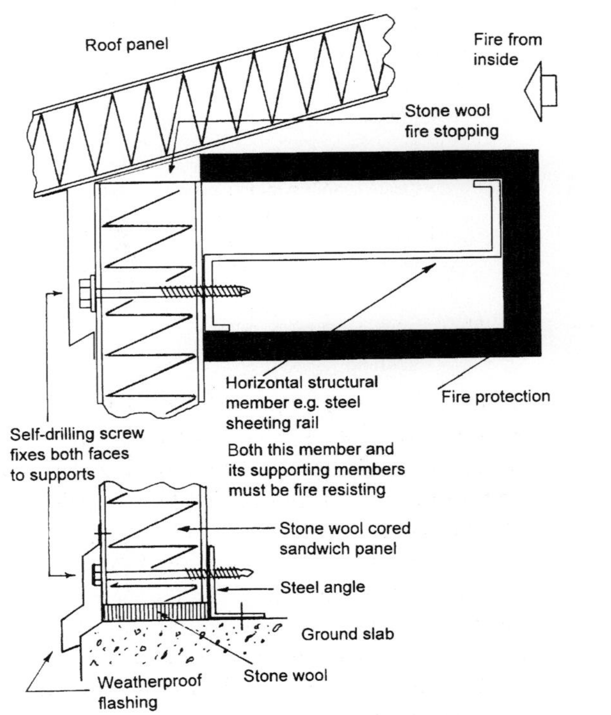

Fig. 5.23 Section through an external wall having fire resistance from the inside.

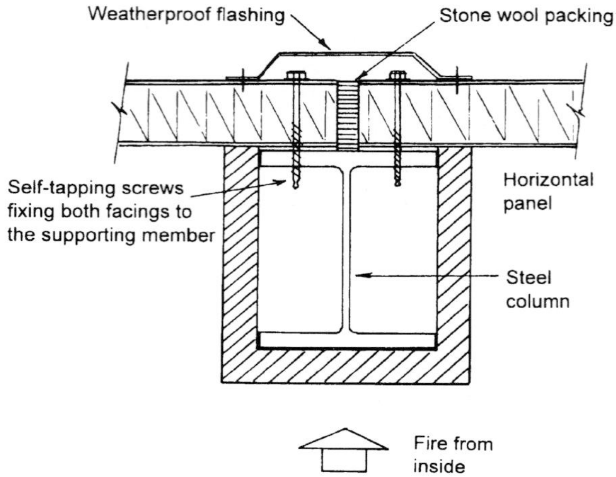

Fig. 5.24 Section through an external wall with fire resistance from the inside.

that the unexposed face of the panel should not increase in temperature by more than 180°C.

The junction formed by an internal fire wall of traditional construction and an external wall of sandwich panels needs to be carefully engineered in order to prevent fire passing from one enclosure to another via the sandwich panels. Figure 5.26 shows several possible routes of fire spread for panels with both combustible and non-combustible cores. These routes can be avoided quite simply for a non-combustible cored panel of stone wool, as shown in Fig. 5.27, but it is more difficult in the case of combustible cored panels, as shown in Fig. 5.28(a) and (b). If the fire wall is thick, and the fire is not severe, it is possible that a PIR or phenolic

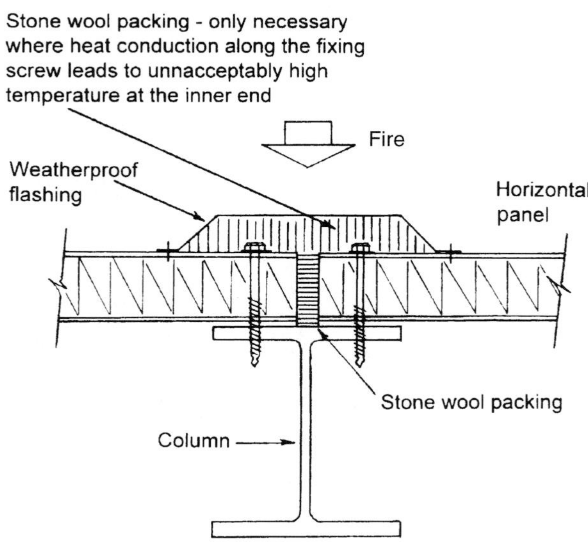

Fig. 5.25 Section through an external wall with fire resistance from the outside.

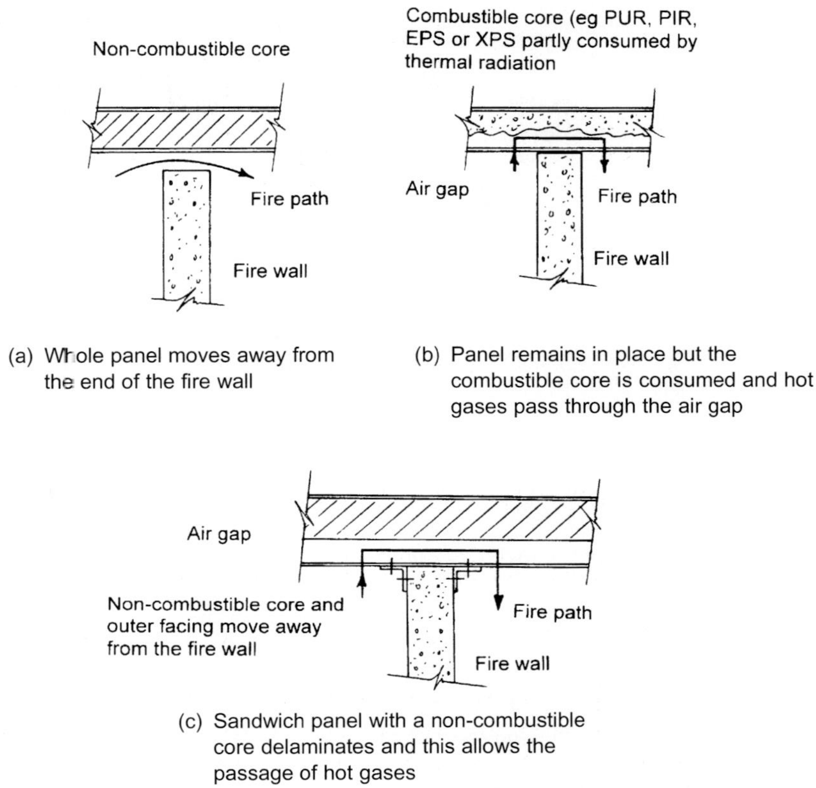

Fig. 5.26 Possible routes for fire spread past the end of a fire wall.

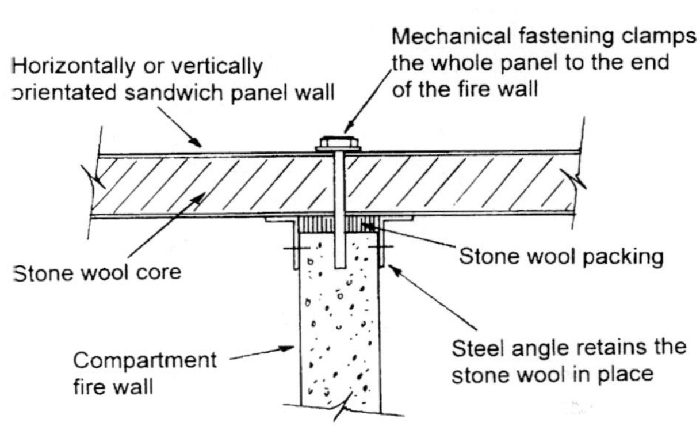

Fig. 5.27 Section through the junction between a fire wall and a stone wool cored panel.

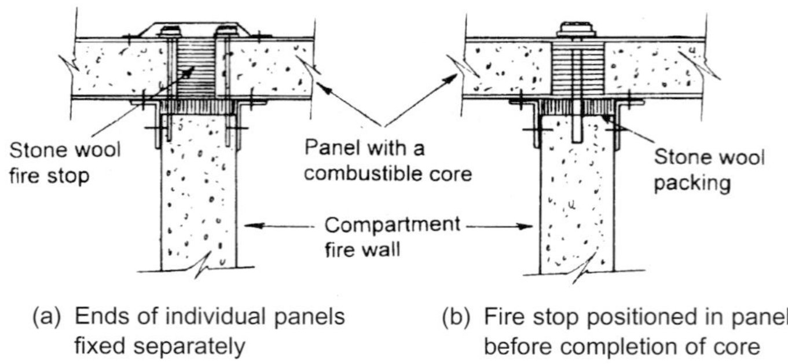

Fig. 5.28 Section through the junction between a fire wall and a panel with a combustible core.

foam core would not decompose to such an extent that fire can be transmitted by radiation or conduction from the inner facing in the non-fire affected enclosure.

The junction between a traditional fire compartment floor and an external wall of sandwich panels needs careful engineering in order to prevent fire spread to the enclosure above the floor. The possible routes of fire spread are similar to those illustrated for the junction of an internal fire wall as shown in Fig. 5.26. The problem is, however, made more difficult by the thermal bowing of the fire resisting floor as illustrated in Fig. 5.29, which means that any connection between edge of floor and the wall panels must accept a vertical deformation which can be more than 200 mm for reinforced concrete floor spans greater than 4.5 metres. The thermal bowing displacements of concrete floors of 4.5-metres span for different thicknesses and types of concrete and different standard fire test exposures are

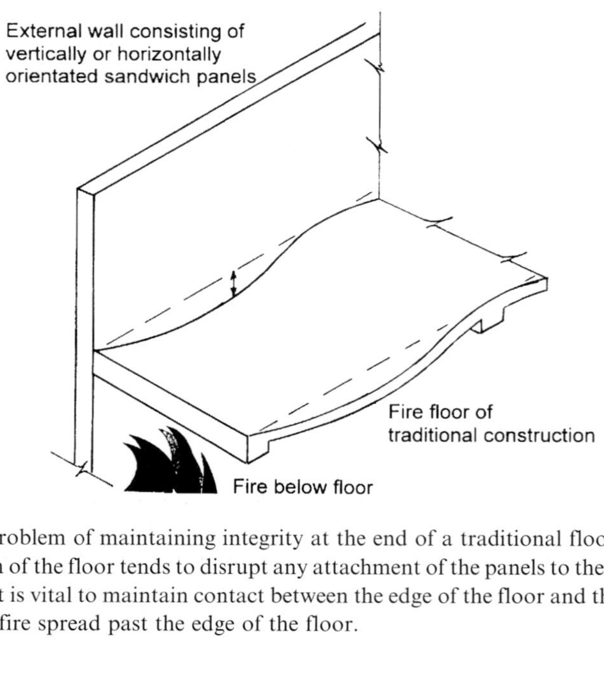

Fig. 5.29 The problem of maintaining integrity at the end of a traditional floor. Thermal bowing deflection of the floor tends to disrupt any attachment of the panels to the edge of the floor. However, it is vital to maintain contact between the edge of the floor and the panels in order to prevent fire spread past the edge of the floor.

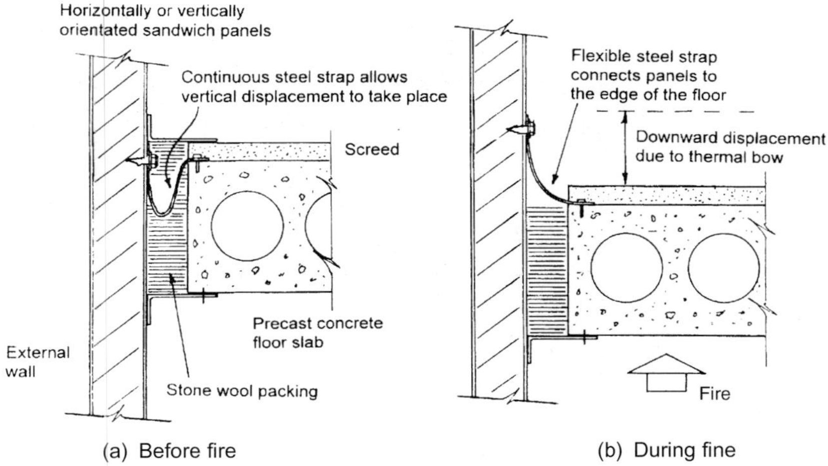

Fig. 5.30 A method of fire stopping the junction between a wall and a floor subject to thermal bow.

available together with a method for calculating the displacements for different thicknesses and spans of floor.[5.43] The relative deformations between the floor and the wall can be accommodated by the detail shown in Fig. 5.30.

5.5.2 Internal walls

Internal walls may need to possess fire resistance when exposed to fire from either side. Unlike an external wall, an internal wall can be laterally restrained at the top by the structure of the enclosure which is unaffected by fire. Figure 5.31 shows the scenario.

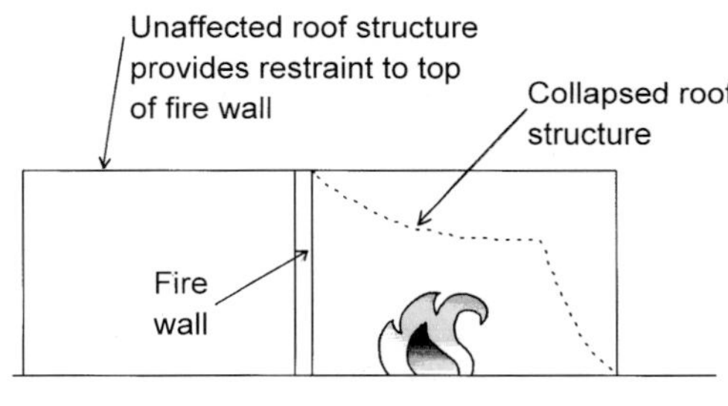

Fig. 5.31 The internal fire wall scenario.

Figure 5.32 shows a vertical section through the wall incorporating vertical sandwich panels. The panel edges at the top of the panel are attached to the structure above by means of special steel cleats which are sufficiently flexible to allow the structure above to move up and down in normal use but which become suspension devices to prevent the facings falling down when fire occurs. The fire

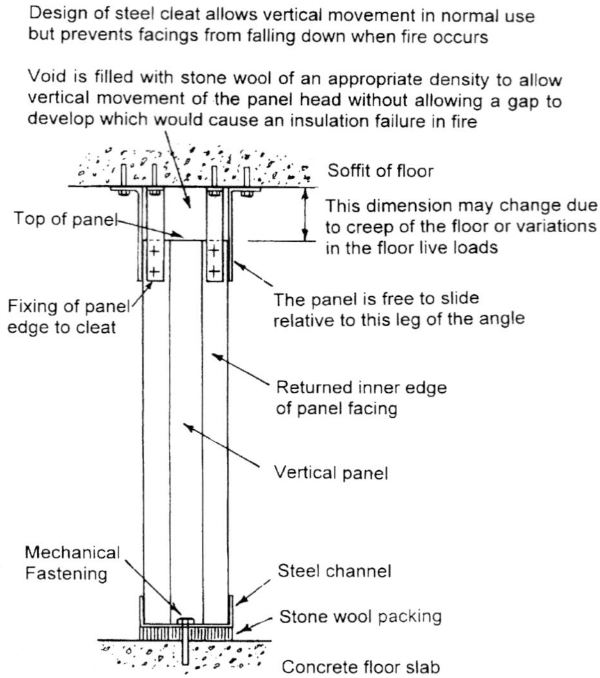

Fig. 5.32 Section through an internal wall with fire resistance from each side.

insulation of stone wool, which fills the void between the top of the panel and the underside of the structure above, needs to be of an appropriate density such that it will expand after compression so that the void is always completely filled, which is necessary if insulation or integrity failure is not to occur.

Where two fire resisting walls form a corner it is essential that the junction does not open up to allow the passage of fire. Figure 5.33 shows a horizontal section through a corner formed by two internal walls. The panels are not directly joined together but are fixed to a structural member, in this case a steel I-section column, which gives the necessary support over their full height.

5.5.3 Roofs

Under fire safety building regulations, roofs rarely need to possess fire resistance. The exceptions are when the roof forms part of a means of escape or when the roof is connected to a higher multi-storey building where fire breaking through the roof could set fire to the higher building.

Where fire resistance is required, the roof support structure has to have at least the same fire resistance as the sandwich panelled roof, but common sense suggests that the stringent ISO 834 criteria for integrity and insulation could be partly relaxed.

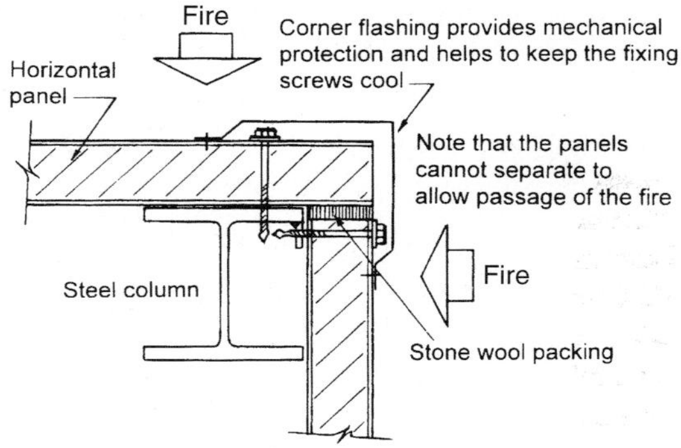

Fig. 5.33 Section through the corner of an internal wall with fire resistance from one side.

5.5.4 Ceilings

Fire-rated ceilings are normally designed to resist fire from below. However, where there is a cavity between the top of the ceiling and the underside of the roof or floor, fire attack can be from either side. In some cases the ceiling may be required to make a contribution only to the fire resistance of the roof or floor. In other cases the ceiling acts as a separate component and is required to possess the full period of fire resistance in itself.

Where the ceiling contributes to the fire resistance of the floor or roof, as in Fig. 5.34, the ceiling has to possess good integrity in its assembled form but the ISO 834 insulation criterion does not apply. The ceiling needs to limit the temperature of the air in the cavity between the ceiling and floor only to a level which keeps the unprotected steel members (steel beams, steel cross-member and hangers) below

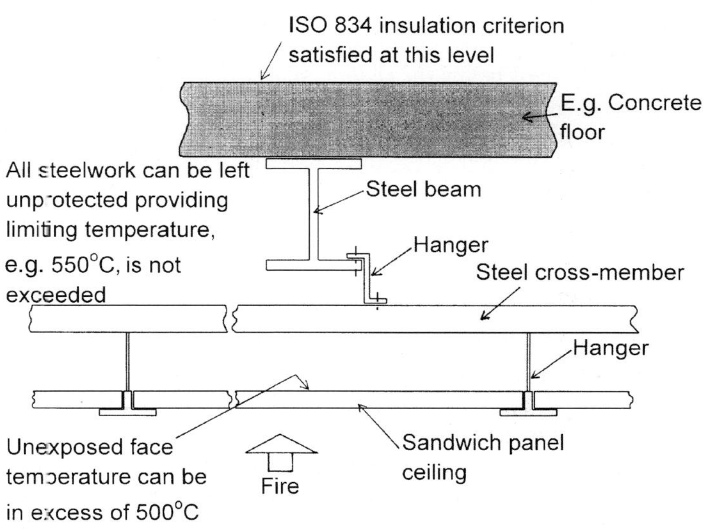

Fig. 5.34 Ceiling which contributes to the fire resistance of a floor.

their limiting temperatures, which will depend on the load level in those members. The member which controls the fire resistance of the whole construction is usually the steel beam because, if properly designed, its load level is high and its limiting temperature to resist runaway deflection is correspondingly low.

Where the ceiling provides all the fire resistance needed, the ISO 834 insulation criterion has to be satisfied on the unexposed (upper) face of the ceiling. With a steel-faced sandwich panel, it is important, as emphasised earlier, to prevent the facings from falling down and this is done by forming two separate membranes (a membrane of the upper facings and a membrane of the lower facings) using T-sections and straps as shown in Fig. 5.35 which shows a vertical section through a typical T-section support.

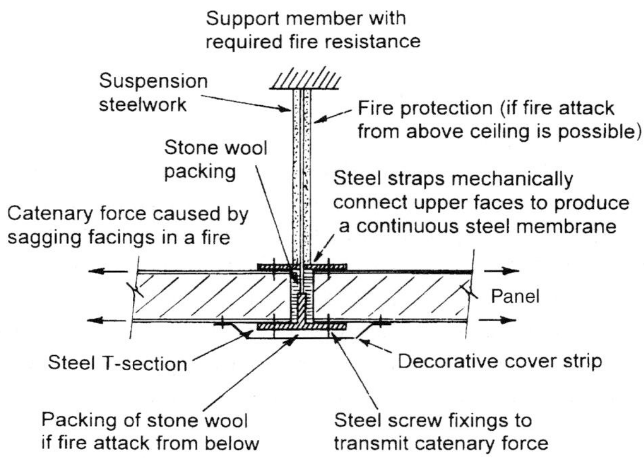

Fig. 5.35 Section through a sandwich panel ceiling showing the T-section support.

It should be noted in Fig. 5.35 that the details provide for resistance to fire attack from above or below the ceiling. With fire attack from above it is necessary for the support member and the suspension steelwork to possess fire resistance. For fire attack from below it may be necessary to include a packing of stone wool above the decorative cover strip in order to limit the temperature attained by the T-section and the screws joining the lower facing to the T-section.

Where the ceiling stops at an external wall and the wall at that level is not robust enough to transmit the catenary force in the panel facings, it is necessary to incorporate a diagonal bracing member, as shown in Fig. 5.36, which is capable of transmitting the catenary force back to the main support structure. The downward leg of the braced member is fixed to the leg of the flange of the angle section and the upper facing is also attached to the braced member. Failure to provide the diagonal bracing may allow integrity failure at the junction of wall and ceiling.

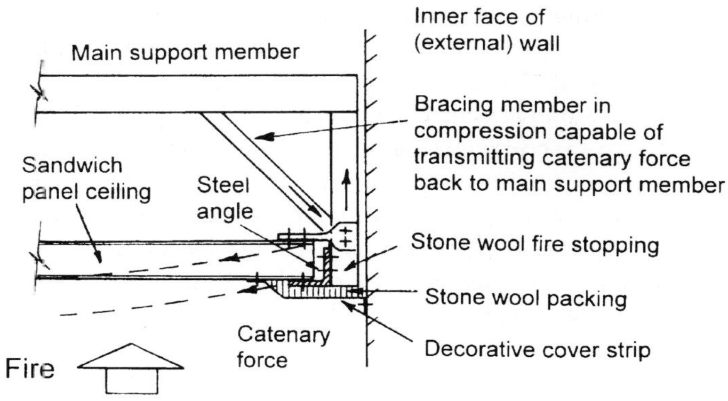

Fig. 5.36 Method of resisting the catenary force at the perimeter of the ceiling.

5.6 REPARABILITY AFTER FIRE

Fire-damaged panels can usually be easily replaced. The extent of panels to be replaced may be expected to be less for panels with non-combustible cores since it is impossible to get fire spread within the core beyond the fire affected zone. However, there is ample evidence from actual fires that this is also true for panels with good-quality polyurethane core material.

If the damage in fire is only to the surface coatings of the sheet metal facings, repair can be accomplished by washing and/or repainting. If, however, the heat has caused debond of the fire-exposed facing, the panel can easily be replaced.

5.7 CHECK LIST FOR DESIGNERS

The following check list is provided for the designers and specifiers of sandwich panels and is intended to help to ensure that all of the relevant factors are taken into consideration.

(1) Consult, at an early stage, the building and fire authorities and the insurer in order to establish the desired fire performance in terms of life safety, property protection and environmental protection.
(2) Identify the relevant fire regulations, the relevant fire test methods and the classification system.
(3) Select a panel system with adequate fire resistance properties when these are required.
(4) Wherever possible, choose a panel system in which both faces are firmly secured to the supporting structure.
(5) Select panels in which the combustible content of all materials in the panel has been minimised. This will reduce the amount of smoke, toxic and irritant products produced in a fire. Try to minimise the use of materials (e.g. PVC) which give off HCl when involved in fire which can damage the structure, finishes and services in the rest of the building. Avoid using panel materials which present a toxic and obscuration hazard to fire fighters.

(6) Minimise the use of materials which can melt and form flaming droplets to cause new outbreaks of fire. This is especially important in ceiling and roofing applications.
(7) In panels required to possess fire resistance, ensure that all fasteners are steel, preferably stainless steel. Remember that fasteners proved acceptable in the fire resistance test may be inadequate for panel assemblies much larger than that tested due to increased stresses and thermal movement.
(8) Remember that fire resistance tests are made on specimen panel assemblies typically 3 m × 3 m for a wall and 3 m × 4.5 m for a ceiling or roof and that larger assemblies may behave differently. If in doubt have an assessment made by a suitably qualified and experience fire safety engineer.
(9) Ensure compatibility of fire resistance requirements. There is little point in using a panel assembly that has fire resistance if the panel support structure has less or zero fire resistance. This is especially important when using the panel suspension concept.
(10) If in doubt about the fire insulation performance of a panel design, consider undertaking an indicative fire resistance test (a test on a 1-m square specimen) on a specimen which incorporates a panel joint. Measurement of temperatures on the unexposed face will indicate if the insulation criterion can be met. However, be careful when trying to predict integrity performance as this is very difficult – only under very clearly defined conditions can integrity be predicted.[5.44]
(11) Remember that, in the fire resistance test, only a very small positive pressure is developed in the furnace, whereas, in the building, pressure acting on an internal or external wall due to wind or internal air pressure may be greater and may have an adverse effect. However, this need not be considered in most projects.
(12) For an enclosure to be a fire compartment, it is essential that there is continuity of fire resistance. Special attention is needed where fire resisting elements connect, for example where a fire wall connects to a roof, if passage of fire between them is not to occur.
(13) Make a calculation of the fire load of the panels and compare this with the expected fire load of the building contents in order to determine whether the fire load of the panels is important, see Section 5.1.5.2.
(14) Fire stopping may need careful consideration if extensive spread of fire through combustible cores is to be prevented. Spread of fire behind panel facings is often difficult to extinguish and may lead to unnecessary fire damage.
(15) Sandwich panels used as ceilings can carry imposed loads if a profiled sheet is used as the upper facing.
(16) When considering the use of sandwich panels, or any other construction, think about the fire scenarios that can occur (see Section 5.5) and decide if the construction will meet the challenge.
(17) Ask for a copy of the full report on a fire resistance test and check that the panel support and restraint conditions in the building are at least as good as those that were used in the fire test.
(18) Consider using wall panels with an effectively fixed fire-protecting inner facing such as plasterboard or other fire-protecting board which does not

fail early in fire. This is important where cores are combustible and capable of producing toxic and irritant combustion products. The use of poorly-protected flammable plastic cores is less objectionable if escape routes and exits are good and easily seen, few people occupy the building, the building has a large smoke reservoir or a system of smoke vents, and a fire detection and alarm system is installed, so that the occupants can vacate the building before the atmosphere becomes untenable.

(19) Before insisting that the combustible core or, if appropriate, the inner facing is treated with a flame retardant, check with the manufacturer to determine the range of fire size and levels of incident radiation for which the retardant works. A retardant which effectively quenches a flame after removal of a small ignition source may not be effective in a real fire. Indeed some flame retardants may increase the fire hazard by producing more smoke of a dense and irritant nature than the equivalent non-flame-retarded product.

(20) Avoid using timber-based facings where it is important to minimise the surface flammability. Small areas of timber, which do not form part of an escape route lining, may not require the use of flame retardants in order to achieve this.

(21) Ensure that both steel facings of a horizontal or vertical sandwich panel are made into continuous membranes and suitably restrained by the boundary structure.

(22) Avoid panel seals which are easily penetrated by fire or give off toxic fumes.

(23) Use fire stopping junctions between sandwich panels and fire separating walls or floors so that fire cannot spread beyond the fire separating element via the combustible core of the panel. Intermediate fire stops around the perimeter of each panel, and within large panels, are also highly desirable. This can reduce the life hazard and property loss.

(24) Avoid extrapolation of integrity data gained from standard fire resistance test results when designing tall wall panels supported at the lower edge since the increased dead load may cause collapse or loss of integrity. Remember that the fire resistance test for walls is conducted on specimens only 3-m high. Suspend tall panels from the upper edge making due allowance for vertical expansion.

(25) Consider the safety of firemen and people in or near the building. For example, avoid external wall or roof panel designs which, in fire, allow the outer facing to become detached and act as a missile hazard and also allow unexpected high levels of thermal radiation to be emitted to people and property nearby. Avoid sandwich panel ceilings in which the lower facing can fall down when exposed to fire from above or below. See item (21) of this check list.

(26) Avoid using sandwich panels with plastic foam cores in rooms where people may be sleeping and where there is no early warning such as an automatic smoke detection and alarm system. The exception might be where the panels can be modified to provide a delayed contribution to the production of fire effluent in the room by, for instance, cladding the fire risk side of the panel with a fire protecting board material or by ensuring that the effluent escapes harmlessly into the outside air.

5.8 REFERENCES

5.1 Davies, J.M. (1996) Fire and cladding in places of public assembly. *New Steel Construction*, (April/May), 24–26.

5.2 European Communities (1989) Council Directive of 21 December 1988 on the approximation of laws, regulations and administrative provisions of the member states relating to construction products (89/106/EEC). *Official Journal of the European Communities*, **L40** (32).

5.3 International Organisation for Standardisation (1993) ISO 9705. Fire tests – Full size room tests for surface products. Switzerland.

5.4 International Organisation for Standardisation (1975) ISO 834, Fire resistance tests – Elements of building construction. Switzerland (including Amendments).

5.5 European Committee for Standardisation (CEN) (1994) Eurocode 1: Basis of design and actions on structures, Part 2.2: Actions on structures exposed to fire, ENV 1991-2-2: 1994, Brussels.

5.6 Purser, D.A. (1995) Toxicity assessment of combustion products. In *SFPE Handbook of Fire Protection Engineering*, National Fire Protection Association, Quincy, Massachusetts, 2nd edn., Chapters 2–8.

5.7 Purser, D.A. (1985) Behavioral impairment in smoke environments. In *International Colloquium on Advances in Combustion Technology*, FAA Civil Aeromedical Institute, Oklahoma City, Oklahoma, USA, April 11–13, 1995.

5.8 Purser, D.A. (1990) The development of toxic hazard in fires from polyurethane foams and the effects of fire retardants. *Flame Retardants*, 17–18 January, London.

5.9 International Organisation for Standardisation ISO 13344 Determination of the lethal toxic potency of fire effluents. Geneva.

5.10 Babrauskas, V. (1995) Toxic fire hazard comparison of fire insulations: the realism of full scale testing contrasted with assessments from bench-scale toxic potency data alone. In *Proceedings of the 1st International ASIAFLAM Conference*, Interscience Communications Ltd., London, pp. 435–452.

5.11 Woolley, W.D. (1985) Are foams a fire hazard?. *Cellular Polymers*, **4**, 81–115.

5.12 British Standards Institution (1996) Fire safety engineering in buildings, BSI DD 240.

5.13 Cooke, G.M.E. (1994) The severity of fire in a large compartment with restricted ventilation. In *IMAS Conference: Fire Safety on Ships – Developments into the 21st Century*, 26–27 May 1994, London.

5.14 Babrauskas, V. and Peacock, R.D. (1992) Heat release rate: the single most important variable in fire hazard. *Fire Safety Journal*, **18**, 255–272.

5.15 Babrauskas, V. and Grayson, S.J. (eds) (1992) *Heat release in fires*, Elsevier Applied Science, 644 pp.

5.16 Tewarson, A. (1995) Generation of heat and chemical compounds in fire. In *SFPE Handbook of Fire Protection Engineering*, National Fire Protection Association, 2nd edn, Quincy, USA, pp. 3-102–3-106.

5.17 British Standards Institution (1987) Fire tests on building materials and structures, BS476: Part 20: Method of determination of the fire resistance of elements of construction (general principles), London, (including Amendments).

5.18 Boughey, R. (1996) Fire safety in buildings. *Fire Safety Engineering*, CEN/TC127, Paramount Publishing, London, pp. 9–11.

5.19 Ball, G.W. *et al.* (1971) Fire performance data for urethane cellular plastics. *Journal of Cellular Plastics*, (Sept/Oct), 241–264.

5.20 Warrington Research Centre (1974) Special investigation made on 3rd April 1974 for Messrs H.H. Robertson (UK) Ltd, Ellesmere Port. Report No WRCSI 15232.

5.21 Jagfeld, P. (1988) Brandschutz: Verhalten von Stahl-PUR-Sandwichelementen im Naturbrandversuch. *IFBSINFO, Industrieverband zur Forderung des Bauens mit Stahlblech* [in German].

5.22 Cope, B. (undated) A comparison of the fire performance of polyurethane insulated products produced with CFC11 and pentane, ISOPA Report.

5.23 Troitzsch, J.H. (1986) How do foams perform under fire conditions? In *Proceedings of the Fire and Cellular Polymers*, Queen Mary College, October 1984, Elsevier Applied Science, pp. 77–91.

5.24 Warrington Fire Research (1993) Investigation into the effects of water cooling on fires in polystyrene cored sandwich panel constructions. Test Report WARRES No. 60802.

5.25 Warrington Fire Research (1994) Large scale investigation into the effects of water cooling internal walls and ceilings in polystyrene cored sandwich panel constructions. Test Report WARRES No. 61696, Test carried out on 16 March 1994.

5.26 Polystyrene gets put to the test (1998) *FMRC Update*, **12** (2), Factory Mutual Research Corporation.

5.27 Babrauskas, V. (1996) Wall insulation products: full scale tests versus evaluation from bench scale toxic potency data. *Interflam 96, Proceedings of the 7th International Fire Science and Engineering Conference*, St John's College, Cambridge, March 1996, Interscience Communications Ltd.

5.28 Karst, H.F. (1987) Das Brandverhalten von Stahl-PUR-Sandwichelementen (Behaviour in fire of steel-polyurethane sandwich panels). *Proceedings of Polyurethanes World Congress* [in German with English summary].

5.29 Karst, H.F. (1995) Steel PUR composite panel performance in fire. In *Sandwich Construction 3, Proceedings of the 3rd International Conference on Sandwich Construction*, Southampton, September 1995, EMAS, UK.

5.30 Wittbecker, F.W., Walter, R. and Sommerfeld, C. (1996) Fire performance of facades and roofs insulated with rigid polyurethane foam – a review of full scale tests. *Interflam 96, Proceedings of the 7th International Fire Science and Engineering Conference*, St John's College, Cambridge, March 1996, Interscience Communications Ltd.

5.31 Jagfeld, P. (1988) Brandschutz: Verhalten von Stahl-PUR-Sandwichelementen im Naturbrandversuch (Performance of steel-PUR-sandwich elements in natural fire tests). *IFBSINFO, Industrieverband zur Forderung des Bauens mit Stahlblech* [in German].

5.32 Blachere, G. Tephany, H., Trottein, Y. and March, J. (1989) *Fire reaction tests in the EEC – Can a material be accepted in a Member State on the basis of the results of tests performed in another Member State?* Commission of the European Communities, Report 111/3197/88-EN, Brussels.

5.33 European Communities (1999) Commission Decision implementing Council

Directive 89/106/EEC as regards the classification of the reaction to fire performance of construction products, CONSTRUCT 98/319 Rev 3 (Final), Brussels, 12 April 1999.

5.34 Cooke, G.M.E. (1994) *The use of plate thermometers for standardising fire resistance furnaces*, BRE Occasional Paper 58, Building Research Establishment, UK.

5.35 Davies, J.M. and Wang, H.-B. (1996) A numerical and experimental study of GRP panels subject to standard cellulosic and hydrocarbon fire tests. In *Interflam 96, Proceedings of the 7th International Fire Science and Engineering Conference*, St John's College, Cambridge, March 1996, Interscience Communications Ltd.

5.36 Gibson, A.G. *et al.* (1995) A model for the thermal performance of thick composite laminates in hydrocarbon fires. *Composite Materials in the Petroleum Industry, Revue de l'Institut Français du Pétrol* (Special Issue), **50**(1), 69–74.

5.37 Davies, J.M. (1994) Core materials for sandwich cladding panels. In *Proceedings of ICBEST-94, International Conference on Building Envelope Systems and Technology*, Singapore, 7–8 December 1994.

5.38 Socrates, G. (1988) Flammability of materials. *Construction and Building Materials*, **2** (3).

5.39 Briggs, P.J. (1986) Fire behaviour of rigid foam insulation boards. In *Proceedings of the Conference of Fire and Cellular Polymers*, Queen Mary College, October 1984, Elsevier Applied Science, pp. 117–133.

5.40 Requirements and tests for LPCB approval of wall and ceiling lining products and composite cladding products. *LPS 1181*, Issue 2, Loss Prevention Council, UK, March 1996.

5.41 Approval Standard for Class 1 insulated wall or wall and roof/ceiling panels, Class No. 4880, Factory Mutual Research, Norwood, Mass., August 1994.

5.42 Newman, G.M. (1990) Fire and steel construction: the behaviour of steel portal frames in boundary conditions. Report No. 087, 2nd edn, The Steel Construction Institute, UK.

5.43 Cooke, G.M.E. (1988) Thermal bowing in fire and how it affects building design. BRE Information Paper IP 21/88, Building Research Establishment, UK.

5.44 Cooke, G.M.E. (1993) Can integrity in fire resistance tests be predicted? In *Proceedings of the 6th Interflam 93 Conference*, Oxford, 30 March–1 April 1993, Interscience Communications Ltd, pp. 321–330.

5.8.1 Additional references for wider reading

5.45 Society of Fire Protection Engineers (1995) *SFPE Handbook of Fire Protection Engineering*, 2nd edn, Boston, Mass.

5.47 Babrauskas, V. and Grayson, S.J. (1992) *Heat Release in Fires* (eds), Elsevier Applied Science, London and New York.

5.47 Drysdale, D. (1985) *An Introduction to Fire Dynamics*, Wiley, Chichester.

5.48 The International Association of Cold Store Contractors (European Division) (1999) *IACSC: Guide on the Design, Construction, Maintenance,*

Specification and Fire Management of Insulated Envelopes in Temperature Controlled Environments.

5.49 Cooke, G.M.E. When are sandwich panels safe in fire? Part 1: An overview; Part 2: Avoiding collapse; Part 3: Fire scenarios. *Fire Engineers Journal* (July 1998), 37–41; (Sept. 1998), 25–33; (Jan. 1999), 18–25.

Chapter 6
Durability

6.1 GENERAL ASPECTS OF DURABILITY

The sandwich panel is one of the most modern components used in the building industry in which each constituent material has to perform several functions. Sandwich panels may act as external wall panels, roof panels, ceiling panels or as partitioning panels. In these applications, they may have to carry wind, snow, imposed or dead loads, provide thermal insulation and act as weather protection. They have to meet large variations in external temperature, they can act as fire insulating barriers and, if the panels act as facades, they may have to meet stringent aesthetic requirements.

The panels therefore have to maintain their strength properties, maintain their insulating properties, to remain weather resistant and to keep the surface finish in good condition over the long period of their expected lifetime. Their durability is defined in terms of their ability to fulfil these requirements throughout this lifetime. The requirements for durability are, therefore, complex.

The required lifetime may vary from one application to another. For industrial buildings, a lifetime of 25 years or less can be sufficient as changes in the industrial activity necessitate frequent rebuilding. In other cases, the requirement can be much higher and may be as much as 50 years or more. The loads that are used in design calculations are usually predicted to appear once every 30 to 50 years. In a proposal[6.1] to CEN regarding guidelines for European technical approvals (for EOTA), it has been proposed that the required lifetime should be 25 years for products that are repairable or replaceable (with some effort) in most types of building. This is the case with sandwich panels. For products that are not replaceable, the requirement for the product is the same as that for the building itself which can vary in the range 10–100 years.

Because every material component in a sandwich panel has to meet several requirements, their possible degradation may lead to significant changes in the performance of the structure. For example, long-term changes in the insulation material may not just affect the thermal insulation properties, they may also have an effect on the strength of the structure.

When developing new structural materials, this leads to a problem. A developer can not wait 50 years to see whether or not the lifetime of the newly-developed material is adequate. There is a need for a faster way to estimate the lifetime so that it is necessary to predict the long-term durability by the use of accelerated ageing methods. There is also a related need for the customers for a new construction product to have some guarantee that the lifetime is sufficient.

It follows that there is a great need for suitable methods to predict and control the durability of structural elements in general and of sandwich panels in particular.

In sandwich construction, the materials that form the insulating core part of the structural element also have an important load-bearing function. The technique of attaching the core material to the faces of a sandwich panel is by bonding (which may be part of a foaming operation or may be by using adhesives) and this raises some new and quite difficult durability problems. In order to achieve a reliable structure, the particular problem to be solved may vary from material to material.

This chapter presents an accelerated ageing procedure for the different components of a sandwich panel which may be used to determine the durability of the panel.

A three-layer sandwich panel, consisting of two thin surfaces and a core, has a number of critical components. Figure 6.1 illustrates a typical cross-section of such a panel and shows all of the material components. With the exception of the external coatings, every layer makes a contribution to the strength of the panel and also, therefore, has an influence on its durability. The main structural components are the faces and the core but there are several layers of material between them. It is shown that there may be as many as fifteen critical layers in a typical sandwich panel.

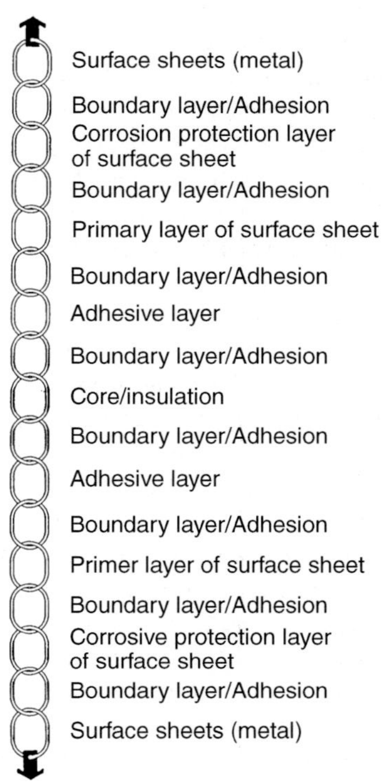

Fig. 6.1 Critical layers in a sandwich panel.

6.2 PREDICTION OF THE LIFETIME OF A STRUCTURE

At the present time, there are no generally accepted methods nor generally accepted criteria for predicting the lifetime of a structure.

The requirements for durability may be stated in two parts as follows.

(1) The durability limit in respect of the *mechanical properties* is reached when the action S_d exceeds the resistance R_d of the structure so that the structure is no longer able to carry the design loads as illustrated by Fig. 6.2. Mathematically, this limit is reached when Equation 6.1 is no longer valid:

$$S_d(t) \leq R_d(t) \tag{6.1}$$

where t is the time period under consideration.

$S_d(t)$ represents the design value of the *load action* (e.g. wind, snow or permanent load) at the time t. Generally the design values of the snow or wind loads are estimated with a 30–50 year return period so that, at time t, the remaining life is this number of years minus t. For natural loads, the maximum load that is predicted to occur usually decreases as the remaining lifetime is reduced. The predicted load is therefore smaller for a short time period than for a longer one, hence the falling curve for S in Fig. 6.2.

$R_d(t)$ represents the design resistance of the structure at time t. This resistance can be affected by *load actions*, by *physical actions* such as temperature, moisture, etc. or by *time actions* which means that the mechanical properties may also decrease with time.

If the resistance R reduces more quickly than the load S, the life of the structure is reduced by considerations of durability as shown in Fig. 6.2.

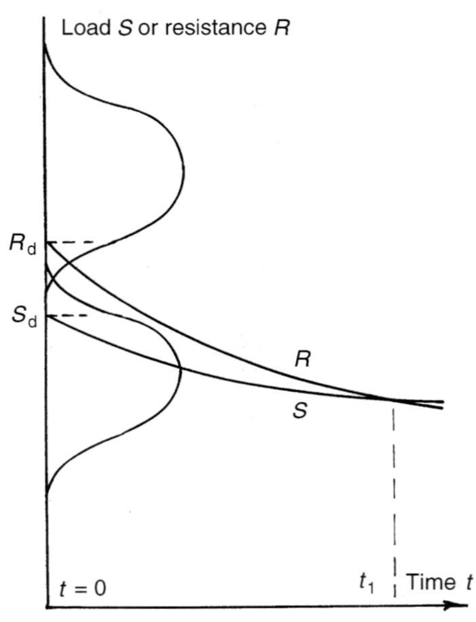

Fig. 6.2 Variation of the actions $S(t)$ and the resistance $R(t)$ of a structure with time.

(2) The durability limit in respect of *physical properties* is reached if:
   ○ the thermal insulation is no longer sufficient
   ○ the weather proofing is no longer adequate
   ○ the aesthetic condition of the structure becomes unacceptable

The main purpose of this chapter is to present methods for predicting the mechanical durability.

In Fig. 6.2, the lifetime of a structure t_1 is reached when $S(t) \geq R(t)$. Normal structural design is carried out at time level zero ($t = 0$) which means that the initial strength divided by a material factor γ_M is compared with the maximum factored value of an action which is estimated to occur over a time period of 30–50 years:

$$S_d = \gamma_q S_k < R_k/\gamma_M = R_d \tag{6.2}$$

where the values with subscript k are characteristic values determined as the 5% fractile values. The safety factors γ are determined so that the probability for a failure, when Equation 6.2 is no longer satisfied, is about 10^{-6}.

The theoretically correct way to calculate the expected lifetime of a structure should be to solve Expression 6.1 in such a way that, for example, a certain defined population of the structure does not fulfil Expression 6.2. If, as a simple example, this proportion is 50% so that the characteristic value is equal to the mean, Expression 6.2 becomes:

$$S_m(t) < R_m(t) \tag{6.3}$$

The problem is that the changes of the actions $S(t)$ with time are not very well understood and that there are hardly any methods available with which to determine the change of the resistance $R(S, t)$ with time and with respect to different actions. Therefore it is not possible, at the present time, to solve Equation 6.1. Neither, of course, is it possible to gather usable results on the basis of experience alone. The only remaining possibility is to develop some appropriate accelerated laboratory testing methods and to use these to estimate a relationship between resistance and action over a reasonably short time. The resulting accelerated testing methods are more or less empirical methods which can be used to compare newly developed structures with older known structures in order to make them fulfil certain agreed criteria.

6.3 ACTIONS ON SANDWICH PANELS WHICH MAY CAUSE DEGRADATION

The actions, against which the panels should be checked in order to assess possible degradation of their structure, are:

- long-term loading (e.g. self-weight, snow)
- repeated loading (e.g. wind, foot traffic on ceilings)
- movements and forces caused by temperature difference over the panel
- temperature variations over the panel which may also cause internal moisture variations within the panel or lead to blistering
- moisture gradient over the panel
- chemical degradation (e.g. migration of the blowing agent)
- ultra-violet radiation on the surface of the panel

- corrosion of both the surface and the fasteners
- degradation of adhesives and sealants

6.3.1 Actions causing mechanical degradation

- *Long-term loading* can cause an increase in deflections and a reduction of strength. The reduction of strength may be determined from tests with a long-term static load. For roof panels the tests should last at least 1000 hours (see Sections 13.9.3.5 and 13.9.4.3).
- *Repeated loading*, which can be caused by snow load, wind load or forces caused by restrained temperature movements, can cause a successive degradation of the strength properties. In order to check that this does not happen, tests with repeated short-term and long-term loading may be carried out. The European Recommendations[6.2] offer a test with which to investigate the 'walkability' of ceiling panels and this is described later in this chapter. Section 13.9.5.4 describes a repeated bending test on a screw fastener which investigates the influence of cycles of temperature changes on fasteners passing through a panel.
- *Internal moisture variations* in the panel, caused by a temperature difference across the width of the panel, can cause condensation on the internal sides of the faces of the panel. This can lead to internal oxidation (corrosion) on the faces and can result in a loss of the bond between the core and the face.
- *A combination of high moisture content and large changes in temperature* can also cause degradation of the core material. In order to check that this is not critical, suitable tests are proposed in the following parts of this chapter.
- *High temperature* can, especially for foams, cause a degradation of the core material which may be checked using the methods described later in this chapter.
- *High temperature* can also, especially in panels produced using foaming techniques, cause so-called 'blistering' effects. These are a result of a combination of rapidly increasing temperature and low bonding strength of the face material to the core material or low tensile strength of the core material. Blistering is initiated by rapidly increasing internal cell gas pressure in the core material which leads to migration of the core gases to locations where the cell structure is flawed followed by local delamination of the panel at points of weakness. Blistering in panels blown with the traditional blowing agent, CFC11, may be said to be well enough understood. However, since the phasing out of CFC11 according to the Montreal Protocol, there is relatively little experience with the alternative blowing agents which are now used or proposed.

6.3.2 Actions causing physical degradation

- *Moisture gradient* over the panel can cause penetration of moisture into the panel if the surface or the joints are not sufficiently tight. This can lead to a reduction of the thermal insulation properties.

- *Sun radiation* can cause changes in the colours of the face material which may not be acceptable for the panels acting as a facade.
- *High humidity, high temperature and an aggressive atmosphere* can cause corrosion in the surface layer of the panel or in the fasteners.

6.4 ACCELERATED TEST METHODS

This section describes methods for studying the effect of long-term loading and the conditions of use on the durability of sandwich panels. These procedures can act as guidelines for the development of new products and when assessing the durability properties of existing products.

6.4.1 Durability of the panel

In this part of the chapter, methods are presented with which the durability of the mechanical properties of sandwich panels may be determined. Their resistance to variable and long-term loading are studied together with their resistance to ageing effects caused by temperature and moisture variations. Because it is not possible to investigate long-term effects by calculation, the methods are entirely experimental and, therefore, the appropriate test procedures are also described.

There is, however, a general requirement for a durable panel which is based on experience. The procedures for design for mechanical resistance, which are described in Chapters 9, 10 and 11, do not include any performance requirements which require a check of the tensile strength of the core and the tensile bond between the faces and the core. However, panels with a low tensile bond strength have been found to be unsatisfactory in service so that this quantity is now seen to offer an important durability criterion. The European Recommendations,[6.2] therefore, require the tensile bond to have a minimum *characteristic* value of 0.075 N/mm^2.

This argument is then taken a stage further, as described later, by defining the *degradation* of the strength of a panel due to ageing in terms of the tensile strength of the core and its bond to the faces.

6.4.1.2 Durability with respect to variable and long-term loading

The following tests are to be carried out on full width panels of realistic span, especially in the case of panels with a core of a discontinuous structure (e.g. using slabstock) or with new foam types. The panels are to be simply supported and subject to uniformly distributed loads. The length of the span is chosen so that shear failure is the critical failure mode.

These tests are appropriate for all types of sandwich panels and are particularly important when new materials are proposed.

Tests for long-term loading

For panels that are used to carry long-term loads, such as snow load in northern latitudes or permanent loads, long-term tests are carried out with a loading history as shown in Fig. 6.3. From these tests the creep factor and the influence of the long-term load on the strength are determined.

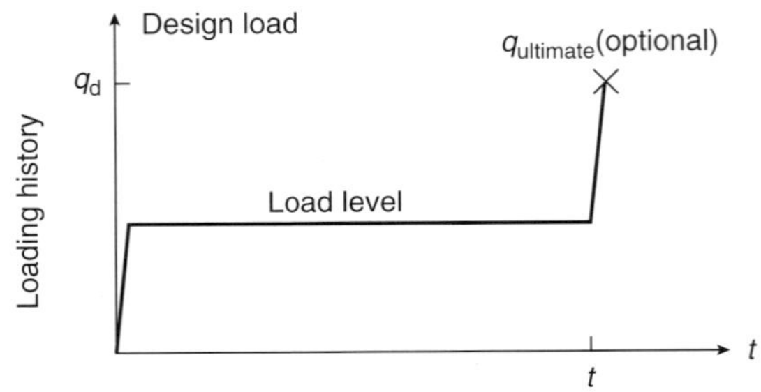

Fig. 6.3 Long-term loading history.

In order to determine the creep factor, it is customary to test simply supported panels of typical span subject to uniformly distributed load. The load level used in this test is of the order of 30% of the mean load for shear failure at the ultimate limit state and t is at least 1000 hours. Deflection readings are taken when the load is first applied and at various times during the test so that a graph of deflection versus time can be plotted. When plotted to a log–log scale, sufficient points should be obtained on the linear part of the curve to allow reliable extrapolation. This procedure is suitable for self-weight and snow but should not be used for very high permanent loads.

In order to determine the long-term shear strength, it is usually considered sufficient to test small beam specimens subject to four point loading. A number of long-term load tests are carried out with a range of load levels such that at least ten of the samples fail within the time interval 0.1 hour $\leq t \leq$ 1000 hours. Ideally, they should be more or less equally spread out within this range so that a graph of failure stress against time can be drawn as shown in Fig. 6.4. The long-term shear strength is the shear strength at a time to failure of 1000 hours.

With good quality polyurethane foam or mineral wool, the long-term shear strength determined in this way is usually of the order of 80% of the short-term strength. However, if long-term tests are not carried out, the characteristic value of short-term strength should be multiplied by a factor of 0.5 in order to give a safe value of the long-term strength for design purposes.

Detailed procedures for these tests are given in Sections 13.9.3.5 and 13.9.4.3.

Tests for walkability (taken from Reference 6.2)

Unless proper walkways are installed, for panels which are to be used as roofs or walk-on ceilings, it is necessary to perform tests which demonstrate that the panels can sustain foot traffic safely and without damage.

Foot traffic is caused mainly by operatives walking on the panels for the purposes of installation or maintenance. There are two main cases to consider.

(1) *Temporary walking caused by people mounting the panels during installation or when carrying out occasional maintenance* In such cases, the total intensity over the life of the panel is of the order of 10–100 steps per square metre of the panel. Generally, sandwich panels with thin metal faces can sustain this kind of traffic without difficulties.

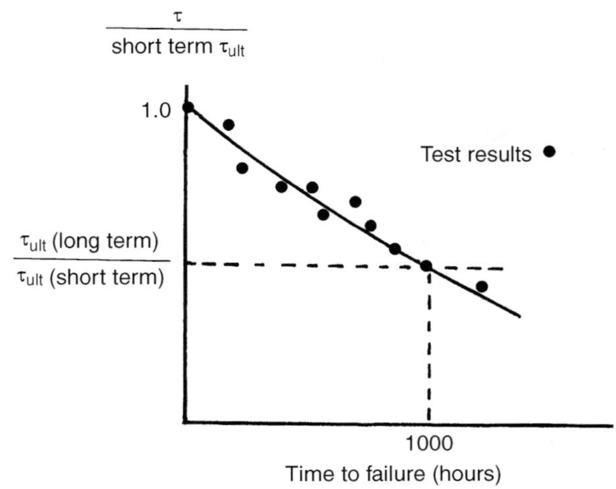

Fig. 6.4 Method to determine the long-term strength from a series of test results.

(2) *Repeated walking caused by traffic on walking routes* This type of traffic may arise during construction when operatives are moving to and from the place where the panels are stored or being placed, or when they are carrying out other installation work with the panels forming a working platform. It can also arise during operation if the panels form a maintenance access route. The total traffic intensity can vary considerably and may be as large as several thousand steps per square metre of the panel. In most cases, sandwich panels require protection against this kind of action.

It is important to note that sandwich panels with thin flat faces are usually only suitable for temporary walking during installation. For repeated walking, some form of protection is generally required.

The tests are divided into two categories as follows.

Test for temporary walking

The test is carried out by loading the panel with a point load of 1.2 kN. The point load is placed in the middle of the longest allowable span and either on the edge rib, if the upper face is profiled, or at edge of the panel if the upper face is flat or lightly profiled. The load is applied through a 100 mm × 100 mm timber block with an optional soft 10-mm thick layer between the plate and the panel. The load is held in place for at least 5 minutes.

There are three possible outcomes from this test:

(1) If the panel carries the applied load without permanent visible damage, access to the roof or ceiling is permitted, one person per panel on an occasional basis, either during or after erection.
(2) If the panel supports the load, but with permanent visible damage, then measures must be taken to avoid damage during erection (e.g. by the use of

walking boards). Furthermore, there must be no provision for access to the roof or ceiling after the building work is completed.

(3) If the panel fails to support the load, then it shall not be used in practice.

For multi-span panels, the maximum allowable span indicated by the above test may be increased by 25%.

Test for repeated walking

The following procedure is proposed in Reference 6.2. This test requires a minimum of ten tensile specimens with a size of 100 mm × 100 mm with the faces intact. 50% of the specimens are tested to failure in tension in the standard way which is described in Section 13.9.3.2. The remaining 50% of the specimens are first subject to 250 cycles of compressive loading with a stress level varying between zero and 0.08 N/mm^2 applied as a central load of 0.45 kN over a loading area of 75 mm × 75 mm. The rate of testing is not critical but shall not exceed 1 Hz. These specimens are then tested to failure in tension, as previously.

The reduction in the tensile strength below the datum value gives an indication of the susceptibility of the panel to degradation caused by repeated walking. If the average of the tensile strength results obtained after cyclic loading falls below 80% of the average value obtained without cyclic loading, the panels may be considered to be unsuitable for regular access without added protection.

Note: The walkability is dependent on the compressive strength of the material and, in order to ensure adequate walkability, it may be necessary to define a minimum level of compressive strength, e.g. 0.1 N/mm^2.

Repeated long-term loading

For panels used to carry variable long-term loads, such as snow load, tests with repeated loading according to Fig. 6.5 may optionally be carried out in order to study the long-term stability of the structure. If the long-term loading test indicates a poor creep performance, this test may be appropriate in order to examine why. When investigating new materials such as adhesives, this test also provides an indication of whether the creep is visco-elastic or visco-plastic.

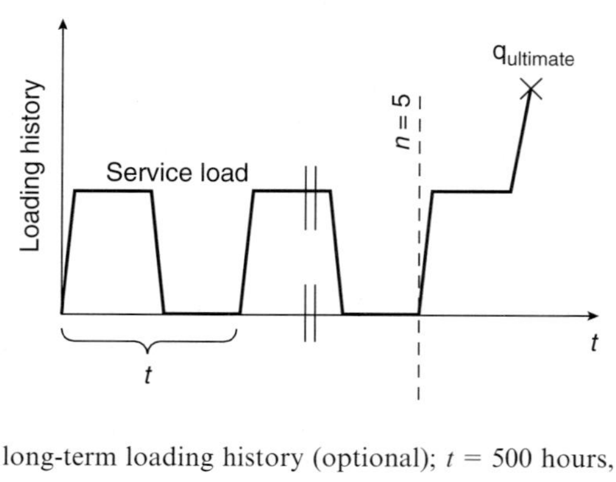

Fig. 6.5 Repeated long-term loading history (optional); t = 500 hours, n = 5 times.

This is essentially a research and development test which requires expert interpretation of the load–time curve.

Repeated short-term loading

This test, with repeated cycling loading at the level of the service shear load, tests the durability of the bonding action and the core strength with respect to frequently repeated loads, such as temperature impact loads in multi-span structures and wind loads. The loading is carried out with a cycling loading regime according to the scheme shown in Fig. 6.6.

For a satisfactory outcome of this test, the reduction in the strength after repeated loading should be less than 30%. There should not be any significant increase in deflection from one cycle to the next.

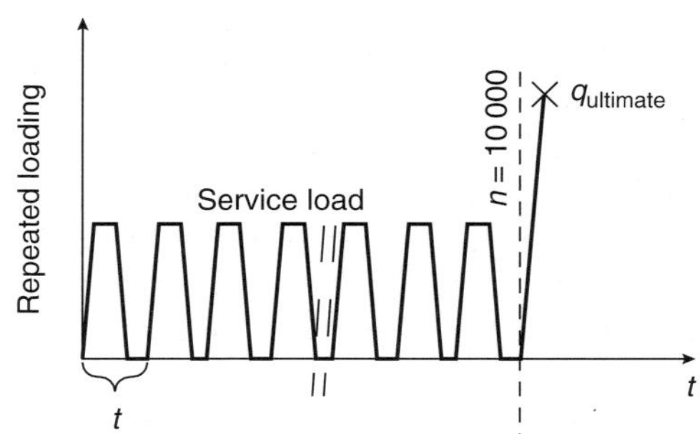

Fig. 6.6 Repeated short-term loading history; $t = 2\,\mathrm{s}$, $n =$ up to 10 000 times.

6.4.2 Effect of temperature variations and moisture on strength properties

In order to study the ageing properties of a sandwich panel, some novel laboratory methods for accelerated ageing under combinations of humidity and temperature have recently been developed. A common feature of these methods is that samples cut from a panel are kept in a closed box where both the temperature and the humidity can be varied under controlled conditions in order to accelerate natural ageing. It is first necessary to study what combinations of the real conditions are causing degradation. Then real ageing may be simulated by keeping the samples in critical conditions which are known to be more severe than the real ones.

In order to determine the durability characteristics for core materials and foam systems, tests with different combinations of temperature and relative humidity have to be carried out in order to achieve structures with controlled durability properties. The nature of the problem is illustrated in Fig. 6.7 (see Reference 6.3). Examples of good durability, controlled durability and unacceptably poor durability can all be found in current sandwich panel production.

For materials with a known risk of possible degradation as, for example, the blistering of panels with a polyurethane foam core or adhesive bonding failure in panels fabricated using structural slabstock (e.g. structural rock wool), the ageing properties may be checked with an appropriate purpose-designed test (the blistering test and the wedge test for adhesive bond performance). In order to

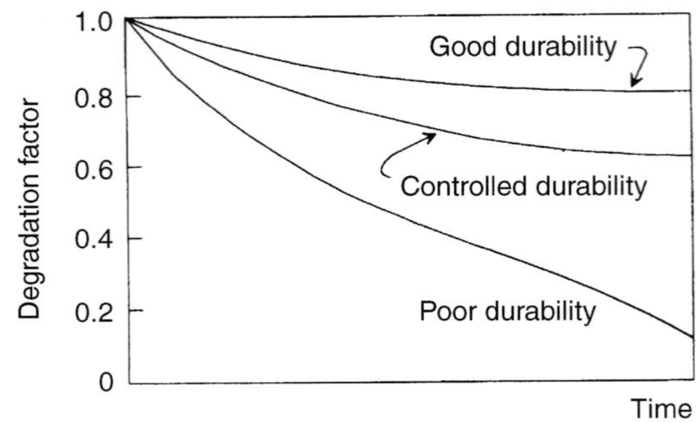

Fig. 6.7 Degradation factor and range of durability of currently available panels.

check for the effect of rapidly increasing temperature, all panels with foam cores should undergo the so-called blistering test.

For materials with a less well-defined risk of progressive degradation, the ageing effect is measured by comparing the strengths before ageing and after ageing. The durability can be readily defined in terms of a 'durability factor' or 'degradation factor' which is defined as the ratio of the tensile strength after ageing to the short-term unaged tensile strength (see Section 13.9.3.2).

6.4.3 Tests for durability

Blistering test

Blistering has been described in Section 6.3.1. The blistering test is carried out on full-scale panels in order to establish whether they may be susceptible to blistering. The exposed face of the panel is heated up rapidly to +90°C and kept at this temperature for two hours while the other face is maintained close to ambient temperature. The panel should then be examined carefully and the size and position of any 'blisters' noted before it is allowed to cool down to the room temperature. A good quality panel should not exhibit any discernible blisters.

This process should then be repeated but this time the panel is cooled down rapidly by spraying with cold water. In a good quality panel, this 'thermal shock' should not cause any significant delamination.

Determination of the durability or degradation factor

In References 6.2 and 6.3, the term 'degradation factor' was used. In the American literature, the term 'durability factor' may be found. For the purpose or our discussion, the two terms will be considered to be synonymous. In order to check for progressive degradation the following methods can be used.

One relevant process for accelerated ageing is described in the American standard ASTM C 481. In this procedure, the test specimen is subject to each of six different ageing conditions in succession and the cycle is repeated six times. Two different combinations of conditions are given, Type A and Type B and the more critical one is selected. A combination of both types may also be used.

Type A

Condition	Temperature (°C)	Time (h)
Water	49	1
Spraying steam	93	3
Cold air	−12	20
Hot dry air	99	3
Spraying steam	93	3
Hot dry air	99	18

Type B

Condition	Temperature (8C)	Time (h)
Water	49	1
Spraying water	71	3
Cold air	−40	20
Dry air	71	3
Spraying water	71	3
Dry air	71	18

After subjecting the specimens to the appropriate accelerated ageing regime, they are re-acclimatised to ambient conditions and tested in tension. The durability factor is the ratio of the strength after ageing to the unaged strength.

This test procedure has a duration of about 14 days. To the best of the authors' knowledge, it has not been widely used and they have no experience of it.

From References 6.3–6.6, or other appropriate methods, the durability in accordance with Fig. 6.7 can be found. Based on this information it is evident that the durability characteristics can be determined in a relatively simple way for the main core materials that are in current use. For foams, high temperature rather than high humidity is generally the degrading factor, whereas for mineral wools, conversely, high humidity rather than high temperature is the degrading factor. With foams there is an additional problem that the hardening process of fresh foams continues under elevated high temperature and a degradation effect can be only be seen after a relatively long time.

The ageing process should be determined for each core material in turn commencing with those considered to be the most critical. A significantly long test period should be used with different valid combinations of values of temperature and humidity in order to establish a critical degradation regime. Then, life-cycle simulation can be commenced according to References 6.5 and 6.6.

For product development purposes, in order to determine the durability factor, and for quality control purposes there is a need for more rapid methods.

It has been shown,[6.3] for example in the case of representative stone wool panels, that a more severe ageing regime *for this class of panel* than that given by ASTM C 481 is simply to maintain the sample for 24 hours under constant conditions of 70°C and 100% relative humidity. These conditions may be obtained by keeping the specimen in a closed box over water heated to a temperature of 70°C, as shown in Fig. 6.8. After keeping the test specimen for 24 hours in a climate box in which this condition is maintained, the strength is measured. The durability factor is the ratio of the strength after ageing to the unaged strength. While the original development work used a time period of 24 hours it has recently been found that this is too short for modified mineral wool and some other material types and an augmented procedure has been evolved.

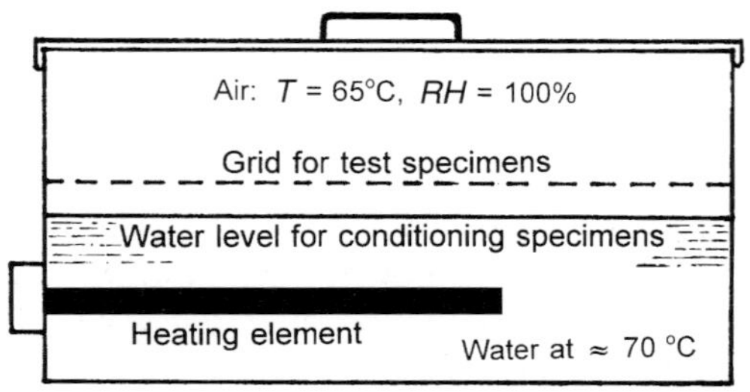

Fig. 6.8 Conditioning chamber for accelerated ageing.

It may be noted that, in the apparatus shown in Fig. 6.8, it is the temperature of the air which is controlled, not the temperature of the water.

The above accelerated ageing methods are suitable for assessing the effect of ageing on the tensile, compressive or shear strength of the core material or the adhesive bonding strength between the core and the faces of the panel. The degradation factor is often found to be almost the same for all strength components. In general, the test specimens are aged with the faces intact and the test specimen includes the faces of the panel, including the protective coating of the face material. Thus the bonding strength of the internal protection layer is also studied. This is very important in order to find an acceptable combination of the adhesive system together with the core material and the internal surface material of the face of the panel.

In all accelerated ageing testing methods, prior to final testing, the samples are returned to room temperature for sufficient time for the ambient conditions to be achieved. This ensures that the test determines the effect of ageing, not the effect of extreme conditions.

When testing the samples in tension after ageing, the failure should take place in the core material not in the adhesive or the surface layer.

During the 1990s, the durability of sandwich panels was extensively researched at Tampere University of Technology in Finland. As a result of this work, a general procedure for the determination of the degradation factor (durability factor) for sandwich panels has been proposed.[6.6] This proposal is based on extensive tests on specimens with a variety of core materials (polyurethanes, polystyrene, stone wool and glass wool). The proposed procedure uses specimens cut from the complete panel with a size of the order of 100 mm × 100 mm the thickness of the panels which are aged according to the following series of conditions.

Cycle C1

- 5 days at +70°C and 90% RH, followed by
- 1 day at −20°C, followed by
- 1 day at +90°C[a] under dry conditions

[a]This temperature may cause problems with some plastic foams and it is only necessary for dark coloured faces exposed to strong sunlight. For lighter colours or internal partitions, a lower value (> 70°C) may be used.

Cycle C2

- 28 days at +65°C and 100% RH

In order to determine the durability characteristics, three points are determined. The first point is the average initial (unaged) tensile strength. The second point is the average tensile strength after one ageing cycle C1 and the third point is the average tensile strength after repeating the cycle C1 five times. The degradation factor is the result of dividing the average aged strength by the average unaged strength. The degradation factor should not be less than 0.6 after one C1 cycle and not less than 0.4 for five C1 cycles.

The durability factor after the C2 cycle should not be less than 0.4. Thus, durability is defined in terms of changes in the tensile bond strength between the panel and the core. This is, of course, not directly representative of the true degradation of the strength of the panel. The tensile test is chosen because it is simple to perform and because experience has shown that a good tensile bond is essential for a durable panel. Furthermore, this procedure is conservative in comparison with the actual degradation of panel strength. There is a relatively good correlation between the degradation of tensile strength and shear strength but a much weaker correlation between the degradation of wrinkling strength and tensile strength. It is here that the procedure is conservative because the reduction in wrinkling strength is of the order of half the reduction in tensile strength.

It has been shown in References 6.5 and 6.6 that panels that fulfil the above criteria can have a predicted lifetime of over 50 years. For panels that have been type tested in this way, it is sufficient to control just the core material quality and the bonding procedure.

6.4.4 Durability of the adhesive

One of the most important requirements for a sandwich panel in use is that the bonding strength of the faces to the core is reliably maintained. The sandwich action, and thus also the strength of the panel, is fully dependent on this bond. However, there is very little publicly available information on the bonding properties of the different types of adhesive, not to mention their many possible combinations with different surface treatments.

With the testing procedures described above, one of the acceptance criteria for a good bonding system is that failure should be obtained in the core material, not in the adhesive layer itself. It follows that the bonding properties of the adhesive material are not tested because they are generally stronger than the core material.

Therefore, the tensile test regimes described above are not sufficient alone and, in the next section, methods are described for the study of the bonding properties of adhesives to surface materials in a separate test.

Tests for the durability of the adhesive bonding system

The test procedure requires that two strips of the face material, with the actual surface treatment to be used in practice, are bonded together with the adhesive that is proposed.

The initial strength properties can be measured in terms of the shear strength (Fig. 6.9), the peel strength (Fig. 6.10) or the resistance to crack formation (Fig. 6.11), the latter being termed the wedge test.

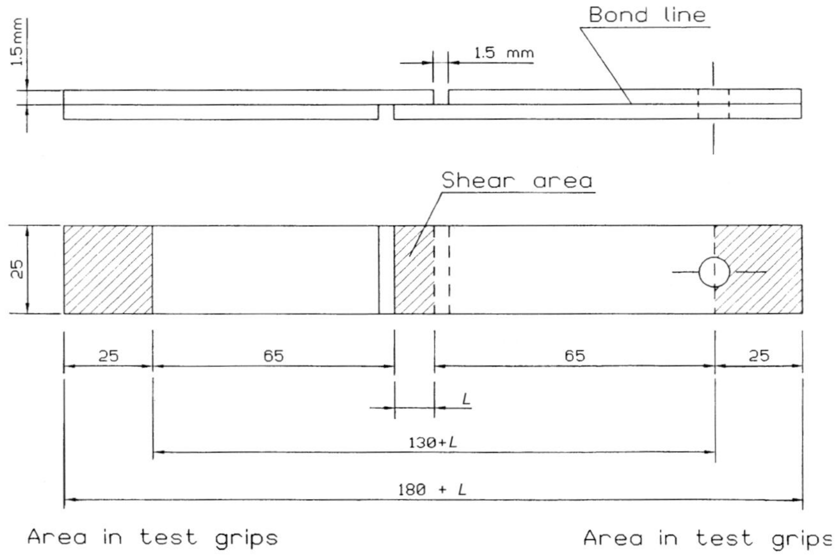

Fig. 6.9 Determination of the shear strength properties of adhesives according to ASTM D 3165 or DIN 53 283.

The remaining specimens may then be aged for 24 hours by immersing them in water heated up to 70°C. This can be conveniently by done by raising the water level in the apparatus shown in Fig. 6.8 to a level above the grid supporting the specimens. The strength is determined after ageing and compared with the initial strength.

However, it has been found that when the adhesive layer is slightly stressed during the ageing procedure, the ageing procedure is much more effective. It is therefore recommended that, whichever test procedure is used, the ageing procedure is carried out with the adhesive layer under a small stress.

These tests on the adhesive layer are largely qualitative and the primary requirement is that the mode of failure should always be a cohesion failure in the adhesive layer itself, not an adhesion failure in the surface between the adhesive and the facing material nor in any of the (primer) surface layer components of the

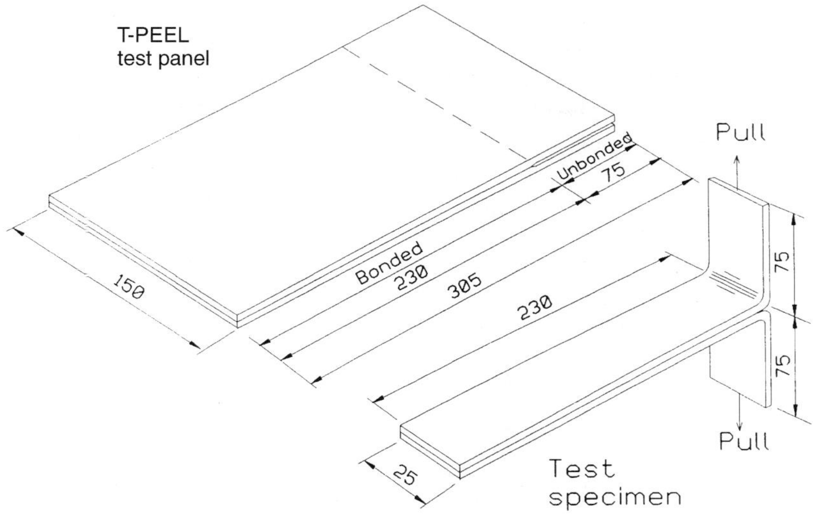

Fig. 6.10 T-peel test in general accordance with ASTM D 1876 or DIN 53282.

facing material. Using testing methods where numerical values can be measured (shear and peel tests), the reduction in strength after ageing should not be more than 30%. Using the wedge method, neither the initial crack nor the crack growth should be more than about 20 mm.

Thus, in the wedge test, the small wedge is pressed into the adhesive layer and this forces the faces apart causing an initial crack and inducing tensile stresses at the end of the crack. During ageing in hot water, the crack grows.

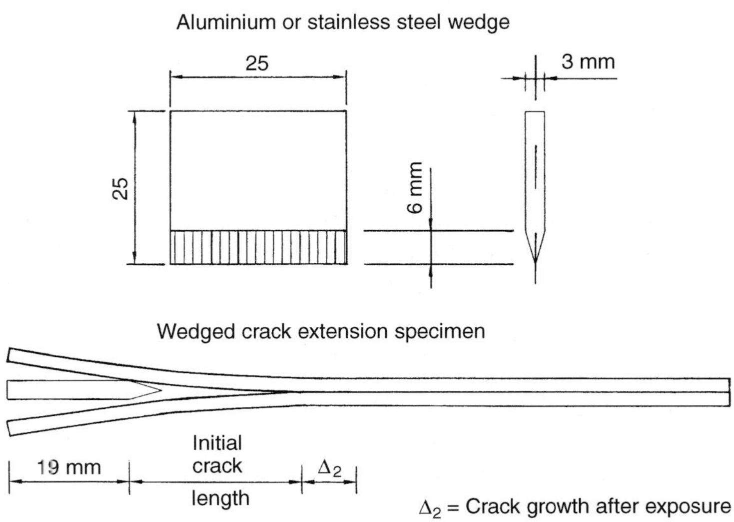

Fig. 6.11 The wedge test in general accordance with ASTM D 3762 or DIN 65448.

The wedge test in accordance with ASTM D 3762 was developed for rather thick aluminium plates. In sandwich panels, the faces are relatively thin and, the thinner the faces, the smaller is the tensile stress component caused by the wedge in the bottom of the crack. When the crack grows, the tensile stresses diminish and so also does the rate of growth of the crack. To apply the ASTM D 3762 method directly would give tensile stresses in the adhesive layer which are too small and it is therefore concluded that a small modification is required for the case under consideration.

Therefore a modified wedge test has been developed for sandwich panels with faces of thin metal sheets (< 1 mm) as shown in Fig. 6.12.

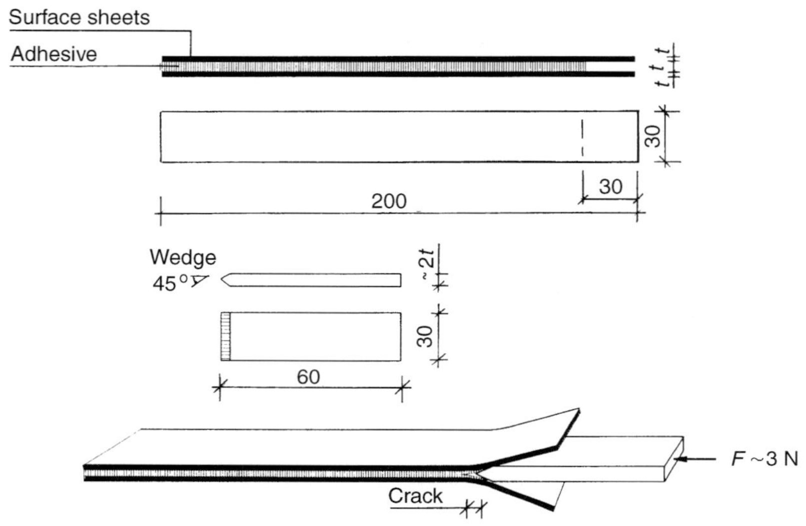

Fig. 6.12 The modified wedge test according to the Paroc Tropic 70 method.

In order to compensate for the effect of using faces with a lower stiffness than those for which ASTM D 3762 was devised, the wedge is chosen to be three times thicker than the thickness of the faces. The wedge also carries a small load during the ageing process in order to induce a greater tensile component in the adhesive.

Note: It is very important that these ageing tests are carried out with every adhesive bonding system to be used. Some bonding systems that have good dry properties have been shown to have poor aged properties!

6.4.5 Durability of facing materials

The choice of external coating materials with regard to their durability is the same as for other usages of thin metal cladding and is unchanged by application to sandwich panel construction. This aspect of durability was considered in Chapter 2.

The corrosion resistance of the coated sheets is good when coatings are used as recommended in Chapter 2. This is true for undamaged coatings. However, every sheet contains cut edges which, in particular circumstances, can be affected by corrosion. If the steel sheets are coated with a zinc layer of 250 g/m^2, this layer will also create a protective layer on the cut steel edges. This zinc layer gives adequate protection against corrosion of vertical surfaces such as walls where the wet time is relatively small.

For flat roofs the wet time for normal overlapping joints is longer because the water tends to remain trapped between the two sheets. There are some means available with which to reduce the corrosion risk in this situation as follows.

(1) The end of the sheet may be bent through 180° in such a way that the cut end is no longer exposed to the atmosphere as shown in Fig 6.13. In the right-hand diagram, the sealant shown in the fold is optional.

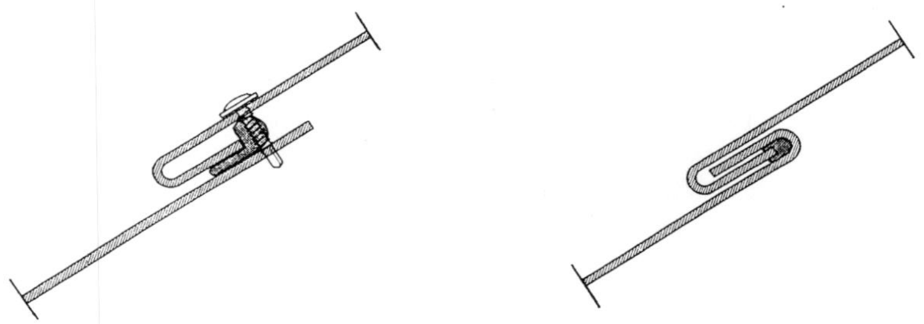

Fig. 6.13 Lap joints in flat sheets incorporating 180° bends.

(2) The space between the two sheets may be increased by using thicker sealants. In this way, the wet time is reduced by increasing the dry-out effect as illustrated in Fig 6.14.

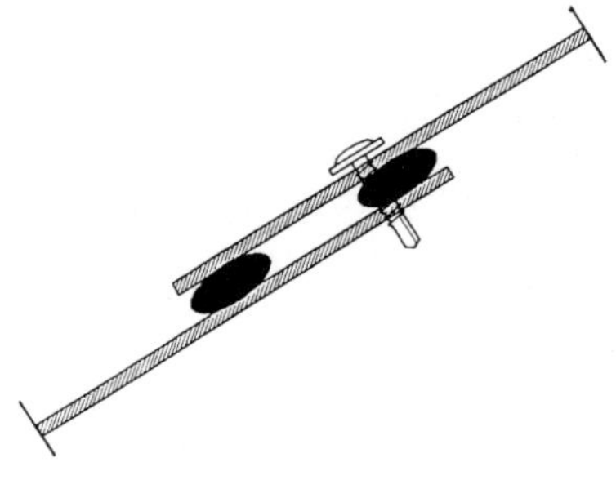

Fig. 6.14 Lap joint with thicker sealant suitable for use with profiled sheets.

(3) The reverse side of the sheet may have a double-primer layer.
(4) The slope of the roof may be increased.

The reverse side coating layer suitable for bonding and durability purposes should be chosen after passing tests in accordance with Section 6.4.4.

6.5 GUIDELINES FOR THE CHOICE OF MATERIALS AND FASTENERS

6.5.1 Core material

All core materials should undergo the tests described in this chapter before being adopted for use. Several of these tests are specifically applicable to new materials. In addition, all mechanical properties (shear, tensile and compression) have to be determined for design and quality control purposes.

Some general considerations when choosing core materials with particular regard to durability are as follows.

- For panels which have to satisfy specific requirements with regard to strength, structural grade core materials should be used.
- With polyurethane foam, special attention should be given to assessing the durability of new chemical formulations. Structural grade foams usually have a density greater than $40\,kg/m^3$.
- For mineral wool cores, only structural grade material which has been specially developed for sandwich panel construction should be used. Classification on the basis of density alone is not valid for mineral wool, the classification should be based on shear strength.
- With expanded polystyrene, care should be taken to ensure that all moisture has been dried out before panel production. The density should preferably not be less than $20\,kg/m^3$.

6.5.2 Adhesives

All adhesives to be used should pass the repeated long-term and/or short-term loading tests and the wedge test. Several types of polyurethane adhesive have proved to be reliable in use.

6.5.3 Face material

External coating
The coating of the outer face of exposed sheets should meet the following requirements where the corrosion class is determined, for example, in accordance with EN 10169:

- protection category 4–5 when used as an external facing
- protection category 3 when used in relatively dry interior climates of normal temperature. For internal conditions with high temperature or humidity, the requirements can be similar to those for external coatings.

Reverse side (primer layer)
As a consequence of its function as a bonding surface, the durability of the primer layer coating is very important. This coating and its adhesive should pass both the wedge test and the shear test in order to check the bond and its durability.

The corrosion protection of the reverse side also has consequences for durability. The interior of a sandwich panel should normally remain dry, in which case there is no problem. However, in order to ensure safe structures, the following recommendations are offered.

- For panels with foamed polyurethane cores, it is recommended that the corrosion protection category on the reverse side should be of the order 3 and, for panels with mineral wool or polystyrene cores, of the order 4.
- The corrosion classification of the reverse side primers is influenced by the number of layers and their composition:
  - one layer (5–10 μm) without corrosion inhibitors is approximately class 1–2
  - one layer (5–10 μm) with corrosion inhibitors is approximately class 2–3
  - two layers (5–10 μm) with corrosion inhibitors is approximately class 3–4 (or alternatively, a single thick layer)

6.5.4 Fasteners

In order to achieve a reliable level of durability with regard to mechanical safety, the fastener should pass the repeated loading test (see Section 13.9.5.4). It is also recommended that stainless steel should be used as the fastener material in order to have adequate safety against corrosion.

6.6 REFERENCES

6.1 Deutsche Institut für Bautechnik (1996) Assumptions for the working life of construction products in guidelines for European technical approvals and harmonised standards. Proposal document TB 96/21/8.

6.2 Preliminary European Recommendations for Sandwich Panels with additional recommendations for panels with mineral wool core material. European Convention for Construction Steelwork, TWG 7.4, International Council for Building Research and Documentation, W56. CIB Publication 148, Reprinted November 1995. Update of this document published as CIB Publication 257 (2000).

6.3 Berner, K. *et al.* (1994) The durability of structural sandwich elements. *Materials and Structures*, **27**, 33–39.

6.4 Hiekkanen, I. (1996) Prediction of the long-term durability of the Paroc element under climatic conditions. Report 28/1996, Tampere University of Technology, Finland.

6.5 Just, M. (1992) Zum Langzeitverhalten von Sandwichplatten mit einem Polyurethan-Hartschaumstoffkern. *Bauingenieur*, **67**, 83–89.

6.6 Kerkkanen, T. and Tianen, T. (1999) Durability of different types of sandwich panels under climatic conditions. Report 15/1999, Institute of Materials Science, Tampere University of Technology, Finland.

Chapter 7
Refrigerated Warehouses, Cold Stores and Chill Rooms

7.1 INTRODUCTION

Refrigerated warehouses and cold stores are used for storing goods, such as foodstuffs, medicines, ice cream, etc., that are perishable at normal temperatures and which will only keep if stored at temperatures as low as $-30°C$. Since maintaining inventories of refrigerated goods imposes severe demands on administrative logistics (special computerised transport systems for moving goods into and out of storage are required), the current trend is towards building large-scale refrigerated warehousing facilities. Several freezer facilities measuring approximately 70-m wide by 170-m long by 35-m high, equipped with high-rise storage racking have recently been constructed in Germany, see Fig. 7.1. In order to serve their purpose, the interiors of such refrigerated warehouses must be maintained at low temperatures for extended periods, which, in turn, implies consistently high energy consumption. Employing outer-wall cladding and roofing with good insulating properties is the key to minimising their energy consumption, and facilities of this type invariably rely on sandwich panels to obtain the optimal thermal insulation.

Fig. 7.1 Refrigerated warehouse under construction ($70\,m \times 170\,m \times 35\,m$).

Relatively small rooms that are kept at a constant temperature by means of cooling are called either chill or cold rooms. The term 'chill room' is usually used for rooms which have an internal temperature lying between approximately +1°C and +8°C. Such rooms are intended for the short-term storage of meat, vegetables, fruit, fish, etc.

It is generally recognised that sandwich panels that, in addition to the excellent insulating properties of their cores, also provide the additional benefits of rapid installation regardless of local weather conditions and encapsulation of the core against the ingress of moisture and the formation of ice, are particularly well-suited to these applications. Sandwich panels used as the walls and ceilings of cold stores and chill rooms usually have an overall thicknesses of 15–30 cm. When employed as the outer walls, they are usually installed vertically oriented as multi-span 'load-bearing' panels fastened to a supporting substructure. There are a number of proprietary wall systems in use which incorporate cam-operated locking devices for rapid and secure installation.

In large refrigerated warehouses, the latching mechanisms of the outer wall panels directly engage the structure of the high-rise racking, which transmits vertical and lateral strains to the wall panels it supports (see Fig. 7.2).

Fig. 7.2 External sandwich panels fastened directly to high-rise pallet racking.

This racking structure rests on a relatively thick (approximately 60 cm) concrete slab which, in turn, floats on a roughly 20-cm thick insulating layer with an intervening layer of PVC sheet. The concrete slab is sealed on all sides by insulating strips. Inserted between this insulating layer and the slab foundation of the building, which rests on a foundation of specially prepared soil strata, are a vapour barrier and a heating system (in order to prevent freezing of the ground, see Fig. 7.3).

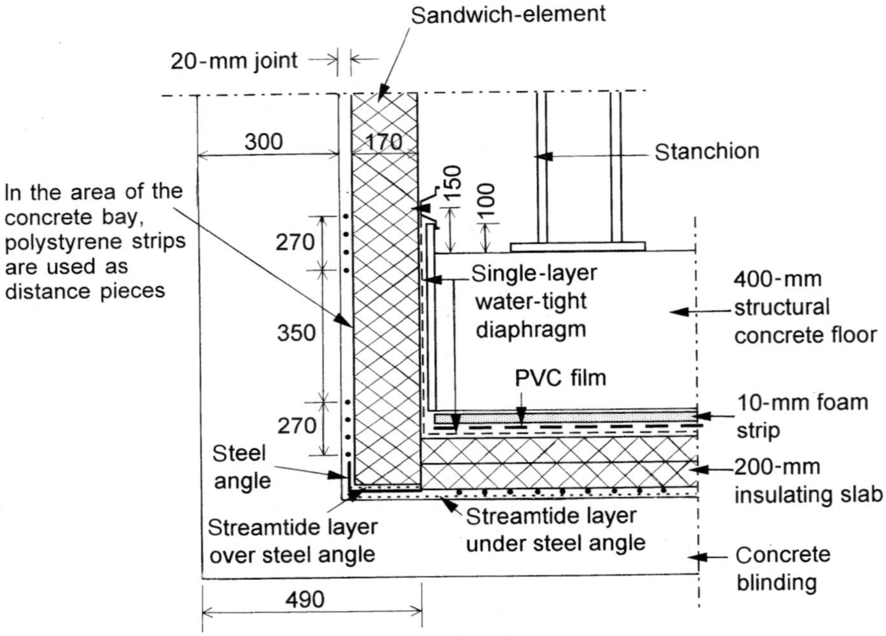

Fig. 7.3 Refrigerated warehouse: detail of connection of outer wall to base.

7.1.1 Cold stores and chill rooms inside the building

Chill rooms usually have an internal temperature lying in the approximate range +1°C to +8°C. Cold rooms and stores are normally designed for temperatures between −18°C and −30°C. In order to obtain a satisfactory performance in these conditions, the following requirements have to be met:

- A very well insulated structure. The operating costs for chill rooms and cold stores are high and therefore a good thermal insulating capacity is required of the building parts in order to keep the maintenance costs at a tolerable level.
- An extremely air-tight and moisture-tight structure. The temperature conditions prevailing in a cold room constantly cause a strong tendency for moisture to migrate from the outside into the cold room, partly by convection and party by diffusion. This influence is of a long-term nature without the possibility of any intermediate drying.

Taking these two very important conditions into account, sandwich panels, with their high thermal insulating capacity and tight structural integrity, offer a particularly suitable choice for both walls and ceilings. However, for optimal cold storage, this is not sufficient and accurate project design together with well-functioning details and skilled panel installation are also required.

7.2 THERMAL INSULATION

The selection of the correct insulation thickness is determined by several factors: the difference between the temperature of the interior and that of the outside environment, the type of core material in the sandwich panel; the energy price and building costs as well as the design of the cooling system.

Cold rooms should preferably be located inside cool buildings in order to avoid the extreme temperature loads caused by solar radiation. This gives a consequential reduction in both the energy consumption and the moisture load. In huge cold stores, where this may not be possible, the insulating panels can be externally protected by, for example, additional profiled steel sheets.

The applicable panel thicknesses usually lie between 200 and 300 mm. These panel thicknesses give sufficiently high surface temperatures on the panel exterior in order to avoid surface condensation.

The use of sandwich panels results in a minimum of cold bridges. These generally only appear at panel fixing screws and corner joints. Panel screws with a good thermal insulation capacity, such as stainless steel screws or nylon spikes, result in a minimum influence on the thermal insulation performance. In details where the panel inner steel sheets extend through to the outer surface, for instance in corner details, the cold bridge can be broken by making a cut in the inner steel sheet.

A type of cold bridge that may cause major long-term problems by seriously reducing the thermal insulation arises when the panel structure is not diffusion-tight and moisture penetrates into the walls and ceilings and then freezes. This often results in significantly increased energy consumption.

7.3 AIR-TIGHTNESS

Cold room structures have to be air-tight for reasons of both energy economy and moisture technology. In general, sandwich construction results in very tight structures. The panel itself is completely air-tight but the panel joints and connections have to be carefully considered. If they are not completely air-tight, warm and moist air may penetrate from the outside into the panel structure and form ice. However, air only flows when there is a pressure difference through the structure. The high differences between the inside and outside temperatures, and the generally large building heights create a pressure difference through the structure, the so-called chimney or stack effect. This pressure difference (in Pa) due to the weight of the column of cold air can be calculated using the approximate expression:

$$\Delta p = 0.043\,\Delta T h$$

where ΔT is the temperature difference between the warm and the cold sides (°C)
h is the height difference between the floor and the ceiling (m)

For example, a temperature of −20°C inside a cold room and 20°C outside with a building height of 7 metres results in a pressure difference of 12 Pa, as shown in Fig. 7.4. Though not a significant structural load, this pressure difference is sufficient to cause a significant flow of air and indicates that large amounts of moisture can pass through untight construction.

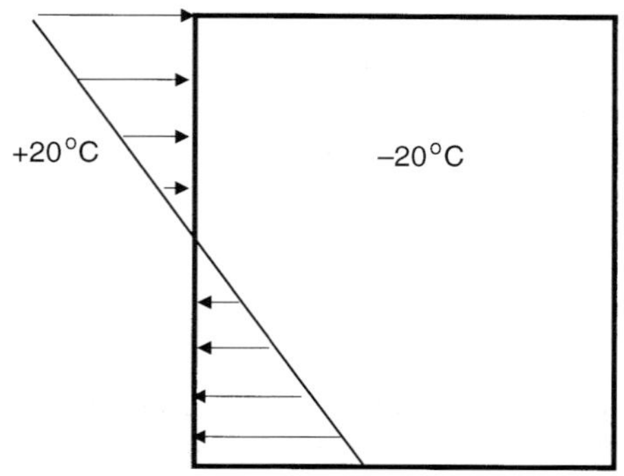

Fig. 7.4 Pressure difference over the height of a cold room.

It should be noted that the pressure difference is inverted relative to the normal case in a building. In a cold store, therefore, a pressure equaliser should be placed near the floor level so that cold, dry air passes out.

In addition, pressure differences caused by the influence of wind have to be added to those caused by temperature difference. A wind speed of 7 m/s may give a pressure difference of approximately 17 Pa over the wall exposed to the wind.

It is also important to check that the ventilation system does not create any underpressure in the cold room.

7.4 MOISTURE BARRIERS

In contrast to normal buildings, in a cold room the walls, ceiling and floor structures are subject to a *constant* inwardly directed vapour pressure. Therefore, diffusion tightness is an essential requirement of the building because the moisture flow never reverses and there is no possibility of drying. In this case, the diffusion barrier and its associated sealants must be located in the outside of the wall (the warm side). The sandwich panel itself, with its metal sheet facings, is diffusion-tight but here, as in the case of air-tightness, diffusion can appear in the panel joints and connections. As for the sealants their vapour resistance should have a minimum value of 2000×10^3 s/m (seconds per metre) which corresponds to a minimum thickness of plastic foil of about 0.2 mm. The sealants should also be non-ageing and able to resist possible movements in the structure. Their adhesion to the connecting surfaces is also important. When the cold room temperature is suddenly reduced from, for example, +20°C to −25°C and no valves or doors are open, very high pressure differences can arise through the structure. There is then a high risk of damage to the panels and vapour barrier with accelerating moisture penetration into the panels.

7.5 FIRE

Where the walls and ceilings are required to act as fire-resisting partitions, sandwich panels with a stone wool core can be used. In this case it should be observed that the air permeability is higher in stone wool than in cellular plastic foams such as polyurethane and polystyrene. This usually means that any open surfaces, which may arise on the short side of panels with stone wool core, need to be protected with a diffusion-tight membrane in order to prevent moisture from penetrating into the panels.

7.6 DETAILS

As emphasised earlier, the detailed design solutions used in chill and cold room applications are important. In addition, the installation must be carried out by skilled operatives. Often there are companies who specialise in cold room installation and these are to be preferred. Some detailed solutions are presented in Figs 7.3 and 7.5–7.8. Special care must be taken with penetrations through the walls. The basic principle is to ensure that there is a perfect air- and vapour-tight barrier on the warm side.

Wall panels can be mounted either vertically or horizontally.

- *Vertical* mounting is the more frequently used because this has the advantage of avoiding end joints within the cold region of the wall, thus reducing the risk of ice formation. Vertical mounting generally leads to multi-span solutions where the spans are reduced in order to reduce the large forces at the supports caused by temperature differences of up to 100°C.

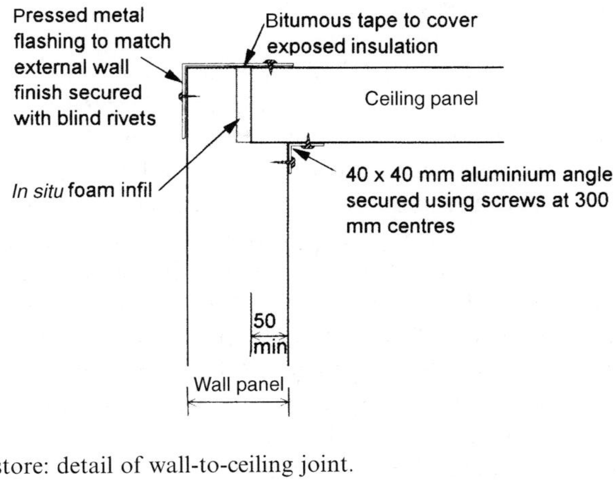

Fig. 7.5 Cold store: detail of wall-to-ceiling joint.

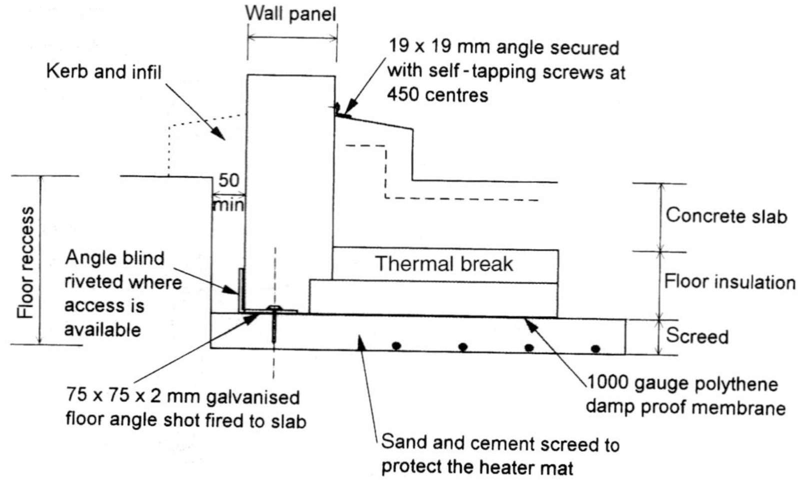

Fig. 7.6 Cold store: detail of connection of internal wall to base.

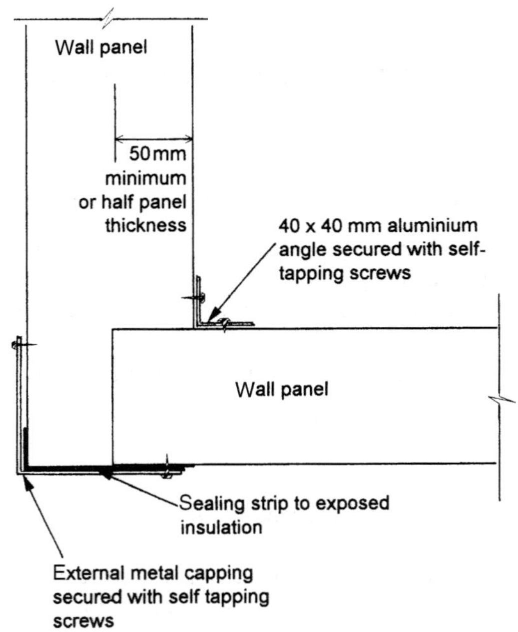

Fig. 7.7 Cold store: detail of wall-to-wall joint.

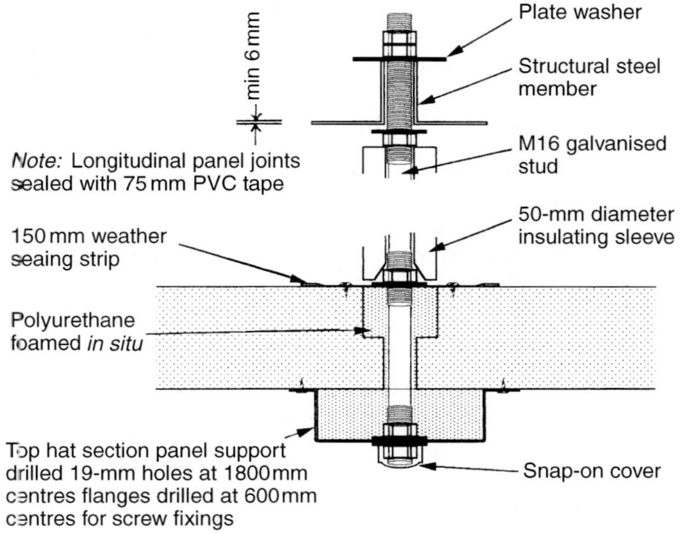

Fig. 7.8 Coldstore: ceiling support detail.

- *Horizontal* mounting has the advantage of allowing simply-supported panels which give rise zero temperature forces. The disadvantage is the added risk of moisture penetration at the end joints.

Figure 7.9 gives typical examples of vertical and horizontal mounting arrangements.

7.7 SPECIAL CONSIDERATIONS APPLYING TO SANDWICH PANEL OUTER WALLS

There are three fundamental problems to be considered when using sandwich panels as the outer walls of refrigerated warehouses and cold stores:

(1) the shortening of the internal structure of the building relative to the panels as it is cooled down
(2) designing for the stresses and deflections associated with thermal bowing of the panels
(3) preventing ingress of water and the formation of ice within the core

These will now be considered in turn.

7.7.1 Allowing for reactions to mechanical and thermal loads

The temperature load cases are major factors to be considered when selecting sandwich panels for use as the outer walls of refrigerated warehouses. These panels are subject to very high thermal loadings due to the extremely large differences in the temperatures between their outer faces (which might reach 55°C–80°C when

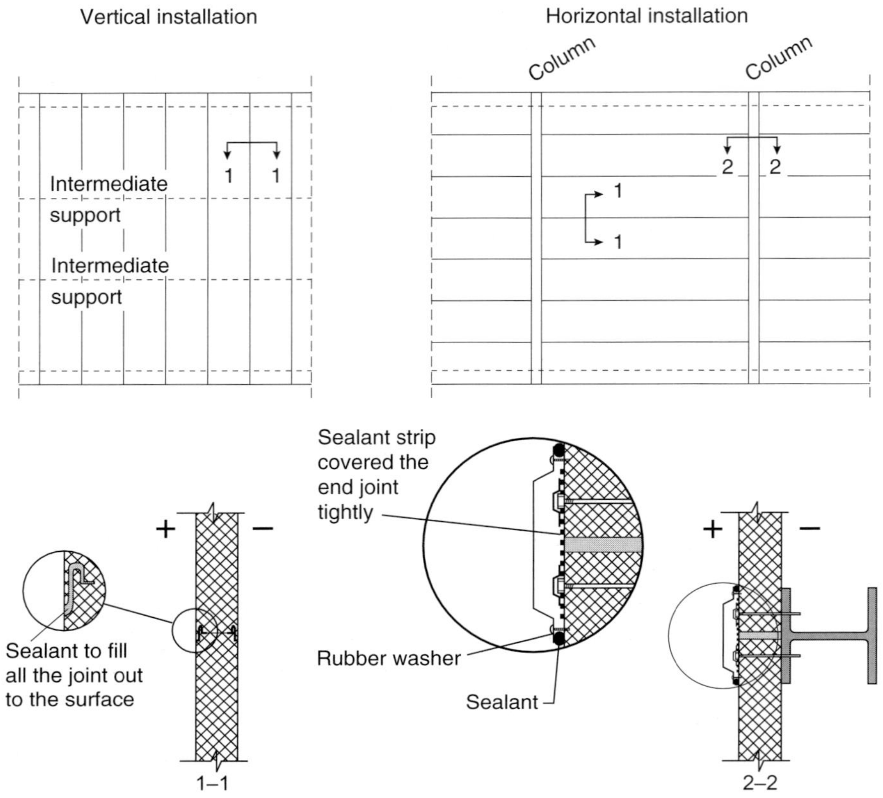

Fig. 7.9 Vertical and horizontal wall mounting arrangements.

exposed to strong sunlight) and their inner faces (−30°C), so that wind loading plays a relatively minor role. In the case of panels which are continuous over two or more spans, a peculiarity of the load cases involving extremely large temperature differences is that, contrary to the case for normal loading, reducing the spans of the panels (the separation between wall-panel latching mechanisms) will not yield correspondingly large reductions in the forces caused by thermal loading. There is, of course, no comparable problem in the case of the single span panels which are used in the construction of relatively small cold stores.

It follows that the only practical approaches remaining available which will allow the use of reasonable support member spacings (say 3.0–5.0 m between wall-panel latching mechanisms depending upon the internal construction of the wall panels and their supporting structures) require the incorporation of structural modifications in the design. These may take the form of flexible panel supports having lateral degrees of freedom (i.e. free to move normal to the panel) or may require the use of prestressed panel faces (see Section 7.4). In any event, the architect must accept some visible curvature in the wall.

A specially constituted air mixture is often employed in order to minimise the energy consumption in large freezer facilities. This air mixture must be continually replaced and this can lead to underpressures (which experience indicates may reach approximately 0.1 kN/m^2) occurring within such facilities and thus to further loads on the panels which must be taken into account in the engineering design. It may be noted here that an overpressure within the storage area is more favourable with respect to moisture movement but underpressure is more usual in large cold stores.

7.7.2 Thermal expansion/contraction and panel deformation

The engineering design of refrigerated warehouses must take into account the thermal contractions of all structural elements that occur during cooling. Assuming that erection takes place at ambient temperatures of about 20°C, and that the building interiors are to be cooled to −30°C, the temperature difference to be employed in computing the thermal expansions and contractions of the structural elements and the resulting deformations of the outer wall panels is −50°C.

A major factor to be considered in the engineering design of sandwich panel walls is that they will generally bow outwards when their inner faces are cooled. This is because the thermal contraction of their outer faces in the plane of the panel is small whereas the inner faces and all internal structural elements will undergo uniform linear thermal contraction. This statement is confirmed by measurements taken while large freezer facilities are cooled down to their operating temperatures.

For buildings with a height greater than about 15 metres, the contraction of the internal supporting structure invariably necessitates that the supports of sandwich panels which span vertically should be designed so that they are free to move vertically, since, for a wall of say 30-m height, a total vertical contraction of $v = 1.2 \times 10^{-5} \times (-50°\mathrm{C}) \times 3000\,\mathrm{cm} = 18\,\mathrm{mm}$ will have to be accommodated. Securing-bolts will severely stress the panels if movable or flexible panel supports have not been incorporated. Supports free to move laterally in the plane of the panel must also be incorporated if severe panel stresses and deformations due to the thermal contraction of steel substructures are to be avoided.

One possible means of meeting this requirement is the use of the specially designed clamps which will be described in detail in Section 7.9. An alternative way of avoiding large thermal compression forces is for the wall panels to span horizontally and for the longitudinal joints between the panels to have provision for the relatively small thermal compression that take place within the cover width of the panel.

7.8 THE ENGINEERING DESIGN OF SANDWICH PANELS

The engineering design of sandwich panels should be based on procedures similar to those described in Chapters 9–11, paying particular attention to the temperature load cases. In the following sections, we will outline the procedures involved in the engineering design of prestressed panels hung on supports having lateral degrees of freedom using, as an example, a solution that has been employed in two completed freezer facilities.

7.8.1 Using panel supports having lateral degrees of freedom

Figure 7.10 depicts an outer wall panel fastened to a steel storage racking substructure by latching mechanisms laterally spaced at intervals of 4.44 m. The use of special supporting clamps allows support displacements totalling as much as 45 mm over intermediate supports 2, 3 and 4. Allowing for both nominal summertime load-case temperatures (55°C outdoors and −30°C indoors) and wind-suction loading, we find that sandwich panels of overall length 18.45 m will satisfy the major criteria regarding resistance to wrinkling at the intermediate supports.

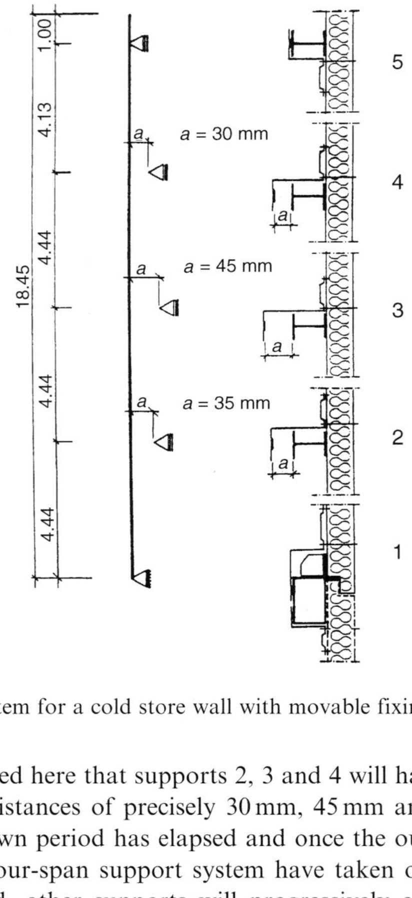

Fig. 7.10 Statical system for a cold store wall with movable fixings at the supports.

It has been assumed here that supports 2, 3 and 4 will have shifted their lateral positions through distances of precisely 30 mm, 45 mm and 35 mm respectively, once the cooling down period has elapsed and once the outward bulges in panel claddings and the four-span support system have taken on their final forms. If cooling is continued, other supports will progressively contact the panel as it continues to deform, generating further intermediate support systems (see Fig. 7.11).

Calculations of panel deformations should take account of the designs of all intermediate support systems due to the relevant load cases. In a final calculation stage, the resultant bending moments transmitted by supports may be computed by superimposing bending moments due to wind loads on the thermally induced bending moments for the supports involved.

When conducting final stress computations, it is critically essential that the

Plate 1. Horizontally-mounted panels forming the wall of an industrial building (courtesy of Paroc, Finland)

Plate 2. Sandwich panels forming an internal partition wall (courtesy of Paroc, Finland)

Plate 3. A walk-on ceiling in a hygienic environment conceals technical installations (courtesy of Paroc, Finland)

Plate 4. Prefabricated curved corner panels with a micro-profiled surface (courtesy of IFBS)

Plate 5. Horizontally-mounted panels with a curved corner flashing (courtesy of IFBS)

Plate 6. Corporate headquarters for Esso (UK) plc, Leatherhead, UK

Plate 7. Compaq Computer Corporation, Erskine, Scotland

Plate 8. Toray Textiles, Mansfield, England

Plate 9. An office building with architectural PUR panels, the Netherlands (courtesy of IFBS)

Plate 10. A multi-purpose office building with PUR panels, the Netherlands (courtesy of IFBS)

Plate 11. A sports building with smooth, horizontally-mounted panels, Norway (courtesy of Paroc, Finland)

Plate 12. Vertically-mounted panels, Mercedes Benz, Germany (courtesy of Paroc, Finland)

Plate 13. Horizontally-mounted panels, Nokia Mobile Phones, Salo, Finland (courtesy of Paroc, Finland)

Plate 14. Horizontally-mounted panels, Katri Antell Bakery, Rovaniemi, Finland (courtesy of Paroc Finland)

Plate 15. Vertically-mounted architectural PUR panels on a waste materials plant, the Netherlands (courtesy of IFBS)

Plate 16. Horizontally-mounted PUR panels on a commercial centre, the Netherlands (courtesy of IFBS)

Plate 17. Vertically- and diagonally-mounted panels, Helsinki Airport, Finland (courtesy of Paroc, Finland)

Plate 18. Wall panels, Midtermolen, Denmark

Plate 19. Horizontally-mounted panels, Naestved, Denmark (courtesy of Paroc, Finland)

Plate 20. Copper-coated, horizontally-mounted panels, Vantaa, Finland (courtesy of Paroc, Finland)

Plate 21. Horizontally-mounted panels, Rovaniemi Power Plant, Finland (courtesy of Paroc, Finland)

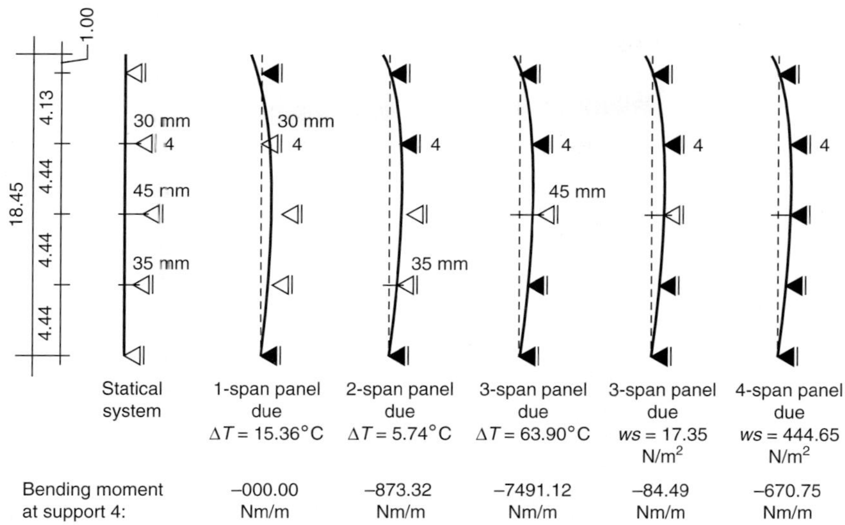

$$M(4) = -9119.68\ \text{N m/m} = -9119.68 \times 10^3\ \text{N mm/m}$$

panel width = 1000 mm; distance between centroids $e = 169.44$ mm; thickness of faces $t = 0.56$ mm

$$\text{maximum stress } \sigma_{max} = \frac{9119.68 \times 10^3}{1000.0 \times 169.44 \times 0.56} = -96.11\ \text{N/mm}^2$$

with safety factor $\gamma = 1.1$; design value $\sigma_{max} = -96.11 \times 1.1 = -105.72\ \text{N/mm}^2$

Therefore required wrinkling stress $\sigma_{wr} \geq -105.72\ \text{N/mm}^2$; available $\sigma_{wr} = 105.3\ \text{N/mm}^2$ is just OK

Fig. 7.11 Statical system for sandwich panel wall as cooling continues.

bending stresses calculated at the centre of the span of a single-span beam never reaches the wrinkling threshold (i.e. 105 N/mm^2 for the panel under consideration). For nominal summertime load-case temperatures and nominal wind-suction loads, the compressive loads of interest are those transmitted to outer panel faces by the panel supports, i.e., exerted in the vicinities of the bolt heads, since we are considering the case of highly localised (point) loads. Significant correlations of wrinkling thresholds to externally induced strains must be taken into account here (see Chapter 10). In Germany, a 20% decrease in free-span wrinkling thresholds is currently assumed to occur in the vicinities of bolt heads.

A major precaution related to installation practices is that securing bolts should not be tightened to the point where they produce indentations in panel claddings, since this might create additional flaws that would further reduce wrinkling thresholds.

A general comment regarding the result obtained above for the case of laterally stationary supports, i.e., for a four-span support system, is that under the particular set of assumptions (the type of wall panels, thermal loads, etc.) involved, computations indicate that thermally induced stresses sufficient to cause wrinkling of panel outer faces may occur during cool-down cycles (see Fig. 7.12, which depicts wrinkling of the outer face of outer wall panels of a large freezer facility for the case where panel supports have inadequate lateral travel), and it appears highly likely that this wrinkling will also occur in actual practice.

Fig. 7.12 Damage caused by wrinkling of the outer face during cool down.

Reducing the loadings by decreasing the spacings of the supports (which would be difficult to implement in actual practice since the spacings of the panel latching mechanisms are usually set by the dimensions of the storage racking) would require the use of extremely closely spaced supports (in this case, a spacing of about 2.33 m which would be unacceptable in any practical application).

7.8.2 Pre-flexed panel outer faces

Another prospective means of reducing the stresses in external wall panels due to extremely high thermal loadings involves supporting sandwich wall panels on several supports distributed over their full widths and intentionally employing intermediate supports protruding farther from the substructures than the outermost supports. Forces, which are small and may well be manually applied, will thus have to be exerted on the panel edges during installation in order to bring them into alignment with these outermost supports. This will give the panel outer faces a small predefined curvature, which will be visible externally and must therefore be considered to be acceptable aesthetically, and thus a predefined prestress. This pre-flexing or prestressing will be opposed by the panel supports as the panels cool, yielding net reductions in the panel loads sufficient to allow the required spacings of panel latching mechanisms and supporting members to be maintained.

This type of structural modification has thus far been employed on two large freezer facilities. As an illustration of the basic principles involved in the engineering design of prestressed outer wall claddings, we will now discuss in some detail the various aspects of an order-of-magnitude computation of the loadings of the sandwich wall panels shown in Fig. 7.13.

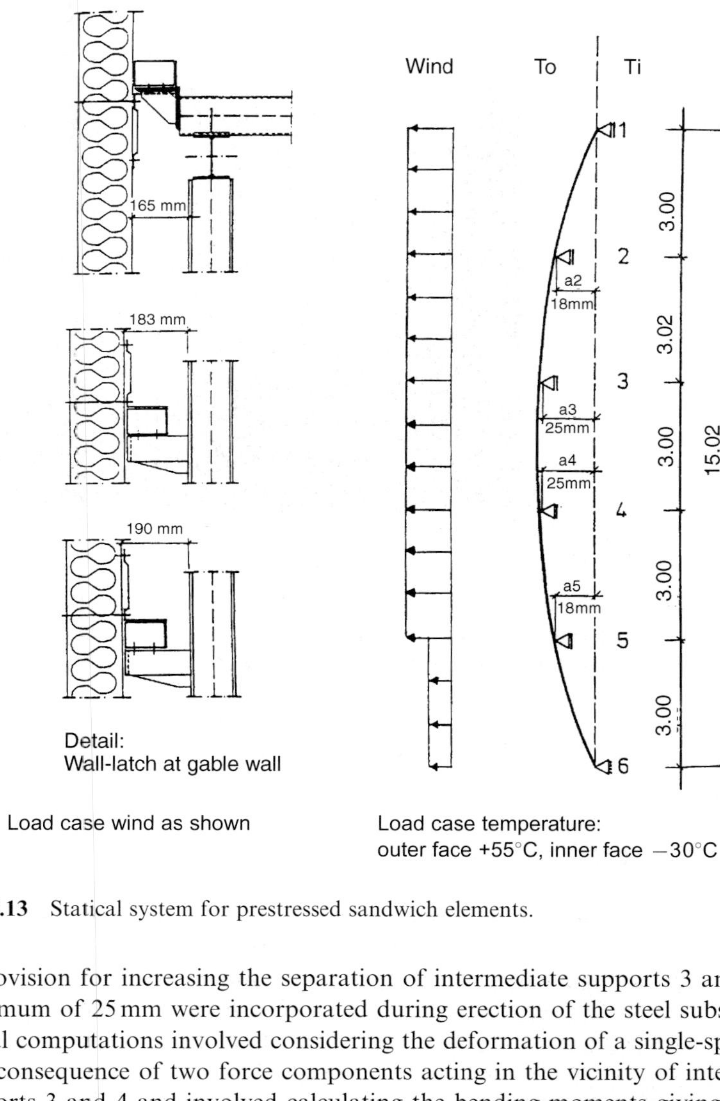

Fig. 7.13 Statical system for prestressed sandwich elements.

Provision for increasing the separation of intermediate supports 3 and 4 by a maximum of 25 mm were incorporated during erection of the steel substructure. Initial computations involved considering the deformation of a single-span panel as a consequence of two force components acting in the vicinity of intermediate supports 3 and 4 and involved calculating the bending moments giving rise to a 25-mm deflection. A shear-creep coefficient of $G_t = 4\,N/mm^2$, a conservative value, was assumed in order to account for shear creeping over extended loading periods.

Nominal thermal loadings for the so-called 'summertime' load case (temperature difference of $-85°C$) and nominal wind-suction loadings are then superimposed upon these loads. The bending moments at Support 2 due to prestress for an outer cladding deflection of 25 mm are as follows. (The calculations are made with the shear modulus, G, for room temperature because small variations of G have little influence on the stresses and it is the initial effects which are important.)

(1) Extended loading periods ignored
$$M_2 = 1.719\,\text{kN m/m}$$
(2) Extended loading periods considered ($G_t = 4.0\,\text{N/mm}^2$)
$$M_2 = 1.113\,\text{kN m/m}$$

The summary of the calculation of the allowable bending moment is as follows – for the summertime load case:

The bending moment at Support 2 due to wind suction, $\Delta T = -85°C$, and prestressing is:

$$M_2(\text{max.}) = 0.409 + 9.671 - 1.113 = 8.967\,\text{kN m/m}$$

Therefore, the requirement for the maximum stress in the outer face is:

$$\sigma_{\max} = \frac{M_{2.\max}}{eA_F} \leq \frac{\sigma_w}{\gamma_M}$$

i.e.

$$\sigma_{\max} = \frac{8.967 \times 10^6}{168.7 \times 560} = 94.8\,\text{N/mm}^2 \leq \frac{105}{1.1} = 95.4\,\text{N/mm}^2$$

where e = the distance between the centroids of the faces = 168.8 mm
A_F = the area of the steel face = 560 mm^2/m
σ_w = the wrinkling stress of the face = 105 N/mm^2

This indicates that prestressing reduces the loading by 14%, so that the load on the sandwich panels remains within the tolerated limits. In the absence of prestressing, much greater loadings, which would probably result in wrinkling of the outer faces of the panels, would occur for the assumed support spacings.

7.9 STRUCTURAL FEATURES REQUIRING SPECIAL ATTENTION

The utmost caution should be exercised in the engineering design of sandwich panel outer walls and roofs when it comes to taking account of the deformations which occur during the cooling-down cycles referred to in Section 7.7.2. Damage due to excessive straining will inevitably result if the contraction of steel substructures due to declining temperatures is ignored. This statement is particularly true for the engineering design of the supports for sandwich wall panels. As mentioned earlier, both vertical and lateral deformations of the supporting substructures may occur in the planes of wall panels. In addition, where designs employ supports having lateral degrees of freedom, lateral deformations normal to the planes of wall panels may also occur.

A suitable choice of solution here involves the use of specially designed clamps, such as those depicted in Figs 7.10, 7.11 and 7.13, which were employed in a completed project. Such clamps (see Fig. 7.14) should be designed to allow for the

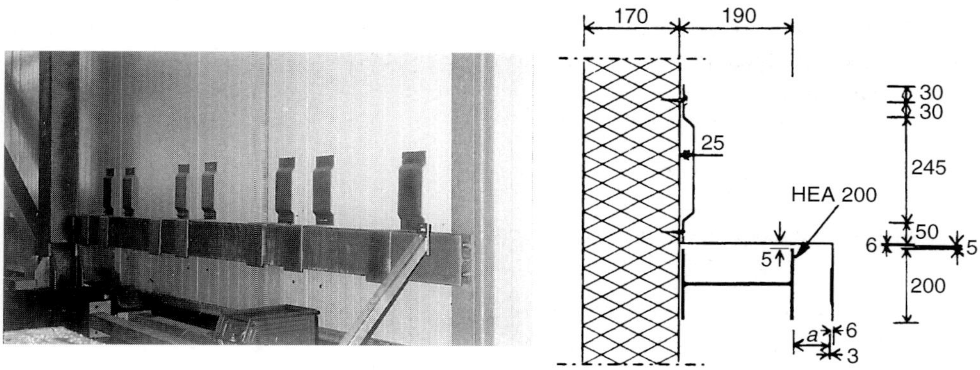

Fig. 7.14 Specially-designed clamp for cold store construction.

effects of bending moments due to the compressive point load (contact clamp/wall) and the tension point load (fastener). Particular attention should be devoted to the compressive forces exerted on sandwich panels by the bending moments.

Deformations occurring during cool-down cycles should also receive due attention when designing wall-panel junctions and intersections, such as corner joints and lateral abutments. Gaps, which are initially sufficiently large to accommodate the contraction of the supporting steel substructure should be invariably be designed in. Simple butt joints, i.e., joints lacking any tongues, grooves or folds, were intentionally employed in order to preclude any mechanical engagement. Gaps should not be permanently caulked until the interior temperatures have stabilised. Corner-joint designs should incorporate cappings fabricated from highly flexible angle stock in order that the cappings will flex in sympathy with the sandwich panels without damaging them (see Fig. 7.15).

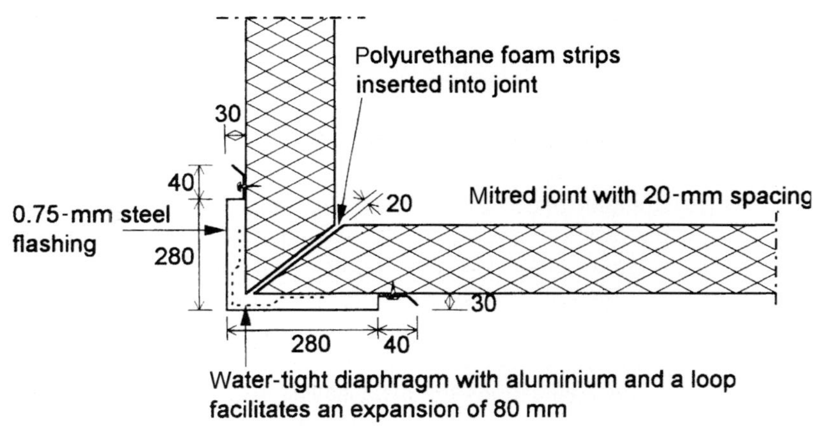

Fig. 7.15 Typical corner construction in a high-rise cold store.

Chapter 8
Special Architectural Applications

8.1 INTRODUCTION

From the point of view of the architect, composite panels fall into two distinct classes. Most of this book is concerned with the sort of sandwich panels that may be produced on a continuous laminating line and used to clad low rise industrial and commercial buildings. Panels of this type are finished on two sides only and use rather rudimentary details at the panel ends and laps. Nevertheless, as will be shown in the examples in this chapter, buildings clad with this type of panel can be made to be aesthetically interesting, and even spectacular.

However, there is another type of panel which is usually smaller, batch-produced and finished on all four sides. This type of panel is produced to a much higher specification and designed for high-technology curtain walling. It usually has a complex fixing system incorporating gaskets which allow individual panels to be mounted or removed without disturbing the remainder of the construction. This book would not be complete without an account of this latter type of sandwich panel which will be considered in the second part of this chapter.

8.2 CLADDING DETAILS

Plates 1–5 show some typical details of the application of sandwich panels in different situations.

8.3 INDUSTRIAL AND COMMERCIAL BUILDINGS

The examples shown in Plates 6–21 are of some typical buildings clad with automatically produced sandwich panels which are either continuously-foamed polyurethane panels with profiled faces or panels with flat faces and with a core of mineral wool. However, the panels could equally have a core of polystyrene without any change to the appearance. The panels are mostly fastened with penetrating screws which are either covered with flashings or else the heads are left visible. Some have proprietary concealed fixings in the sidelaps. The panels generally have colour-coated steel sheets as the facing material.

Typical applications for these panels are commercial buildings such as trade centres, storage warehouses, office and public buildings and a wide range of industrial buildings.

Most of the examples in Plates 6–21 are offered without detailed comment in order to demonstrate what can be achieved. In such cases, the caption to the plate is considered to be sufficient. Some projects are described in a little more detail in the following section.

8.3.1 Esso UK plc corporate headquarters, Leatherhead, UK

The three-storey complex, part of which is shown in Plate 6, consists of two long parallel buildings linked by a full-height atrium. The offices are housed in two blocks which have stepped sections to give sunshading to the south-facing aspects as well as providing circulation galleries to the atrium and north-facing offices. The roof of this outstanding office complex has been clad in 8570 m^2 of 50-mm thick sandwich panel with a profiled upper face giving a U-value of 0.45 $W/m^2 K$. The silver-grey wall panels have a flat outer face with a solar reflective finish in order to reduce solar gain in the building. They also have a 50-mm thick core and provide a U-value of 0 36 $W/m^2 K$.

The architect for this project was the Percy Thomas Partnership and the photo was by Jeremy Cockayne.

8.3.2 Manufacturing facility for Compaq Computer Corporation, Erskine, Scotland

This project was built in two phases, each consisting of 12 100 m^2 and joined at right-angles by a dramatic curved connecting link. The walls were clad with 9-metre long horizontal sandwich panels with flat faces. The square corners of the building were constructed to extremely tight tolerances in order to allow the neat detailing shown in Plate 7. Curved sandwich panels were used in the connecting link.

The American architect for this project was Spencer Herolz and the photograph was by Williams Photography Ltd.

8.4 HIGH-TECHNOLOGY ARCHITECTURAL WALL PANELS

The Sainsbury Arts Centre shown in Fig. 8.1, which was designed by Foster Associates in 1977, is one of the first non-commercial, non-industrial buildings to use an interchangeable panel system. Identical 1.8×1.2 m panels were used for both the walls and roof, with the weatherproofing joint detail dependent upon the neoprene gasket being mounted back against an aluminium carrier system. Full details of this system are given as Case study 22 in Reference 8.1.

The original mounting and sealing system failed for a number of reasons which are now fully understood and this cladding system has subsequently been replaced by a flat composite panel. The neoprene gasket at the Sainsbury Centre was formed as a net, transported to site and site vulcanised at the mid-points prior to fitting the panels. In this way an interchangeable system of wall and roof panels was achieved. A similar principle of continuous neoprene gaskets was later used at Gatwick Airport North Piers.

Composite panels were also used by Foster at the Renault Centre where 75-mm thick panels span 4 metres between the vertical cladding support as shown in Fig. 8.2. Further details of this development are given as Case study 21 in Reference 8.1. The panel system which was bolted to the vertical cantilevered columns by three fixings at each end of the panel provided a very low cost elevation. A large neoprene skirt takes up the deflection at the top of the assembly.

Fig. 8.1 Sainsbury Arts Centre, University of East Anglia, designed by Foster Associates, 1977.

Fig. 8.2 Renault Centre.

Sandwich panels for this type of 'architectural' application are usually made on a discontinuous (batch) process using platen presses or vacuum presses. The advantage of this process is that the sheet-foam materials, typically expanded polystyrene or extruded polystyrene, can be inspected for their surface quality prior to lamination and that the core materials can be changed, within the same process, to polyurethane slabstock, mineral wool lamellas, or expanded and extruded polystyrene. This market is currently estimated as 400 000 square metres in the UK and a similar quantity for Holland. This type of panel, with the possibility of a formed edge on all four sides, offers the opportunity for a more sophisticated integrated arrangement of panels, windows and louvres. This can, of course, then be sold at a much higher cost than continuous foamed panel systems, say £150–200 per m^2 compared with £25–35 per m^2.

Laminated panels using sheet core materials of expanded or extruded polystyrene are being increasingly used on more sophisticated 'architectural' buildings. Integrated systems, including composite panels, are supplied complete with doors, windows and louvres.

With this need for greater sophistication in mind, the Aspect II system, which was developed by Brookes Stacey Randall Fursdon and is now marketed by Coseley Panel Systems, is an example of a state-of-the-art integrated panel system using sandwich panel technology.

8.4.1 Case study: Aspect Mark II (Reference 8.3)

> 'The idea is to create a flexible framework, a framework which embraces the floor, the walls and the ceiling so that within that framework the plan is completely changeable.'
>
> Norman Foster – *Frontiers of Design* (September 1978)

The joining and fixing of metal composite panels is critical to their performance in use. Additional stiffness can often be achieved by the incorporation of UPVC edge profiles, which also act as joint formers, in order to achieve more complex joining and fixing details and to allow interchangeability with other components, such as doors, windows and louvres.

The search for an improved four-way crossover junction, without the use of ladder gaskets (as used for the Sainsbury Arts Centre, University of East Anglia, and Gatwick North Piers), and the need to avoid the use of site-applied sealants, were fundamental design aims in the design of the new cladding system, Aspect II, developed by Brookes Stacey Randall Fursdon now manufactured and marketed by Coseley Panel Products as shown in Fig. 8.3.

This system of interchangeable facade components offers the opportunity of rearranging or adapting the building envelope at any stage during the construction process or after the building is occupied. Aspect II allows sandwich cladding panels to be interchanged with glazed panels, louvres or doors. Even loading bay doors can be relocated. For a full description of the development of the system, see Reference 8.2.

The method of fixing the panels is by means of an aluminium clamping plate located within the UPVC edge profiles and behind the primary gaskets. As shown

Fig. 8.3 Aspect II: interchangeable facade components.

in Fig. 8.4, the clamping plates are secured with a stainless steel counter-sunk socket head machine screw to a rear fixing block, which slides within the vertical rear aluminium carrier. The Aspect II system is entirely secretly fixed and avoids any problems of aligning fixings, sealing fixings or panel damage by fixings.

The panels can be easily assembled or dismantled by a skilled fixing team. The clamping plate has been carefully engineered with allowance for thermal movement, rotational effects and tolerance. The interface of the clamping plate and the UPVC edge section has been designed to minimise the thermal bowing of the panel. The number of clamping plates is dependent upon the wind loading and panel size.

In Aspect II, the weather seal is formed by the primary gasket framing the perimeter of the panel. The inner seal is formed by a horizontal air seal mounted to the top and bottom edges of the panel. This is clamped against the vertical air seal mounted to the rear aluminium carrier. A still air zone exists between the primary gaskets and the air seals. Drainage continuity is maintained using sealed joints in the rear carrier, normally at 4-metre centres, which allow for thermal expansion. The cill detail is designed to drain moisture from within the assembly to the outside.

All components, including windows, louvres and doors, have a common method of jointing and are secretly fixed. This achieves a system of co-ordinated components, through their means of jointing, resulting in a coherent assembly of diverse elements. Throughout these elements, the finish can be common or highly varied depending on the needs of each project and the requirements of the specifying architect.

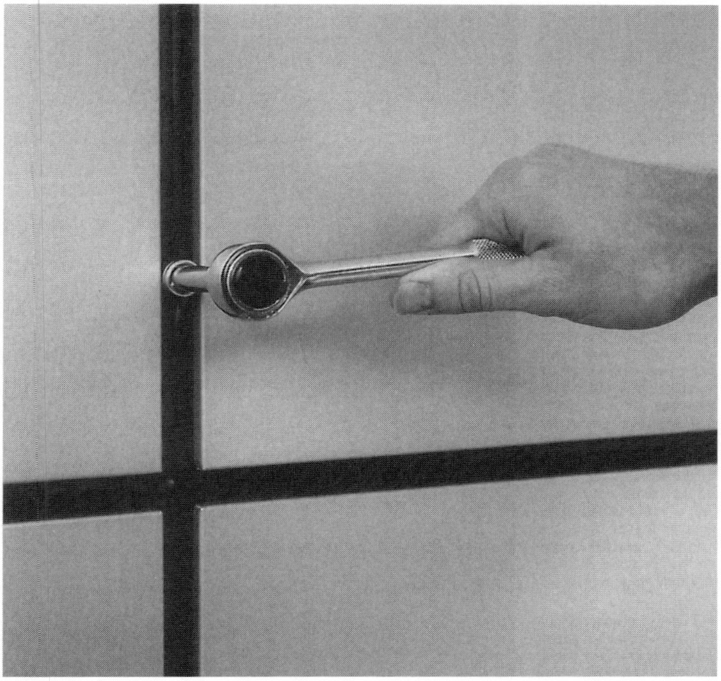

Fig. 8.4 Aspect II: fixing of panels.

The sequence of assembly starts with the aluminium rear carriers, which are fixed back on to the sub-structure of the building, and then aligned, plumbed and shimmed to a flat plane. The fixings and shims will be determined by the nature and tolerance of the sub-structure of each project. The vertical air seals are rolled into the locating channels in the front face of the carriers. These air seals run continuously up the elevation and are initially oversized in order to allow for shrinkage, all as shown in Fig. 8.5.

The first panel, complete with its horizontal air seal and primary framed gasket, is offered up to the carrier and restrained by bolting down the aluminium cladding plate on to the aluminium rear block within the carrier. The clamping plate is then partially backed-off and the adjacent panel offered up to the carrier. The joint between the panels is compressed to 15 mm, see Fig. 8.6, and the panel alignment is checked. The action of tightening the socket head machine screws of the clamping plates compresses the panel and the horizontal air seals against the vertical air seals. It is good practice to assemble a bay of panels vertically.

The weather seal of the horizontal panel to panel joint is formed by the primary framed gaskets, backed up by the panel mounted horizontal air seals. There is no requirement for horizontal rear carriers or sub-frames as shown in Fig. 8.7.

To minimise thermal migration, the rear clamping block of the clamp plate assembly has an extended nose section. This ensures that panels are retained within the gridlines of the panel modules. The assembly is restrained vertically by stop blocks which are located in every third horizontal panel joint and carry the dead load.

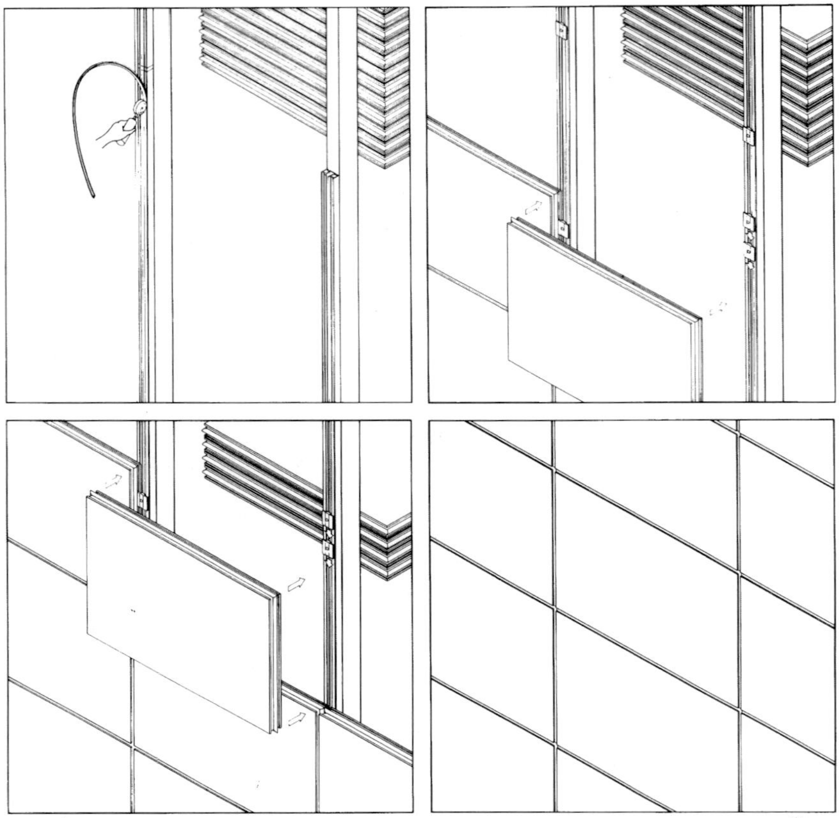

Fig. 8.5 Aspect II: sequence of erection.

The sandwich panel construction of Aspect II can use various core insulation materials to achieve U-values of 0.45–0.35 $W/m^2 K$.

The Aspect II system of panels and its associated components were produced after a design, research and development programme which included the manufacture of mock-ups and prototypes and their subsequent testing. The panel-to-panel joint, with its patented framed gaskets, has been successfully tested according to BS 5368 Parts 1–3 for air penetration, weather resistance and wind loading.

Rotational movements within the panel joints are caused by wind induced deflection and thermal bow. The joints have been designed to accommodate the rotational movement resulting from deflections not greater than span/250.

The panel system allows individual panels, windows, doors or louvres to be removed as required and any part of the assembly may be replaced or relocated as necessary. Replacement, during the contract period, of panels that are inadvertently damaged by other trades is an inevitable requirement on large contracts using lightweight panels. As well as the advantage of interchangeability, the ability to remove individual panels means that, during the life of the building, panels that are damaged can be easily replaced with the minimum of interference to the occupants of the building.

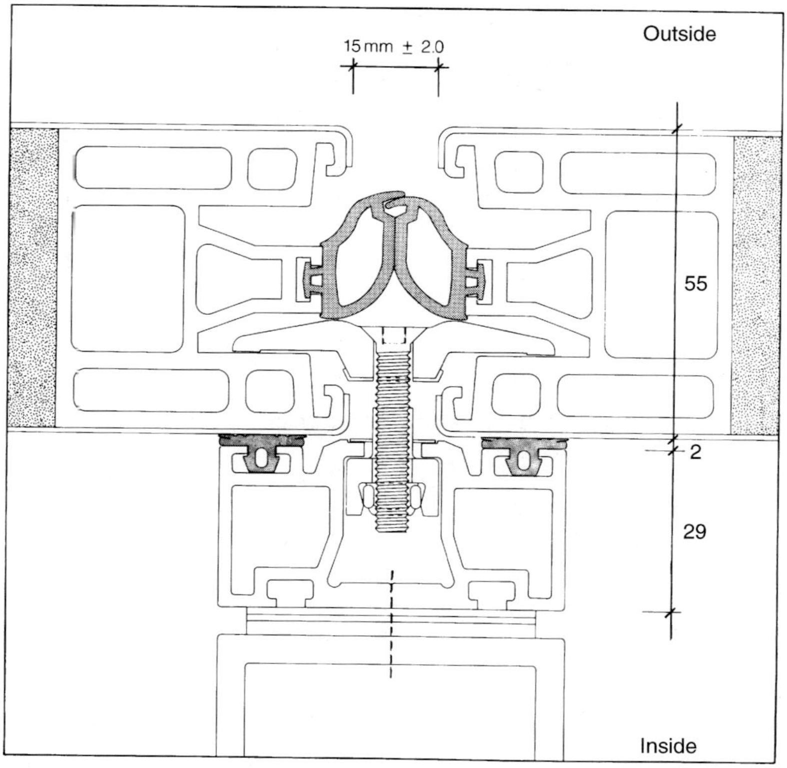

Fig. 8.6 Aspect II: plan view of joint between panels.

The system has been used on a number of projects including John Winter's Mansell Street and was awarded the Aluminium Imagination 'Shapemakers' Award in 1993.

8.5 REFERENCES

8.1 Brookes, A.J. (1985) *Concepts in Cladding*, Construction Press, London.
8.2 Brookes, A.J. (1990) *Architects' Journal*, 24 January.
8.3 Brookes, A.J. and Newman, J. (1994) Performance requirements for composite panels. In *Proceedings of the International Conference on Building Envelope Systems and Technology*, Singapore, December 1994.

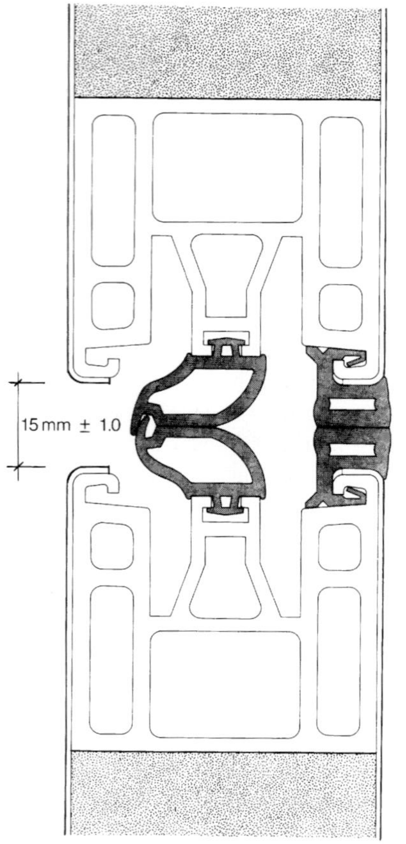

Fig. 8.7 Section through a horizontal joint.

Chapter 9
Principles of Structural Behaviour

9.1 INTRODUCTION

Sandwich panels are relatively simple structural components which can be analysed using the conventional principles of structural mechanics. Provided that a small number of mechanical properties are known, they can be designed on the basis of a structural analysis which predicts the stresses and deflections. This chapter is therefore primarily concerned with the relevant methods of analysis.

However, because of the influence of the relatively flexible core material, sandwich panels cannot be designed on the basis of conventional beam and plate theory. Most sandwich panels span in one direction and therefore require an augmented beam theory which takes into account the shear flexibility of the core. Furthermore, many of the core materials in common use creep under sustained stress. Another important consideration is that core materials are generally chosen because of their thermal insulating properties and, as a consequence, large temperature differences can exist between the faces of the panel. This means that sandwich panels tend to be particularly sensitive to temperature effects. The main factors to be taken into account when considering the structural behaviour are therefore:

- the influence of the shear flexibility of the core on the global behaviour
- the influence of the core in restraining local buckling (wrinkling) of the faces
- the influence of temperature-induced stresses and deflections
- creep of rigid foam core materials with time

This chapter will consider how the stresses and deflections caused by conventional loads, temperature loads and creep are determined analytically. The next chapter will show how the resistance or strength of panels may be calculated with respect to the alternative failure modes. The calculation models for the different failure modes are deterministic and based on either mean or characteristic values of the properties of the face and core materials. Chapter 11, the third in this series, will then consider how the calculated stresses and strengths may be compared under prescribed conditions of safety in order to complete the design. This relationship is illustrated in Fig. 9.1. The safety analysis is based on limit state principles with partial safety factors for stresses (load factors) and resistances (material factors). Because no harmonised procedures for sandwich panel design are available, the safety factors used are based on the published recommendations of the European Convention for Constructional Steelwork (ECCS) and the International Building Council (CIB) which, in turn, arise from the relevant Eurocodes.

The calculation procedures for sandwich panels are predominantly linear elastic. This is essential for the serviceability limit of deflection and also for some of the ultimate limit state modes of failure which exhibit limited ductility. However, as

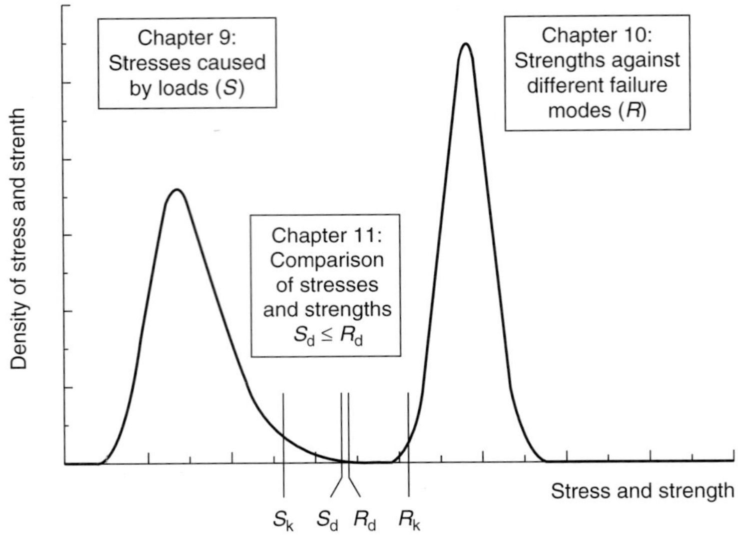

Fig. 9.1 Distributions of stress and strength in a sandwich panel.

will be explained later, a pseudo-plastic design procedure is often used for the ultimate limit state design of continuous panels which fail in a bending mode whereby 'plastic hinges' of zero resistance are assumed at the internal supports. This effectively means that, under appropriate conditions, a continuous panel may be treated as a series of simply supported panels.

Although much of this chapter is concerned with precise analytical procedures based on solutions of the governing differential equations, it should be appreciated at the outset that there are approximate methods and design curves available that are sufficiently accurate for all practical purposes. The results for panels with flat or quasi-flat faces are summarised in Table 9.1. Approximate solutions for panels with profiled faces are given in Section 9.4 together with comprehensive design curves for all of the practical cases. However, because these are all derived from the exact methods, it is necessary to describe these first.

The basic equations given in this chapter have previously been given by a number of authors although the primary sources are now out of print. These sources are given in the references at the end of the chapter.

9.1.1 Types of panels

Each of the metal faces of sandwich panels may be of three types, namely flat, lightly profiled or fully profiled as shown in Fig. 9.2.

For the purposes of global structural analysis (that is the prediction of stresses and deflections), panels in which both faces are flat or lightly profiled must be considered separately from panels in which at least one face is profiled. The former cannot be considered as a special case of the latter because the equations become

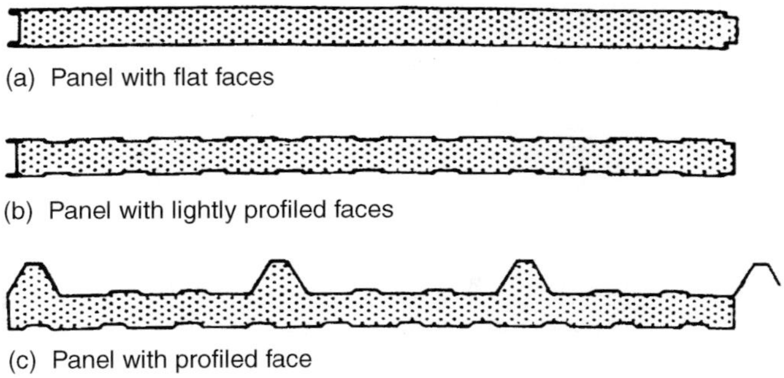

Fig. 9.2 Typical cross-sections of sandwich panels.

ill-conditioned and the analysis breaks down. The global analysis of sandwich panels is therefore considered in two separate parts.

For the conventional types of panels used as the cladding of buildings or in controlled environment rooms, flat face theory may be used for profiling with a depth less than about 4 mm. If flat face theory is used for more strongly profiled panels, the errors introduced by this approximation are more important for stresses than for deflections.

9.2 ANALYSIS OF PANELS WITH FLAT OR LIGHTLY PROFILED FACES

The analysis that is required to predict the stresses and deflections in panels with flat or lightly profiled faces is essentially that of conventional beam theory with the addition of shear deformation. This section includes the derivation of the equations, which is considered to be useful in promoting an understanding of the fundamental behaviour. However, the less mathematical reader may simply make use of resulting design expressions for the cases most frequently met in practice.

Figure 9.3 shows the stress resultants and deformations in a small element of a sandwich panel.

The relationships between the stress resultants and deformations are:

$$M_S = B_S(\gamma_2' + \theta) = B_S(\gamma' - w'' + \theta) \tag{9.1}$$

$$V_S = A_C G_{eff} \gamma \tag{9.2}$$

where, in addition to the quantities defined in Fig. 9.3, a prime denotes differentiation with respect to x which is measured along the length of the panel and

B_S $= E_{F1} A_{F1} E_{F2} A_{F2} e^2/(E_{F1} A_{F1} + E_{F2} A_{F2})$ = bending stiffness of the sandwich
A_{F1}, A_{F2} = cross-sectional areas of the faces
E_{F1}, E_{F2} = Young's moduli of the faces
A_C $= be$ = effective area of the foam core
d_C = depth of core
G_{eff} $= G_C e/d_C$ = effective shear modulus of the core
G_C = shear modulus of the core
w = total deflection

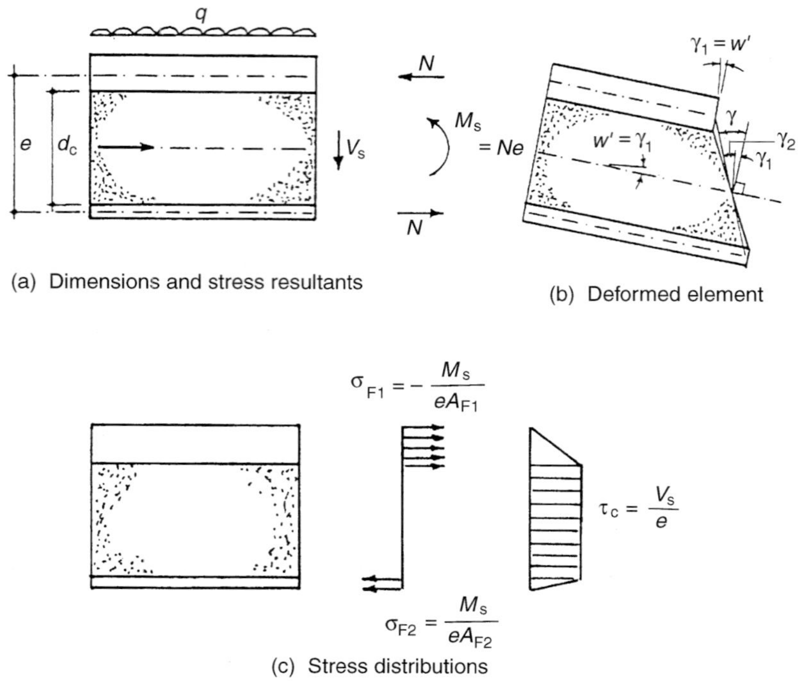

Fig. 9.3 Forces and deformations in a typical sandwich element with thin faces.

γ	= shear strain in core (divergence of the normal from the horizontal axis of the section)
θ	$= (\alpha_{F2} T_2 - \alpha_{F1} T_1)/e$
	= curvature resulting from a temperature difference between the faces
α_{F1}, α_{F2}	= coefficient of thermal expansion of the faces
T_1, T_2	= temperature of the faces
q	= distributed load per unit length
V_S	= shear force
N	= axial force in face

The equilibrium equations are those for the familiar engineer's theory of bending:

$$\frac{dM_S}{dx} - V_S = 0 \tag{9.3}$$

$$\frac{dV_S}{dx} + q = 0 \tag{9.4}$$

and substituting the relationship between the stress resultants and deformation into these gives:

$$B_S(\gamma'' - w''') - A_C G_{eff}\gamma = 0 \tag{9.5}$$

$$A_C G_{eff}\gamma' = -q \tag{9.6}$$

Uncoupling these equations finally yields:

$$B_S w'''' = q - \frac{B_S}{A_C G_{eff}} q'' \tag{9.7}$$

and

$$\gamma'' = -\frac{q'}{A_C G_{eff}} \tag{9.8}$$

In many practical cases, the bending moments and shear forces are statically determinate and it is merely necessary to use the above equations to determine the deflections. In such cases, the following alternative form is more convenient:

$$w'' = -\frac{M_S}{B_S} + \frac{V'_S}{A_C G_{eff}} - \theta \tag{9.9}$$

$$\gamma = \frac{V_S}{A_C G_{eff}} \tag{9.10}$$

If the temperature term θ is considered to give rise to a special case, the two remaining terms on the right hand side of Equation 9.9 are the bending and shear terms respectively. From the practical point of view, it is often convenient to consider these two terms separately and to consider the deflection to be the sum of two components. The first of these is the bending component which is given by the familiar 'Engineer's Theory of Bending'. The second is the shear component which requires a separate calculation according to shear theory.

It may be observed here that the bending moments are carried entirely by the faces and the shear forces by the core. For *statically determinate* panels, the stress resultants (and therefore the stresses) are identical to those given by the Engineer's Theory of Bending. For these panels, the flexibility of the core serves merely to enhance the deflections.

For *statically indeterminate* panels (e.g. panels continuous over two or more spans), the stress resultants are also influenced by the flexibility of the core.

We now need to consider solutions of these equations for different load and support conditions. This will be done for the cases most frequently met in practice.

9.2.1 Case 1: Simply supported panel with a uniformly distributed load

This case is shown in Fig. 9.4. It is a case where the bending moment and shear force are statically determinate and can be written down without calculation. Thus, using the non-dimensional longitudinal coordinate $\xi = x/L$, we have

$$M_S = \frac{qL^2}{2}(\xi - \xi^2) \tag{9.11}$$

$$V_S = \frac{qL}{2}(1 - 2\xi) \tag{9.12}$$

In Equation (9.9), the bending and shear terms are separate and it is convenient to integrate them separately. Thus, the bending component of deflection, w_B, is given by

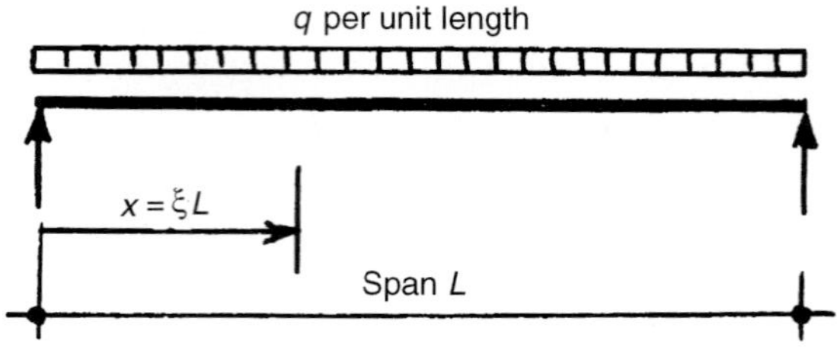

Fig. 9.4 Simply supported panel with a uniformly distributed load.

$$w_B'' = -\frac{M_S}{B_S} = -\frac{qL^2}{2B_S}(\xi - \xi^2) \tag{9.13}$$

Integrating twice and inserting constants of integration A and B,

$$w_B = -\frac{qL^4}{2B_S}\left(\frac{\xi^3}{6} - \frac{\xi^4}{12} + A\xi + B\right) \tag{9.14}$$

The boundary conditions are $w_B = 0$ when $\xi = 0$ and $\xi = 1$, thus $A = -1/12$ and $B = 0$, i.e.

$$w_B = \frac{qL^4\xi}{24B_S}(\xi^3 - 2\xi^2 + 1) = \frac{qL^4\xi}{24B_S}(1+\xi)(1+\xi-\xi^2) \tag{9.15}$$

The shear component of deflection, w_S, is given by

$$w_S'' = \frac{V_S'}{A_C G_{eff}} = -\frac{q}{A_C G_{eff}} \tag{9.16}$$

and integrating twice, as before,

$$w_S = -\frac{qL^2}{A_C G_{eff}}\left(\frac{\xi^2}{2} + A\xi + B\right) \tag{9.17}$$

The boundary conditions are again $w_B = 0$ when $\xi = 0$ and $\xi = 1$ so that $A = -0.5$ and $B = 0$, giving

$$w_S = \frac{qL^2\xi}{2A_C G_{eff}}(1 - \xi) \tag{9.18}$$

The total deflection is, of course, given by $w = w_B + w_S$ and, expressing

$$k = \frac{3B_S}{A_C G_{eff} L^2} \tag{9.19}$$

we have

$$w = \frac{qL^4}{24B_S}\xi(1-\xi)(1 + 4k + \xi - \xi^2) \tag{9.20}$$

The maximum stress $\sigma_{F.max}$ in the faces is at mid-span, where $\xi = 0.5$, i.e.

$$M_{S.max} = \frac{qL^2}{8} \text{ and } \sigma_{F.max} = \pm\frac{qL^2}{8eA_F} \tag{9.21}$$

where A_F is the cross-sectional area of either the top face or the bottom face as required.

The maximum shear stress $\tau_{C.max}$ in the core is at the support so that, if B is the width of the panel and d_C is the depth of the core:

$$V_{S.max} = \pm\frac{qL}{2} \text{ and } \tau_{C.max} = \frac{qL}{2Bd_C} \tag{9.22}$$

The maximum deflection, w_{max}, is also at mid-span and

$$w_{max} = \frac{5qL^4}{384B_S}(1 + 3.2k) \tag{9.23}$$

9.2.2 Case 2: Simply supported panel with a point load

This case is shown in Fig. 9.5 and again the stress resultants are statically determinate.

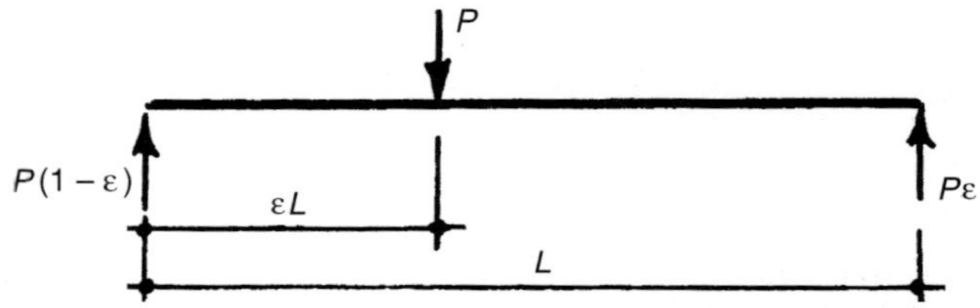

Fig. 9.5 Simply supported panel with a point load.

Using the same notation as in the previous problem and adopting Macauley's convention, whereby quantities in curly brackets are only considered when the contents of the bracket are positive, we have

$$M_S = PL\,(1 - \varepsilon)\xi - PL\{\xi - \varepsilon\} \tag{9.24}$$

$$V_S = P(1 - \varepsilon) - P\{\xi - \varepsilon\} \tag{9.25}$$

Separating the bending and shear components of deflection, as before,

$$w''_B = -\frac{M_S}{B_S} = -\frac{PL}{B_S}[(1 - \varepsilon)\xi - \{\xi - \varepsilon\}] \tag{9.26}$$

and then integrating twice with constants of integration A and B,

$$w_B = -\frac{PL^3}{B_S}\left[(1 - \varepsilon)\frac{\xi^3}{6} - \frac{\{\xi - \varepsilon\}^3}{6} + A\xi + B\right] \tag{9.27}$$

the boundary conditions are $w_B = 0$ when $\xi = 0$ and $\xi = 1$ so that $B = 0$ and

$$A = -\frac{1 - \varepsilon}{6}(2\varepsilon - \varepsilon^2) \tag{9.28}$$

giving

$$w_B = \frac{PL^3}{6B_S}[-(1 - \varepsilon)\xi^3 + \{\xi - \varepsilon\}^3 + \xi\varepsilon(1 - \varepsilon)(2 - \varepsilon)] \tag{9.29}$$

It follows that, to the left of the load,

$$w_{B1} = \frac{PL^3}{6B_S}(1 - \varepsilon)\xi(2\varepsilon - \varepsilon^2 - \xi^2) \tag{9.30}$$

and, after a little rearrangement, to the right of the load,

$$w_{B2} = \frac{PL^3}{6B_S}\varepsilon(1-\xi)(2\xi - \xi^2 - \varepsilon^2) \tag{9.31}$$

Similarly, for the shear component of deflection:

$$w_S = \iint \frac{V'_S L^2}{A_C G_{eff}} d\xi\, d\xi = \frac{PL}{A_C G_{eff}}[\xi(1-\varepsilon) - \{\xi - \varepsilon\} + A\xi + B] \tag{9.32}$$

Here, the boundary conditions result in $A = B = 0$ so that

$$w_S = \frac{PL}{A_C G_{eff}}[\xi(1-\varepsilon) - \{\xi - \varepsilon\}] \tag{9.33}$$

Therefore, to the left of the load, we have

$$w_{S1} = \frac{PL}{A_C G_{eff}}\xi(1-\varepsilon) \tag{9.34}$$

and, to the right of the load,

$$w_{S2} = \frac{PL}{A_C G_{eff}}\varepsilon(1-\xi) \tag{9.35}$$

Summing the components of deflection to give the total deflection, we have to the left and right of the load,

$$w_1 = \frac{PL^3}{6B_S}(1-\varepsilon)\,\xi\,(2k + 2\varepsilon - \varepsilon^2 - \xi^2) \tag{9.36}$$

$$w_2 = \frac{PL^3}{6B_S}\varepsilon(1-\xi)(2k - \varepsilon^2 + 2\xi - \xi^2) \tag{9.37}$$

The deflection below the load is found from either of the above equations by putting $\xi = \varepsilon$.

9.2.2.1 *Special case of a point load at the centre of the panel*

When the point load is at the centre of the panel, $\varepsilon = 0.5$ and we have a particular case of the above equations which occurs frequently in practical calculations. The maximum bending stresses and deflections occur at mid-span where $\xi = 0.5$ giving:

$$M_{max} = \frac{PL}{4} \text{ and } \sigma_{F.max} = \pm\frac{PL}{4eA_F} \tag{9.38}$$

$$w_{max} = \frac{PL^3}{48\,B_S}(1+4k) \tag{9.39}$$

where k is according to Equation 9.19.

9.2.3 Case 3: Simply supported panel with a temperature difference

For this case, the bending moments and shear forces, and therefore the stresses, are all zero. If the temperatures of the faces are T_1 and T_2 and the coefficients of thermal expansion are α_{F1} and α_{F2} respectively, then

$$\theta = \frac{\alpha_{F2}T_2 - \alpha_{F1}T_1}{e} \tag{9.40}$$

and from Equation 9.9,

$$w'' = -\theta \tag{9.41}$$

so that integrating twice and setting the deflection to zero at the ends of the beam, gives

$$w = \frac{\theta L^2}{2}\xi(1-\xi) \tag{9.42}$$

with a maximum deflection at the centre of the panel of $\theta L^2/8$.

9.2.4 Solutions for two- and three-span panels

The results obtained for Cases 1, 2 and 3 can be combined to give the stress resultants and displacements of multi-span panels. Figure 9.6 shows the procedure for a two-span panel subject to a uniformly distributed load. In order to obtain the solution for the two-span case, it is merely necessary to add together the solutions for a single span of $2L$ subject to the uniformly distributed load (Fig. 9.6(b)) and an upward point load at mid-span (Fig. 9.6(c)) where P is chosen so that the two deflections Δ are equal.

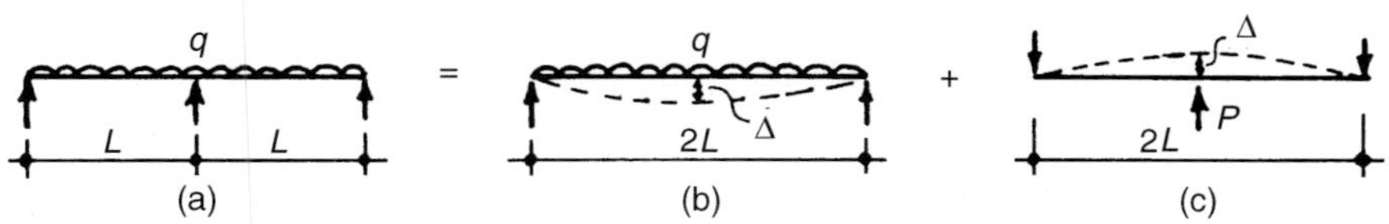

Fig. 9.6 Solution for a two-span panel.

Solutions for two-span panels subject to a temperature difference between the faces may be obtained similarly as a combination of Cases 2 and 3.

Panels with three equal spans may also be solved as a combination of the solutions for a simply supported span of $3L$ subject to the applied uniformly distributed (or temperature) load (Fig. 9.7(b)) and a point load P at each third point (Figs 9.7(c) and (d)). Figures 9.7(c) and (d) are, of course, mirror images and P is chosen so that $\Delta_1 + \Delta_2 = \Delta$.

It follows that the majority of cases of interest to the designers of continuous wall and roof panels can all be obtained from simple combinations of the solutions for single span panels given as Cases 1 to 3.

This has been done, and the results are summarised in Table 9.1. In this table, the symbols are as previously defined and the deflections for two- and three-span panels are approximate.

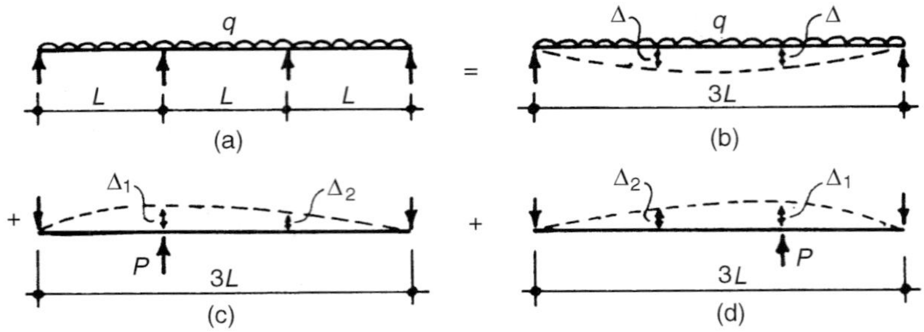

Fig. 9.7 Solution for a three-span panel.

9.2.5 More general cases of loading and support conditions

The above methods should be capable of dealing with the vast majority of load and support conditions likely to arise in practice. Occasionally, a special case arises that is outside the scope of the methods described above. It may be possible to adapt the procedures that have been described but it is more likely that the designer will have to resort to numerical methods.

There are two main possibilities here, namely the finite-element method and the finite-difference method and both have their advocates. The finite-element method will be described here because it has two advantages.

(1) Sandwich panels are one of the class of structure for which the finite-element method is exact. This means that it is only necessary to have a single element between each of the natural nodes of the problem and the calculations are much simpler than in more general finite-element analysis.
(2) Boundary conditions can be treated more simply in finite-element analysis.

The subject element is shown in Fig. 9.8. It has two degrees of freedom per node making four degrees of freedom for the element. There are therefore four displacements available to define the displacement function within the element. As the element is governed by a fourth-order differential equation, the general displacement function contains four arbitrary constants which can therefore be defined exactly.

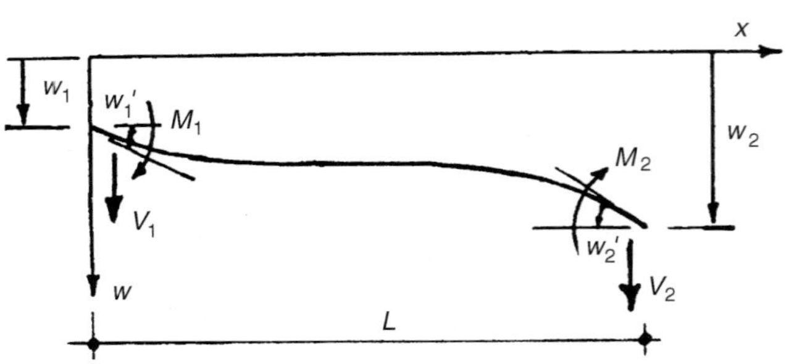

Fig. 9.8 Finite element for the analysis of panels with flat or lightly profiled faces.

Table 9.1 Summary of solutions for one-, two- and three-span panels with flat or lightly profiled faces loaded by a uniformly distributed load q and by a temperature difference between the outer (T_1) and the inner (T_2) face ($k = 3\,B_S/A_C\,GL^2$; $\theta = \alpha_T\,(T_2 - T_1)/e$)

	End support reaction, R_0	Intermediate support reaction, R	Bending moment in end span, M	Bending moment at internal support, M	Maximum deflection in span, w_{max}
Single span of L Uniform load, q	$\frac{qL}{2}$		$\frac{qL^2}{8}$		$\frac{5qL^4}{384B_S}(1+3.2k)$
Temperature difference, $(T_2 - T_1)$					$\frac{\theta L^2}{8}$
Two spans of L Uniform load, q	$\frac{qL}{2}\left[1-\frac{1}{4(1+k)}\right]$	$qL\left[1+\frac{1}{4(1+k)}\right]$	$\frac{qL^2}{8}\left[1-\frac{1}{4(1+k)}\right]$	$-\frac{qL^2}{8}\frac{1}{(1+k)}$	$\frac{qL^4}{48B_S}\frac{(0.26+2.625k+2k^2)}{(1+k)}$
Temperature difference, $(T_2 - T_1)$	$-\frac{3B_S\theta}{2L}\frac{1}{(1+k)}$	$\frac{3B_S\theta}{L}\frac{1}{(1+k)}$	$-\frac{3B_S\theta}{4}\frac{1}{(1+k)}$	$-\frac{3B_S\theta}{2}\frac{1}{(1+k)}$	$\frac{\theta L^2}{32}\frac{(1.089+3.96k)}{(1+k)}$
Three spans of L Uniform load, q	$\frac{qL}{2}\left[1-\frac{1}{(5+2k)}\right]$	$qL\left[1+\frac{1}{2(5+2k)}\right]$	$\frac{qL^2}{8}\left[1-\frac{1}{(5+2k)}\right]$	$-\frac{qL^2}{(10+4k)}$	$\frac{qL^4}{24B_S}\frac{(0.826+5.625k+2k^2)}{(5+2k)}$
Temperature difference, $(T_2 - T_1)$	$-\frac{6B_S\theta}{L}\frac{1}{(5+2k)}$	$\frac{6B_S\theta}{L}\frac{1}{(5+2k)}$	$-3B_S\theta\frac{1}{(5+2k)}$	$-6B_S\theta\frac{1}{(5+2k)}$	$\frac{\theta L^2}{4}\frac{(1.056+k)}{(5+2k)}$

In fact, the element is similar to that for a beam element with shear deformation for which the relevant stiffness matrix has been derived by a number of authors. The derivation is a simplified form of the more general derivation for panels with profiled faces which will be given later. It will not, therefore, be given here in detail but merely quoted:

$$\begin{vmatrix} V_1 \\ M_1 \\ V_2 \\ M_2 \end{vmatrix} = \frac{1}{1+\alpha} \begin{vmatrix} \frac{12B_S}{L^3} & \frac{6B_S}{L^2} & \frac{-12B_S}{L^3} & \frac{6B_S}{L^2} \\ \frac{6B_S}{L^2} & \frac{B_S(4+\alpha)}{L} & \frac{-6B_S}{L^2} & \frac{B_S(2-a)}{L} \\ \frac{-12B_S}{L^3} & \frac{-6B_S}{L^2} & \frac{12B_S}{L^3} & \frac{-6B_S}{L^2} \\ \frac{6B_S}{L^2} & \frac{B_S(2-\alpha)}{L} & \frac{-6B_S}{L^2} & \frac{B_S(4+\alpha)}{L} \end{vmatrix} \begin{vmatrix} w_1 \\ w'_1 \\ w_2 \\ w'_2 \end{vmatrix} \qquad (9.43)$$

where

$$\alpha = 4k = \frac{12\,B_S}{A_C\,G_{\mathrm{eff}}\,L^2} \qquad (9.44)$$

9.3 ANALYSIS OF PANELS WITH PROFILED FACES

When one or both faces of the panel are profiled, a more general analysis is required which takes into account the bending stiffness of the faces. Figure 9.9 shows the stress resultants and deformations for this case. A comparison with Fig. 9.3 reveals that the two are the same except for the addition here of the face bending moments M_{F1} and M_{F2} together with the corresponding shear forces V_{F1} and V_{F2}.

The relationships between the stress resultants and deformations given in Equations 9.1 and 9.2 are unchanged. Further relationships are added as follows:

$$M_{F1} = -B_{F1}w'' \qquad M_{F2} = -B_{F2}w'' \qquad (9.45)$$

$$V_{F1} = -B_{F1}w''' \qquad V_{F2} = -B_{F2}w''' \qquad (9.46)$$

where, in addition to the quantities defined in Fig. 9.9 and below Equations 9.1 and 9.2,

$B_{F1} = E_{F1}\,I_{F1}$ = bending stiffness of upper face
$B_{F2} = E_{F2}\,I_{F2}$ = bending stiffness of lower face

Because the stress resultants in the two faces are proportional to the same deformations, it is convenient to treat them together, so that

$$M_D = M_{F1} + M_{F2} \qquad M = M_D + M_S \qquad (9.47)$$

$$V_D = V_{F1} + V_{F2} \qquad V = V_D + V_S \qquad (9.48)$$

$$B_D = B_{F1} + B_{F2} \qquad B = B_D + B_S \qquad (9.49)$$

The above equations imply the separation of the stress resultants into a sandwich part and a flange part as shown in Fig. 9.10. This separation is fundamental to understanding the behaviour of panels with stiff faces and also turns out to be very useful in practical design calculations. It should therefore be carefully noted.

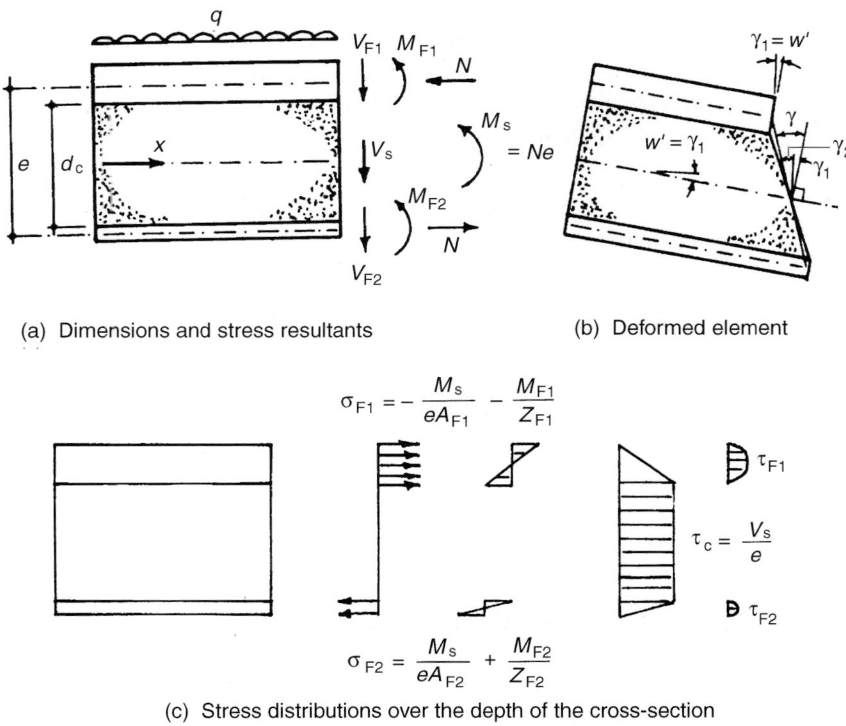

Fig. 9.9 Forces and deformations in a typical sandwich element with profiled faces.

From Equations 9.1, 9.2, 9.45 and 9.46, together with 9.47–9.49, two differential equations are obtained:

$$A_C G_{eff} \gamma - B_D w''' = V \tag{9.50}$$

$$B_S(\gamma' + \vartheta) - Bw'' = M \tag{9.51}$$

Eliminating γ and noting that $V' = -q$, a fourth-order differential equation in w is obtained:

$$w'''' - \left(\frac{\lambda}{L}\right)^2 w'' = \left(\frac{\lambda}{L}\right)^2 \frac{M}{B} + \frac{1+\alpha}{\alpha}\frac{q}{B} - \left(\frac{\lambda}{L}\right)^2 \frac{\theta}{1+\alpha} \tag{9.52}$$

where L is the length of the panel and

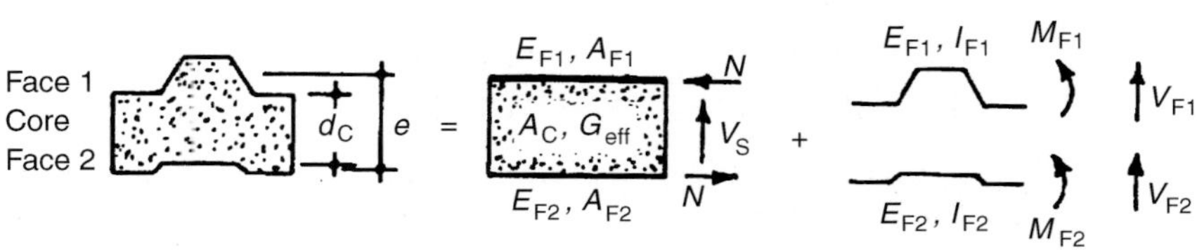

Fig. 9.10 Separation of stress resultants into sandwich part and flange part.

$$\alpha = \frac{B_D}{B_S} \quad \beta = \frac{B_S}{A_C G_{eff} L^2} \quad \lambda^2 = \frac{1+\alpha}{\alpha\beta} \tag{9.53}$$

Similarly, eliminating w from Equations 9.50 and 9.51:

$$\gamma'' - \left(\frac{\lambda}{L}\right)^2 \gamma = -\frac{\beta\lambda^2 V}{B} \tag{9.54}$$

The equations in the above form are particularly useful when the distributions of total bending moment, M, and shear force, V, are known, i.e. for statically determinate systems. For such cases, the general solutions of Equations 9.52 and 9.54 are

$$w = C_1 \cosh\frac{\lambda x}{L} + C_2 \sinh\frac{\lambda x}{L} + C_3 + C_4 x + w_P \tag{9.55}$$

$$\gamma = D_1 \cosh\frac{\lambda x}{L} + D_2 \sinh\frac{\lambda x}{L} + \gamma_P \tag{9.56}$$

where w_P and γ_P are particular integrals which depend primarily on the loading. As these solutions must also satisfy Equation 9.3, it is easy to show that

$$D_1 = (1+\alpha)\frac{\lambda}{L}C_2 \quad D_2 = (1+\alpha)\frac{\lambda}{L}C_1 \tag{9.57}$$

Thus the number of constants of integration reduce to four and these can be determined from the boundary conditions, in particular, for a simply supported panel:

$$w(0) = 0 \quad w''(0) = 0 \quad w(L) = 0 \quad w''(L) = 0 \tag{9.58}$$

As for panels with flat faces, there are three cases which are fundamental, namely simply supported panels subject to:

(1) uniformly distributed load
(2) point load anywhere in the span
(3) uniform temperature difference between the faces

and, from combinations of these, most of the other cases which are important in practice can be derived. The solutions for the significant cases are as follows with $\xi = x/L$ as before.

9.3.1 Simply supported panel with a uniformly distributed load of q per unit length

With

$$M = \frac{qL^2}{2}(\xi - \xi^2) \qquad V = \frac{qL}{2}(1 - 2\xi) \qquad \xi = \frac{x}{L} \tag{9.59}$$

The particular integrals in Equations 9.55 and 9.56 are

$$w_P = \frac{qL^4}{24B}\left(\xi^4 - 2\xi^3 - \frac{12}{\alpha\lambda^2}\xi^2\right) \tag{9.60}$$

$$\gamma_{\mathrm{p}} = \frac{qL^3\beta}{2B}(1 - 2\xi) \tag{9.61}$$

The constants of integration then follow as

$$C_1 = \frac{qL^4}{\alpha\lambda^4 B} \qquad C_2 = -\frac{qL^4}{\alpha\lambda^4 B}\frac{\cosh\lambda - 1}{\sinh\lambda}$$

$$C_3 = -\frac{qL^4}{\alpha\lambda^4 B} \qquad C_4 = -\frac{qL^3}{B}\left(\frac{1}{24} + \frac{1}{2\alpha\lambda^2}\right) \tag{9.62}$$

so that the complete solution is

$$w = \frac{qL^4}{B}\left[\frac{1}{24}\xi(1 - 2\xi^2 + \xi^3) + \frac{\xi(1-\xi)}{2\alpha\lambda^2} - \frac{\cosh\dfrac{\lambda}{2} - \cosh\dfrac{\lambda(1-2\xi)}{2}}{\alpha\lambda^4\cosh\dfrac{\lambda}{2}}\right] \tag{9.63}$$

$$\gamma = \frac{qL^3\beta}{B}\left[\frac{(1-2\xi)}{2} - \frac{\sinh\dfrac{\lambda(1-2\xi)}{2}}{\lambda\cosh\dfrac{\lambda}{2}}\right] \tag{9.64}$$

After a little more calculation, the stress resultants are obtained as

$$M_{\mathrm{S}} = \frac{qL^2}{1+\alpha}\left[\frac{\xi(1-\xi)}{2} - \frac{\cosh\dfrac{\lambda}{2} - \cosh\dfrac{\lambda(1-2\xi)}{2}}{\lambda^2\cosh\dfrac{\lambda}{2}}\right] \tag{9.65}$$

$$M_{\mathrm{D}} = \frac{qL^2\alpha}{1+\alpha}\left[\frac{\xi(1-\xi)}{2} + \frac{\cosh\dfrac{\lambda}{2} - \cosh\dfrac{\lambda(1-2\xi)}{2}}{\alpha\lambda^2\cosh\dfrac{\lambda}{2}}\right] \tag{9.66}$$

$$V_{\mathrm{S}} = \frac{qL}{1+\alpha}\left[\frac{1-2\xi}{2} + \frac{\sinh\dfrac{\lambda(1-2\xi)}{2}}{\lambda\cosh\dfrac{\lambda}{2}}\right] \tag{9.67}$$

$$V_{\mathrm{D}} = \frac{qL\alpha}{1+\alpha}\left[\frac{1-2\xi}{2} + \frac{\sinh\dfrac{\lambda(1-2\xi)}{2}}{\alpha\lambda\cosh\dfrac{\lambda}{2}}\right] \tag{9.68}$$

The important values of stress and deflection at mid-span follow from the above equations with $\xi = 0.5$:

$$M_{\mathrm{S},0.5} = \frac{qL^2}{1+\alpha}\left[\frac{1}{8} - \frac{\cosh\dfrac{\lambda}{2} - 1}{\lambda^2\cosh\dfrac{\xi}{2}}\right] \tag{9.69}$$

$$M_{\mathrm{D},0.5} = \frac{qL^2\alpha}{1+\alpha}\left(\frac{1}{8} + \frac{\cosh\frac{\lambda}{2} - 1}{\alpha\lambda^2\cosh\frac{\lambda}{2}}\right) \tag{9.70}$$

$$\sigma_{F11} = -\frac{M_{S,0.5}}{e\,A_{F1}} - \frac{M_{D,0.5}}{Z_{F1}}\frac{B_{F1}}{B_D}$$

$$\sigma_{F22} = \frac{M_{S,0.5}}{e\,A_{F2}} + \frac{M_{D,0.5}}{Z_{F2}}\frac{B_{F2}}{B_D}$$

where Z_{F1} and Z_{F2} are the relevant section moduli of the upper and lower faces respectively.

$$w_{0.5} = \frac{qL^4}{B}\left(\frac{5}{384} + \frac{1}{8\alpha\lambda^2} - \frac{\cosh\frac{\lambda}{2} - 1}{\alpha\lambda^4\cosh\frac{\lambda}{2}}\right) \tag{9.71}$$

9.3.2 Simply supported panel with a point load anywhere in the span

This case is illustrated by Fig. 9.5, the only difference being that here the panel has faces whose stiffness cannot be neglected. If the point load is applied at a position given by $x = e$, that is $\xi = e/L = \varepsilon$, the bending moment and shear force are given by

$$M = PL(1-\varepsilon)\xi - PL\{1-\varepsilon\} \qquad V = P(1-\varepsilon) - P\{1-\varepsilon\}^0 \tag{9.72}$$

where, as before, according to Macauley's notation, the quantities in curly brackets are set equal to zero when negative. The particular integrals in Equations 9.55 and 9.56 are

$$w_p = \frac{PL^3}{6B}[-(1-\varepsilon)\xi^3 + \{\xi-\varepsilon\}^3] - \frac{PL^3}{B\lambda^2}$$
$$\times\left[(1-\varepsilon)\xi + \frac{1}{\alpha}\frac{1}{2}(\xi-\varepsilon) - \left(\frac{L}{\lambda}\right)\sinh\lambda\,(\xi-\varepsilon)\,\{\xi-\varepsilon\}^0\right] \tag{9.73}$$

$$\gamma_p = \frac{\beta PL^2}{B}[1-\varepsilon-(1-\cosh\lambda(\xi-\varepsilon))\,\{\xi-\varepsilon\}^0] \tag{9.74}$$

giving, with index 1 valid for $0 \leq \varepsilon$ and index 2 for $\varepsilon \leq \xi \leq 1$:

$$w_1 = \frac{PL^3}{B}\left[\frac{1}{6}(1-\varepsilon)\,\xi\,(2\varepsilon-\varepsilon^2-\xi^2) + \frac{1}{\alpha\lambda^2}(1-\varepsilon)\xi - \frac{1}{\alpha\lambda^3}\frac{\sinh\lambda(1-\varepsilon)}{\sinh\lambda}\sinh\lambda\,\xi\right] \tag{9.75}$$

$$w_2 = \frac{PL^3}{B}\left[\frac{1}{6}\varepsilon(1-\xi)(-\varepsilon^2+2\xi-\xi^2) + \frac{1}{\alpha\lambda^2}\varepsilon(1-\xi) - \frac{1}{\alpha\lambda^3}\frac{\sinh\lambda\varepsilon}{\sinh\lambda}\sinh\lambda(1-\xi)\right] \tag{9.76}$$

$$\gamma_1 = \frac{PL^2}{B}\beta\left[1-\varepsilon+\frac{\sinh\lambda(1-\varepsilon)}{\sinh\lambda}\cosh\lambda\,\xi\right] \tag{9.77}$$

$$\gamma_2 = \frac{PL^2}{B}\beta\left[-\varepsilon+\frac{\sinh\lambda\,\varepsilon}{\sinh\lambda}\cosh\lambda(1-\xi)\right] \tag{9.78}$$

$$M_{S1} = PL\frac{1}{1+\alpha}\left[(1-\varepsilon)\xi - \frac{\sinh\lambda(1-\varepsilon)}{\lambda\sinh\lambda}\sinh\lambda\,\xi\right] \tag{9.79}$$

$$M_{S2} = PL\frac{1}{1+\alpha}\left[\varepsilon(1-\xi) - \frac{\sinh\lambda\varepsilon}{\lambda\sinh\lambda}\sinh\lambda(1-\xi)\right] \tag{9.80}$$

$$M_{D1} = PL\frac{\alpha}{1+\alpha}\left[(1-\varepsilon)\xi + \frac{\sinh\lambda(1-\varepsilon)}{\alpha\lambda\sinh\lambda}\sinh\lambda\xi\right] \tag{9.81}$$

$$M_{D2} = PL\frac{\alpha}{1+\alpha}\left[\varepsilon(1-\xi) + \frac{\sinh\lambda\varepsilon}{\alpha\lambda\sinh\lambda}\sinh\lambda(1-\xi)\right] \tag{9.82}$$

$$V_{S1} = P\frac{1}{1+\alpha}\left[1-\varepsilon - \frac{\sinh\lambda(1-\varepsilon)}{\sinh\lambda}\cosh\lambda\xi\right] \tag{9.83}$$

$$V_{S2} = P\frac{1}{1+\alpha}\left[-\varepsilon + \frac{\sinh\lambda\varepsilon}{\sinh\lambda}\cosh\lambda(1-\xi)\right] \tag{9.84}$$

$$V_{D1} = P\frac{\alpha}{1+\alpha}\left[1-\varepsilon + \frac{\sinh\lambda(1-\varepsilon)}{\alpha\sinh\lambda}\cosh\lambda\xi\right] \tag{9.85}$$

$$V_{D2} = P\frac{\alpha}{1+\alpha}\left[-\varepsilon - \frac{\sinh\lambda\varepsilon}{\alpha\sinh\lambda}\cosh\lambda(1-\xi)\right] \tag{9.86}$$

9.3.3 Simply supported panel with a temperature difference between the faces

If the temperatures of the two faces are T_1 and T_2 with coefficients of linear extension α_{F1} and α_{F2}, then in the absence of bending stiffness, the panel would bend into a curvature of

$$\theta = \frac{\alpha_{F2}T_2 - \alpha_{F1}T_1}{e} \tag{9.87}$$

It is convenient to use θ as a parameter in the equations for this case and θ has been included in the derivations of the governing differential Equations 9.52 and 9.54. The particular integrals in Equations 9.55 and 9.56 for this case are

$$w_p = \frac{\theta}{2(1+\alpha)}x^2 \qquad \gamma_p = 0 \tag{9.88}$$

Thus the complete solutions which follow may be obtained:

$$w = \frac{\theta L^2}{1+\alpha}\left[\frac{\xi}{2}(1-\xi) - \frac{1}{\lambda^2}\frac{\cosh\frac{\lambda}{2} - \cosh\lambda\frac{(1-2\zeta)}{2}}{\cosh\frac{\lambda}{2}}\right] \tag{9.89}$$

$$\gamma = \frac{-\theta L}{\lambda}\frac{\sinh\lambda\frac{(1-2\xi)}{2}}{\cosh\frac{\lambda}{2}} \tag{9.90}$$

$$M_S = -\frac{\alpha\theta B_s}{1+\alpha}\frac{\cosh\frac{\lambda}{2} - \cosh\lambda\frac{(1-2\xi)}{2}}{\cosh\frac{\lambda}{2}} \tag{9.91}$$

$$M_D = +\frac{\alpha\theta B_s}{1+\alpha}\frac{\cosh\frac{\lambda}{2} - \cosh\lambda\frac{(1-2\xi)}{2}}{\cosh\frac{\lambda}{2}} \tag{9.92}$$

$$V_S = -\frac{\theta B_S}{\beta\lambda L}\frac{\sinh\lambda\frac{(1-2\xi)}{2}}{\cosh\frac{\lambda}{2}} \tag{9.93}$$

$$V_D = +\frac{\theta B_S}{\beta\lambda L}\frac{\sinh\lambda\frac{(1-2\xi)}{2}}{\cosh\frac{\lambda}{2}} \tag{9.94}$$

9.3.4 Solutions for two- and three-span panels

Solutions for panels with two equal spans L subject to uniformly distributed load follow as a combination of the above cases as shown in Fig. 9.6. It is merely necessary to add together the solutions for a single span of $2L$ subject to the uniformly distributed load (Fig. 9.6(b)) and an upward point load P at mid-span (Fig. 9.6(c)) where P is chosen so that the two deflections Δ are equal.

Solutions for two-span panels subject to a temperature difference between the faces may be obtained similarly.

Panels with three equal spans L may also be solved as a combination of the solutions for a simply supported span of $3L$ for (Fig. 9.7(b)) the applied uniformly distributed or temperature loading and (Fig. 9.7(c) and 9.7(d)) a point load P at the third point. Figures 9.7(c) and 9.7(d) are, of course, mirror images and P is chosen so that

$$\Delta_1 + \Delta_2 = \Delta \tag{9.95}$$

The authors have programmed the complete set of equations given above, together with the extension to panels with two and three equal spans. The solutions are relatively stable numerically and it is considered that these equations represent the best approach to regular situations. The finite-element methods which are discussed later are essentially applicable to irregular situations such as unequal spans or non-uniform loading.

9.4 APPROXIMATE SOLUTIONS FOR SOME SIMPLE CASES

A useful approximate solution can be used for some simple cases of panels that have either one or both faces profiled. This makes the usual assumption that the applied load is shared between two separate load-carrying systems, namely the sandwich part, which includes the influence of core shear, and the flange part which merely involves bending of the flanges. It then makes the further assumption that these two systems are quite independent except that their deflections coincide at some critical point, usually at mid-span. The notation for this method has already been given in Fig. 9.10.

If the bending stiffness of the faces is neglected, we have the 'sandwich part' of the section that carries load as a consequence of axial forces N in the flanges and a shear force V_S in the core. The bending stiffness B_S and shear rigidity S of this part of the panel are given by:

$$B_S = \frac{E_{F1}A_{F1}E_{F2}A_{F2}e^2}{E_{F1}A_{F1} + E_{F2}A_{F2}} \tag{9.96}$$

$$S = \frac{G_C e^2 b}{d_C} = A_C G_{eff} \tag{9.97}$$

where b is the width of the section.

We also have the 'flange part' of the section whose bending stiffness is simply

$$B_D = E_{F1}I_{F1} + E_{F2}I_{F2} \tag{9.98}$$

Considering the case of a panel simply-supported over a span L and carrying a uniformly distributed load q, the deflections at mid-span of the two parts are respectively

$$w_S = \frac{5q_S L^4}{384 B_S}\left(1 + \frac{9.6 B_S}{A_C G_{eff} L^2}\right) = \frac{5q_S L^4}{384 B_S}(1+k) \tag{9.99}$$

$$w_D = \frac{5q_D L^4}{384 B_D} \tag{9.100}$$

where q_S = component of the load q carried by the sandwich part
q_D = component of the load q carried by the flange part, and

$$k = \frac{9.6 B_S}{A_C G_{eff} L^2} = \text{shear factor} \tag{9.101}$$

Equating w_S and w_D and noting that $q_S + q_D = q$ gives

$$q_S = q\frac{B_S}{B_S + (1+k)B_D} = q(1-\beta) \tag{9.102}$$

$$q_D = q\frac{(1+k)B_D}{B_S + (1+k)B_D} = q\beta \tag{9.103}$$

where

$$\beta = \frac{(1+k)B_D}{B_S + (1+k)B_D} \tag{9.104}$$

The stresses and deflections at any point in the panel now follow on the basis of simple beam theory. Thus, the deflection at mid-span is given by

$$w_{max} = \frac{5qL^4}{384 B_S}(1-\beta)(1+k) \tag{9.105}$$

and the maximum stress in the compressed face is given by

$$\sigma_{F11.max} = \frac{qL^2}{8}\left[\frac{\beta}{Z_{F1}} + \frac{(1-\beta)}{eA_{F1}}\right] \tag{9.106}$$

where Z_{F1} is the relevant section modulus of the compressed face.

The treatment of temperature loading is similar in principle as follows.

If a sandwich panel with flat or lightly profiled faces is subject to temperatures of T_1 and T_2 on the upper and lower faces respectively, it is curved into a circular arc though no direct stresses are induced. The deflection at mid-span is

$$w_S = \frac{L^2\theta}{8} \tag{9.107}$$

where θ has been defined previously in Equation 9.87 and α_{F1} and α_{F2} are the coefficients of expansion of faces 1 and 2 respectively.

If at least one face is profiled, this curvature cannot take place without a bending moment distribution being set up and equilibrium requires that

$$M_S = eN = -M_{F1} - M_{F2} = -M_D \tag{9.108}$$

These bending moments imply shear forces near the end of the panel that are assumed to have the distribution shown in Fig. 9.11.

Fig. 9.11 Assumed shear force distribution under temperature difference.

The shear transfer distance e_s is assumed to be small compared with the span so that the bending moments M_S and M_D may be assumed to be uniform over the length of the panel. The deflections at mid-span of the sandwich part and the flange part are, therefore,

$$w_S = \frac{L^2\theta}{8} + \frac{M_S L^2}{8B_S} + \frac{M_S}{A_C G_{eff}} \tag{9.109}$$

$$w_D = \frac{M_D L^2}{8B_D} \tag{9.110}$$

equating these and inserting $M_S = -M_D$ gives

$$M_D = -M_S = \frac{B_S B_D \theta}{B_S + (1+k)B_D} = B_D\theta(1-\beta) \tag{9.111}$$

where

$$k = \frac{8B_S}{A_C G_{eff} L^2} \tag{9.112}$$

and β is given by Equation 9.104.

It follows that the maximum deflection is given by

$$w_{max} = \frac{L^2\theta}{8}(1 - \beta) \qquad (9.113)$$

Equations 9.102 and 9.103 can be used for other loading conditions provided that the appropriate expressions for k is used. Some useful values of k are given in Table 9.2.

Table 9.2 Expressions for the shear factor k

General arrangement	Expression for k
Uniformly distributed; span L	$k_Q = 9.6 \dfrac{B_S}{G_{eff} A_C L^2}$
$q_0 \sin \frac{\pi x}{L}$; span L	$k_{sin} = \pi^2 \dfrac{B_S}{G_{eff} A_C L^2}$
Temperature difference ΔT; span L	$k_T = 8 \dfrac{B_S}{G_{eff} A_C L^2}$
Point load P at midspan; spans $L/2$, $L/2$	$k_P = 12 \dfrac{B_S}{G_{eff} A_C L^2}$

These approximate procedures can be extended, with similar simplicity, to two-span beams but the accuracy is then diminished to the point where their value is questionable. It may be noted that the division of the load-carrying behaviour of a sandwich panel into the sandwich part and the flange part is of fundamental importance and is also used in the development of the finite-element methods which are discussed next.

In an alternative simplification, design curves are given in Reference 9.7 for the important case of panels with one face fully profiled and one face lightly profiled or flat, as shown in Fig. 9.12.

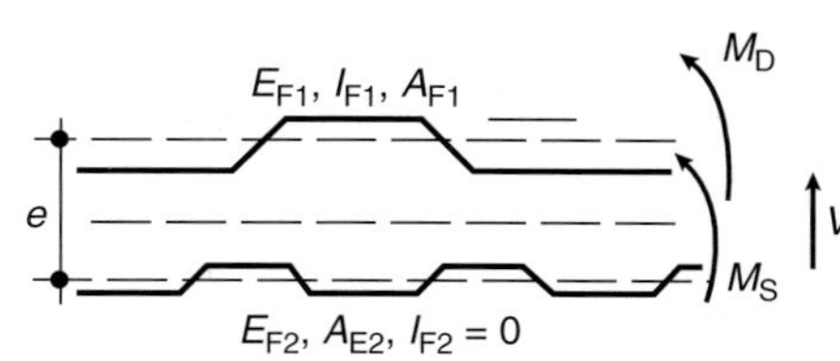

Fig. 9.12 Section used in design curves.

For this case, the basic parameters are:

$$B_S = \frac{E_{F1} A_{F1} E_{F2} A_{F2} e^2}{E_{F1} A_{F1} + E_{F2} A_{F2}} \qquad B_D = E_{F1} I_{F1} \qquad k = \frac{B_S}{A_C G_{eff} L^2} \tag{9.114}$$

The design curves give reduction factors β on the conventional beam bending moments, determined without consideration of the shear flexibility, in terms of the shear factor k and the stiffness ratio B_D/B_S.

For single-span panels, the reduction factors are given in Figs 9.13 and 9.14.

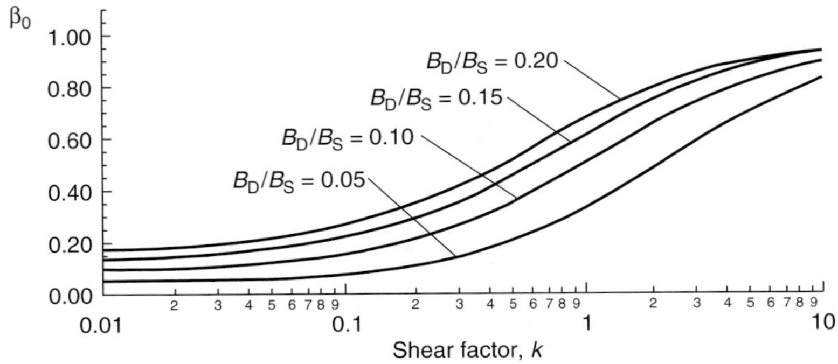

Fig. 9.13 Reduction factor β_0 for single-span panels under uniformly distributed load.

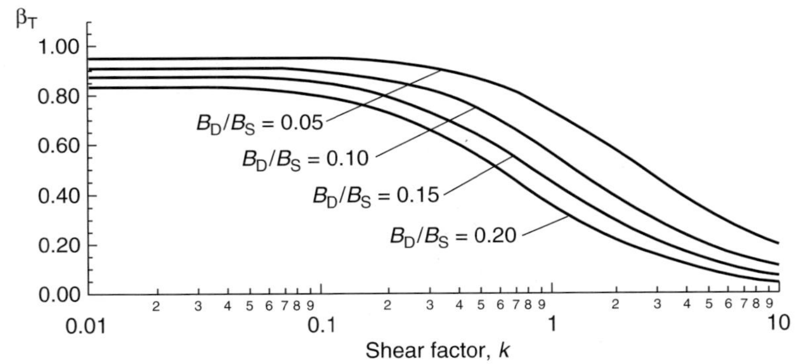

Fig. 9.14 Reduction factor β_T for single-span panels with a temperature difference between the faces.

Here the bending moment distribution M on the complete panel is statically determinate. The factors β indicate how this bending moment is distributed between the sandwich part and the flange part. Thus, for a uniformly distributed load we use β_0:

Flange part: $M_D = \beta_0 M$ (9.115)

Sandwich part: $M_S = (1 - \beta_0)M$ (9.116)

and for a temperature difference we use β_T:

Flange part: $M_D = \theta B_D \beta_T$ (9.117)

Sandwich part: $M_S = -M_D$ (9.118)

For two-span panels, the reduction factors are given in Figs 9.15–9.18.

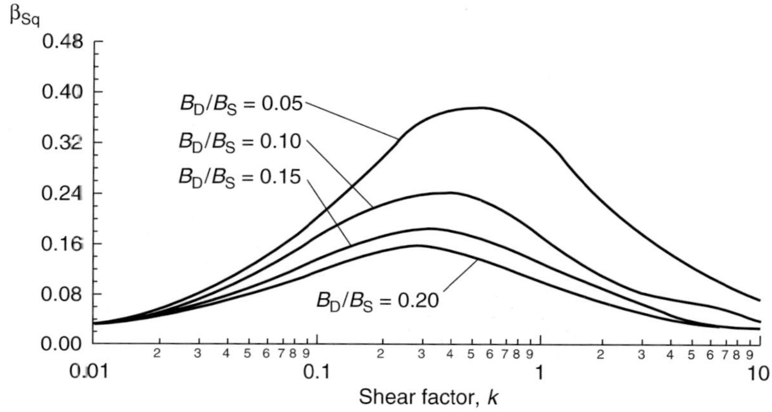

Fig. 9.15 Reduction factor β_{Sq} for two-span panels under uniformly distributed load.

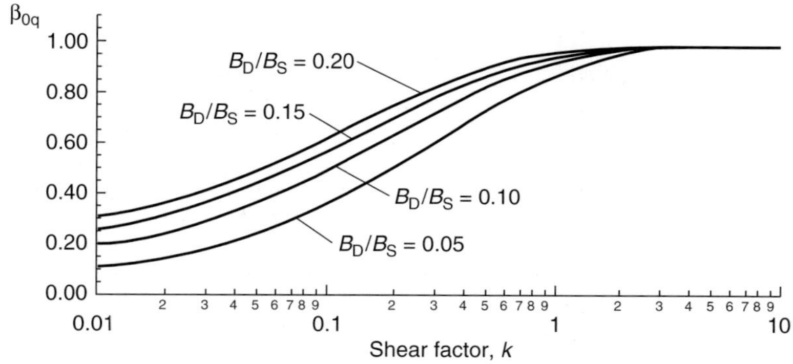

Fig. 9.16 Reduction factor β_{0q} for two-span panels under uniformly distributed load.

Here the bending moment distribution M on the complete panel is no longer statically determinate and depends on the relative stiffness of the parts of the panel. The procedure starts with the bending moments M given by conventional beam theory. The modification of M is in two stages. The factors β_S consider the change in the global bending moment as a consequence of sandwich action. The factors β_0 then indicate how this bending moment is distributed between the sandwich part and the flange part.

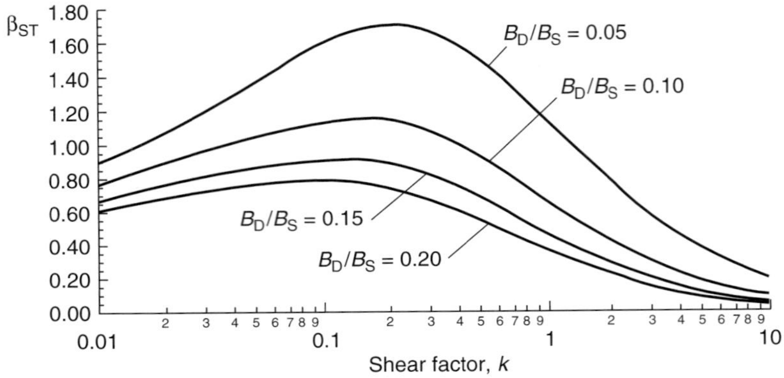

Fig. 9.17 Reduction factor β_{ST} for two-span panels with a temperature difference between the faces.

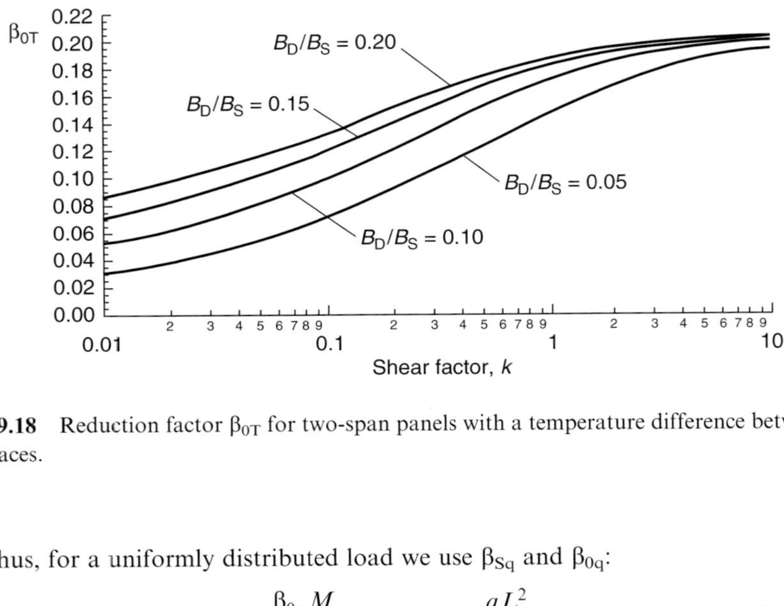

Fig. 9.18 Reduction factor β_{0T} for two-span panels with a temperature difference between the faces.

Thus, for a uniformly distributed load we use β_{Sq} and β_{0q}:

$$\text{Flange part:} \quad M_D = \frac{\beta_{0q} M}{1 + \beta_{Sq}} \quad \text{with } M = \frac{qL^2}{8} \tag{9.119}$$

$$\text{Sandwich part:} \quad M_S = \frac{1 - \beta_{0q})M}{1 + \beta_{Sq}} \tag{9.120}$$

and for a temperature difference we use β_{ST} and β_{0T}:

$$\text{Flange part:} \quad M_D = -\theta B_D \beta_{ST} \tag{9.121}$$

$$\text{Sandwich part:} \quad M_S = M_D\left(\frac{1}{\beta_{0T}} - 1\right) \tag{9.122}$$

For three-span panels, the procedure and equations for the bending moments at an internal support are the same as for two-span panels using the reduction factors given in Figs 9.19–9.22.

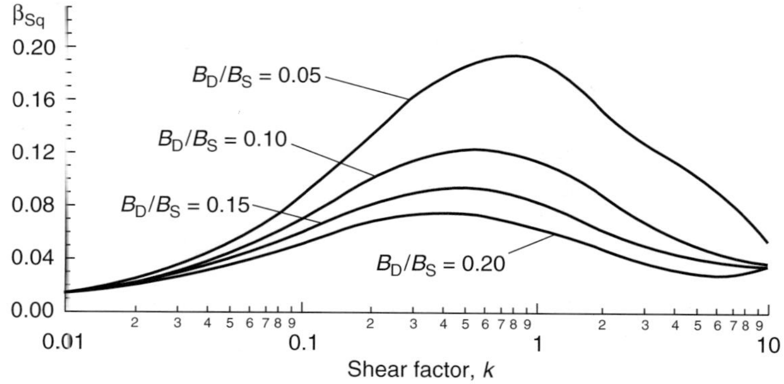

Fig. 9.19 Reduction factor β_{Sq} for three-span panels under uniformly distributed load.

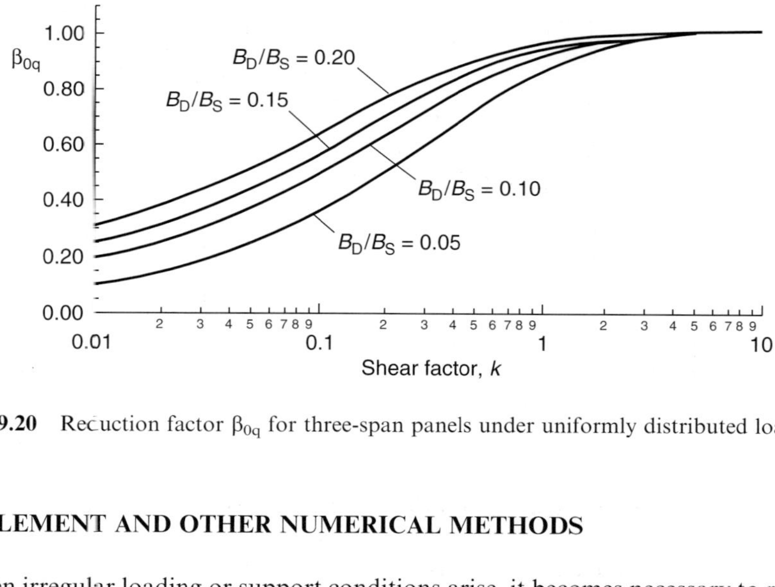

Fig. 9.20 Reduction factor β_{0q} for three-span panels under uniformly distributed load.

9.5 FINITE-ELEMENT AND OTHER NUMERICAL METHODS

When irregular loading or support conditions arise, it becomes necessary to resort to numerical methods of analysis and, for general purposes, the authors believe that the conventional finite-element method offers the best approach.

In many applications, the finite-element method is approximate and it is necessary to use a large number of elements in order to obtain accurate solutions. For three-layered sandwich beams, the solutions are exact and the minimum number of elements necessary to model the problem will give a precise solution.

The general solution for the bending of panels with profiled faces was given first.[9.1] It was then extended to panels subject to combined axial load and bending,[9.2] giving solutions for panels with both flat and profiled faces. As the former is

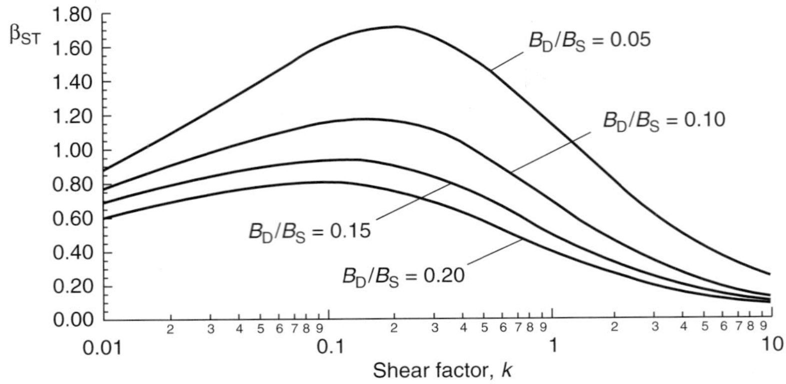

Fig. 9.21 Reduction factor β_{ST} for three-span panels with a temperature gradient.

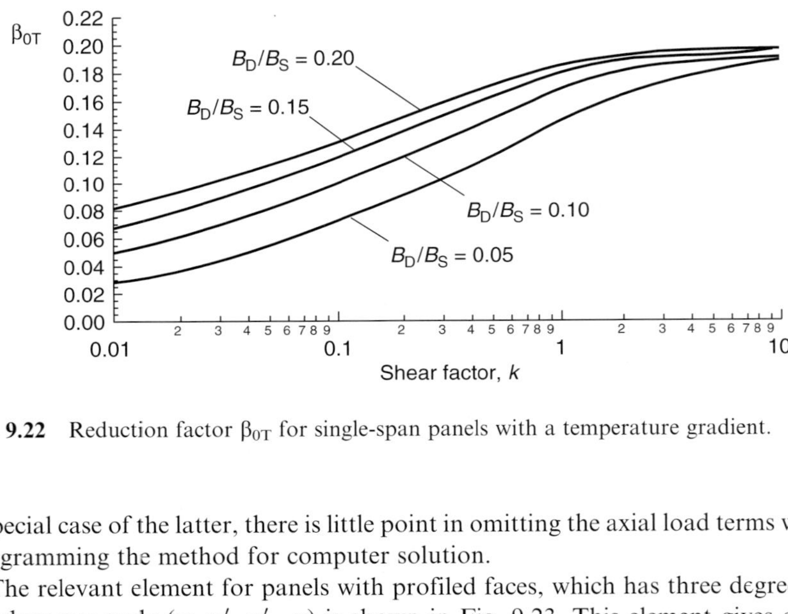

Fig. 9.22 Reduction factor β_{0T} for single-span panels with a temperature gradient.

a special case of the latter, there is little point in omitting the axial load terms when programming the method for computer solution.

The relevant element for panels with profiled faces, which has three degrees of freedom per node (w, w', $w' - \gamma$) is shown in Fig. 9.23. This element gives exact results because the governing differential equation is of sixth order so that its solution incorporates six arbitrary constants. The six displacements at the two nodes are, therefore, sufficient to define exactly the internal displacements in the element.

The element for panels with flat faces that was considered earlier is similar but with the bending moments M_D and the corresponding displacements omitted.

The governing differential equations are derived as follows in terms of the forces and deformations described in Fig. 9.9.

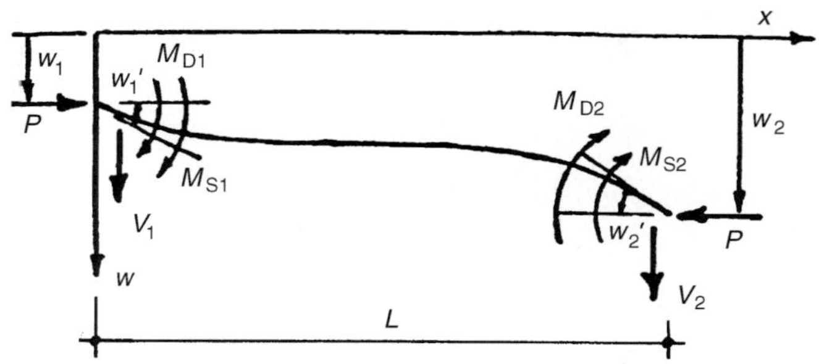

Fig. 9.23 Finite-element for axially loaded sandwich panels.

9.5.1 Force–deformation relationships

$$M_S = B_S(\gamma' - w'') \tag{9.123}$$

$$M_{F1} = -B_{F1}w'' \quad M_{F2} = -B_{F2}w'' \tag{9.124}$$

$$M = B_S\gamma' - Bw'' \tag{9.125}$$

$$V_S = A_C G_{eff}\gamma \tag{9.126}$$

$$V_{F1} = -B_{F1}w''' \quad V_{F2} = -B_{F2}w''' \tag{9.127}$$

$$V = A_C G_{eff}\gamma - B_D w''' \tag{9.128}$$

Equilibrium conditions

$$M' - V_z - Pw' = 0 \tag{9.129}$$

$$V_z' + q_w \tag{9.130}$$

In these equations, a prime denotes differentiation with respect to x and in addition to the quantities defined in Figs 9.9, 9.10 and 9.23:

B_S	= bending stiffness of the sandwich part of the cross-section $= E_{F1}A_{F1}E_{F2}A_{F2}e^2/(E_{F1}A_{F1} + E_{F2}A_{F2})$
B_{F1}	$= E_{F1}I_{F1}$ = bending stiffness of the upper face
B_{F2}	$= E_{F2}I_{F2}$ = bending stiffness of the lower face
A_{F1}, A_{F2}	= areas of the faces
E_{F1}, E_{F2}	= Young's moduli of the faces
A_C	$= be$ = area of the foam core
G_{eff}	= effective shear modulus of the core $= G_{nom}e/d_C$
G_{nom}	= nominal shear modulus of the core
q_w	= uniformly distributed load on the panel
w	= total deflection
γ	= shear strain in the foam core (divergence from the normal to the axis of the cross-section)

Because the stress resultants in the two faces are proportional to the same deformations, it is convenient to treat them together, thus:

$$M_D = M_{F1} + M_{F2}$$
$$M = M_D + M_S$$

$$
\begin{aligned}
B_D &= B_{F1} + B_{F2} \\
B &= B_D + B_S \\
V_D &= V_{F1} + V_{F2} \\
V &= V_D + V_S
\end{aligned}
$$

Combining Equations 9.123–9.130 gives the two differential equations

$$B_S(\gamma'' - w''') - A_C G_{\text{eff}}\gamma = 0 \tag{9.131}$$

$$A_C G_{\text{eff}}\gamma' - Pw'' - B_D w'''' = -q_w \tag{9.132}$$

and if these are uncoupled, we obtain

$$-\frac{B_D}{A_C G_{\text{eff}}} w^{\text{vi}} + \left(\frac{B}{B_S} - \frac{P}{A_C G_{\text{eff}}}\right) w'''' + \frac{P}{B_S} w'' = \frac{q_w}{B_S} \tag{9.133}$$

$$-\frac{B_D}{A_C G_{\text{eff}}} \gamma'''' + \left(\frac{B}{B_S} - \frac{P}{A_C G_{\text{eff}}}\right) \gamma'' + \frac{P}{B_S} \gamma = 0 \tag{9.134}$$

The general solution of these equations has the form

$$w = C_1 \sinh px + C_2 \cosh px + C_3 \sin qx + C_4 \cos qx + C_5 x + C_6 + w_0 \tag{9.135}$$

$$\gamma = D_1 \sinh px + D_2 \cosh px + D_3 \sin qx + D_4 \cos qx \tag{9.136}$$

where

$$p = \sqrt{\frac{A_C G_{\text{eff}}}{2B_D}\left(\frac{B}{B_S} - \frac{P}{A_C G_{\text{eff}}}\right) + \sqrt{\left[\frac{A_C G_{\text{eff}}}{2B_D}\left(\frac{B}{B_S} - \frac{P}{A_C G_{\text{eff}}}\right)\right]^2 + \frac{A_C G_{\text{eff}} P}{B_D B_S}}} \tag{9.137}$$

$$q = \sqrt{\frac{A_C G_{\text{eff}}}{2B_D}\left(\frac{B}{B_S} - \frac{P}{A_C G_{\text{eff}}}\right) - \sqrt{\left[\frac{A_C G_{\text{eff}}}{2B_D} - \left(\frac{P}{A_C G_{\text{eff}}}\right)^2\right] + \frac{A_C G_{\text{eff}} P}{B_D B_S}}} \tag{9.138}$$

and w_0 is the particular integral, namely $q_w x^2/2P$.

The relationship between the coefficients C and D follows from Equations 9.131 and 9.132:

$$D_1 = F_p C_2 \quad D_2 = F_p C_1 \quad D_3 = -F_p C_4 \quad D_4 = F_q C_3 \tag{9.139}$$

where

$$F_p = \frac{B_S p^3}{B_S p^2 - A_C G_{\text{eff}}} = p\frac{(P + B_D p^2)}{A_C G_{\text{eff}}} \tag{9.140}$$

$$F_q = \frac{B_S q^3}{B_S q^2 - A_C G_{\text{eff}}} = q\frac{(P - B_D q^2)}{A_C G_{\text{eff}}} \tag{9.141}$$

The element defined in Fig. 9.23 has three degrees of freedom per node so that the nodal load and displacement vectors are respectively

$$F = \begin{vmatrix} V_z \\ M_D \\ M_S \end{vmatrix} \tag{9.142}$$

$$w = \begin{bmatrix} w \\ w' \\ w' - \gamma \end{bmatrix} \tag{9.143}$$

As the total number of degrees of freedom of the element (six) is the same as the number of unknown coefficients in the general solutions of the differential Equations 9.135 and 9.136, the finite element will be exact. The relationship between the nodal displacement and arbitrary constants for zero applied load q_w is

$$\begin{bmatrix} w_1 \\ w_1' \\ w_1' - \gamma_1 \\ w_2 \\ w_2' \\ w_2' - \gamma_2 \end{bmatrix} = \begin{bmatrix} 0 & 1 & 0 & 0 & 1 & 1 \\ p & 0 & q & 0 & 1 & 0 \\ p - F_p & 0 & q - F_q & 0 & 1 & 0 \\ \sinh pL & \cosh pL & \sin qL & \cos qL & L & 1 \\ p \cosh pL & p \sinh pL & q \cos pL & -q \sin qL & 1 & 0 \\ (p - F_p) \cosh pL & (p - F_p) \sinh pL & (q - F_q) \cos qL & -(q - F_q) \sin qL & 1 & 0 \end{bmatrix} \begin{bmatrix} C_1 \\ C_2 \\ C_3 \\ C_4 \\ C_5 \\ C_6 \end{bmatrix} \tag{9.144}$$

i.e.

$$[w] = [A][C] \tag{9.145}$$

$$[C] = [A]^{-1}[w] \tag{9.146}$$

The nodal force vectors can also be evaluated at each node in terms of the arbitrary constants C. Thus, from Equations 9.123–9.130 using the vertical shear force V_z, rather than the normal shear force V, which was used to derive the fundamental equations:

$$V_z = A_C G_{\text{eff}} \gamma - B_D w''' - Pw' \tag{9.147}$$

$$M_D = B_D w'' \tag{9.148}$$

$$M_S = B_S(\gamma' - w'') \tag{9.149}$$

After changing the sign of the terms in V_1, M_{D2}, M_{S2} to accord with the usual sign convention for finite-element analysis, this gives

$$\begin{bmatrix} V_1 \\ M_{D1} \\ M_{S1} \\ V_2 \\ M_{D2} \\ M_{S2} \end{bmatrix} = \begin{bmatrix} T_1 & 0 & T_4 & 0 & -P & 0 \\ 0 & -T_2 & 0 & T_5 & 0 & 0 \\ 0 & T_3 & 0 & T_6 & 0 & 0 \\ T_1 \cosh pL & T_1 \sinh pL & T_4 \cos qL & -T_4 \sin qL & -P & 0 \\ -T_2 \sinh pL & -T_2 \cosh pL & T_5 \sin qL & T_5 \cos qL & 0 & 0 \\ T_3 \sinh pL & T_3 \cosh pL & -T_6 \sin qL & -T_6 \cos qL & 0 & 0 \end{bmatrix} \begin{bmatrix} C_1 \\ C_2 \\ C_3 \\ C_4 \\ C_5 \\ C_6 \end{bmatrix} \tag{9.150}$$

where

$$T_1 = A_C G_{\text{eff}} F_p - B_D p^3 - P_p \tag{9.151}$$

$$T_2 = B_D p^2 \tag{9.152}$$

$$T_3 = B_S p(F_p - p) \tag{9.153}$$

$$T_4 = A_C G_{\text{eff}} F_q + B_D q^3 - P_q \tag{9.154}$$

$$T_5 = B_D q^2 \tag{9.155}$$

$$T_6 = B_S q(F_q - q) \tag{9.156}$$

The element stiffness matrix is therefore obtained as the product of two 6×6 matrices

$$[F] = [B][C] = [B][A]^{-1}[w] \tag{9.157}$$

where B is the matrix of coefficients in Equation 9.150. For axially loaded sandwich panels in which all other loading is applied as point loads and moments at the nodes, this is all that is required for exact solutions. Element stiffness matrices derived in this way can be assembled to form the global stiffness matrix of the complete sandwich panel according to the usual rules. The global stiffness equations may then be solved to give the displacements at the nodes. The nodal stress resultants then follow from the element stiffness Equations 9.157.

If the values of the displacements or nodal stress resultants are required at points other than the natural nodes, the simplest solution is to insert additional nodes as necessary. Alternatively, Equations 9.144–9.146 can be used to evaluate the arbitrary constants in Equations 9.135 and 9.136 and the complete pattern of deflections and forces obtained from Equations 9.147–9.149.

At a given level of axial load, the equations are linear but, under increasing axial load, they are non-linear and the deflections and stress resultants become infinite at the lowest critical load. This means that the analysis of a given axially loaded panel carrying an axial load generally requires repeated analyses in order to investigate the behaviour under increasing load. Both a stepping procedure and iteration to the critical load using a modified 'Southwell' plot have been successfully used.

When an element is subject to a uniformly distributed load, it is necessary to apply equivalent nodal forces and these require a separate calculation. As shown in Fig. 9.24, the required forces are the 'fixed end moments' and shears reversed. In order to calculate these, it is necessary to return to the original solution of the fundamental differential Equations 9.135 and 9.136 and to insert the appropriate boundary conditions, namely $w=0$; $w'=0$; $\gamma=0$ at $x=0$ and $x=L$. This gives the following equations which may be solved for the arbitrary constants C:

$$\begin{bmatrix} 0 & 1 & 0 & 1 & 0 & 1 \\ p & 0 & q & 0 & 1 & 0 \\ F_\mathrm{p} & 0 & F_\mathrm{q} & 0 & 0 & 0 \\ \sinh pL & \cosh pL & \sin qL & \cos qL & L & 1 \\ p\cosh pL & p\sinh pL & q\cos qL & -q\sin qL & 1 & 0 \\ F_\mathrm{p}\cosh pL & F_\mathrm{p}\sinh pL & F_\mathrm{q}\cos qL & -F_\mathrm{q}\sin qL & 0 & 0 \end{bmatrix} \begin{bmatrix} C_1 \\ C_2 \\ C_3 \\ C_4 \\ C_5 \\ C_6 \end{bmatrix} = \begin{bmatrix} 0 \\ 0 \\ 0 \\ \dfrac{-q_\mathrm{w}L^2}{2P} \\ \dfrac{-q_\mathrm{w}L}{P} \\ 0 \end{bmatrix} \tag{9.158}$$

The full vector of 'fixed end' forces then follows from Equation 9.150 to which the additional vector of terms due to the distributed load q_w must be added, namely

$$\left[0; -B_\mathrm{D}\frac{q_\mathrm{w}}{p}; -B_\mathrm{S}\frac{q_\mathrm{w}}{p}; -q_\mathrm{w}L; -B_\mathrm{D}\frac{q_\mathrm{w}}{p}; -B_\mathrm{S}\frac{q_\mathrm{w}}{p}\right] \tag{9.159}$$

In fact, as a consequence of the symmetry of the problem, only the first three terms in the force vector are required, but it is reassuring to derive the full vector in the manner described and then to note the symmetry. As before, the first, fifth and sixth terms require a reversal of sign in order to comply with the sign convention for finite-element analysis.

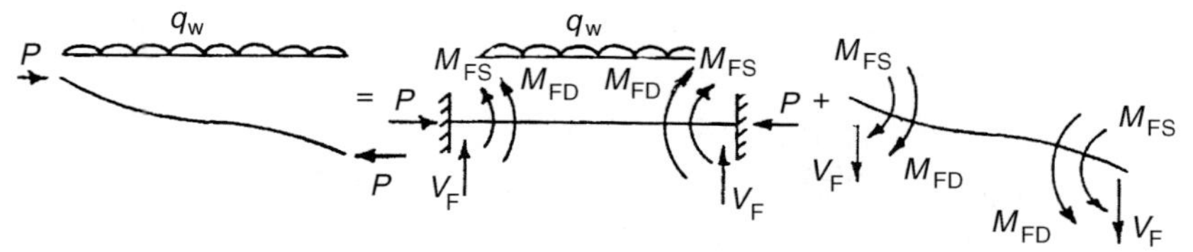

(a) Element with UDL (b) Fixed-ended element with UDL (c) Fixed-end forces reversed

Fig. 9.24 Nodal equivalent forces for a uniformly distributed load.

9.6 AXIALLY LOADED SANDWICH PANELS

Evidently, the detailed finite-element procedure given in Section 9.5 can be used to find solutions for axially loaded sandwich panels. However, in practice, this may not be necessary. There is an explicit solution for the elastic critical load (the equivalent of the 'Euler' bucking load of a pin-ended column) which can be found from the solution in Section 9.5 by setting $q_w = 0$ and inserting the boundary conditions $v = 0$; $w'' = 0$; $\gamma = 0$ at $x = 0$ and $x = L$. This leads to:

$$\sin qL = 0 \tag{9.160}$$

which gives, for the lowest critical load:

$$qL = \pi \tag{9.161}$$

After some manipulation, the resulting expression for the critical load can be expressed as:

$$P_{cr} = \frac{P_E P_{EF} - P_{EF}^2 + P_E P_C}{P_E - P_{EF} + P_C} \tag{9.162}$$

where P_E = Euler critical load of the complete panel $= \pi^{2^2} B/L^2$
P_{EF} = Euler critical load of the flanges only $= \pi^2 B_D/L^2$
P_C = the 'shear critical load' $= A_C G_{eff}$

Figure 9.25 illustrates a typical case of how the critical load varies with the length of the panel. Equation 9.162 represents a transition from the Euler curve $P_{cr} = P_E$ (no shear deformation) for large spans to $P_{cr} = P_{EF}$ (flanges buckling independently) for short lengths. Equation 9.162 can be used to obtain an estimate of the effect of the axial load on the performance of, for example, a wall panel supporting load from above. If P is well below P_{cr}, empirical methods of design may be used, otherwise it is necessary to resort to finite elements. In practical design[9.2] it is usually necessary to take into account the eccentricity of the load application. This tends to cause local bending at the top of the panel, bearing in mind that most of the axial load tends to enter the panel through the inside face.

9.7 SANDWICH PANELS SPANNING IN TWO DIRECTIONS (SANDWICH PLATES)

From the practical point of view, the only two-way panel of interest is the simply supported panel with flat faces. This arrangement is sometimes used, often with

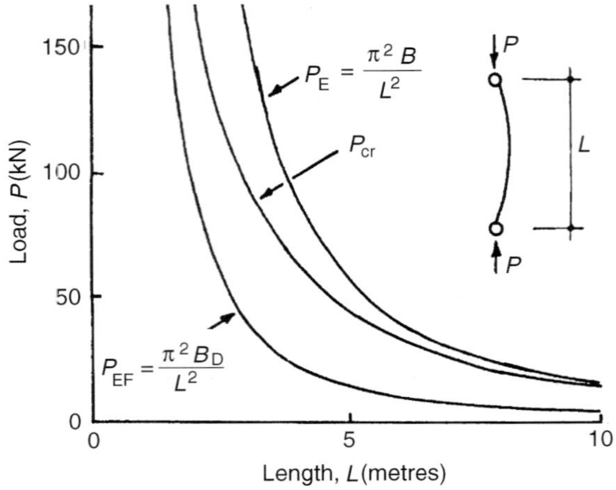

Fig. 9.25 Critical loads of a sandwich panel acting as a pin-ended strut.

highly sophisticated all-round support and sealing, in high-tech facade systems. The equations for this case are not complicated and are given below. For other more complex arrangements, it is necessary to resort to the use of finite elements. General finite-element solutions are available but are outside the scope of this book.

9.7.1 Simply supported panel with flat or quasi-flat faces

Allen[9.3] gives solutions for various cases of two-dimensional plate buckling and bending of sandwich panels, including the case of interest here. He gives the full derivation but, here, for simplicity, only the important results will be quoted.

For an isotropic rectangular panel of length a and width b subject to a uniform pressure q per unit area:

$$\text{Maximum deflection:} \quad w_{max} = \frac{qb^4}{D^2}(\beta_1 + \rho\beta_2) \tag{9.163}$$

$$\text{Maximum bending stress:} \quad \sigma_{F.x} = \frac{qb^2}{dt}(\beta_3 + \nu\beta_4) \tag{9.164}$$

$$\text{Maximum bending stress:} \quad \sigma_{F.y} = \frac{qb^2}{dt}(\beta_4 + \nu\beta_3) \tag{9.165}$$

$$\text{Maximum shear stress in the faces:} \quad \tau_{F.xy} = \frac{qb^2}{dt}(1 - \nu)\beta_5 \tag{9.166}$$

$$\text{Maximum shear stress in the core:} \quad \tau_{C.zx} = \frac{qb}{d}\beta_6 \tag{9.167}$$

$$\text{Maximum shear stress in the core:} \quad \tau_{C.yz} = \frac{qb}{d}\beta_7 \tag{9.168}$$

where

$$\beta_1 = \frac{16}{\pi^6}\sum\sum\frac{(-1)^{(m-1)/2}\,(-1)^{(n-1)/2}}{mn\,\Omega^2} \tag{9.169}$$

$$\beta_2 = \frac{16}{\pi^6}\sum\sum\frac{(-1)^{m-1)/2}\,(-1)^{(n-1)/2}}{mn\,\Omega} \tag{9.170}$$

$$\beta_3 = \frac{16}{\pi^4}\sum\sum\frac{(-1)^{(m-1)/2}\,(-1)^{(n-1)/2}}{\Omega^2}\,\frac{m}{n}\,\frac{b^2}{a^2} \tag{9.171}$$

$$\beta_4 = \frac{16}{\pi^4}\sum\sum\frac{(-1)^{(m-1)/2}\,(-1)^{(n-1)/2}}{\Omega^2}\,\frac{n}{m} \tag{9.172}$$

$$\beta_5 = \frac{16}{\pi^4}\sum\sum\frac{b}{a\,\Omega^2} \tag{9.173}$$

$$\beta_6 = \frac{16}{\pi^3}\sum\sum\frac{(-1)^{(n-1)/2}}{n\,\Omega}\,\frac{b}{a} \tag{9.174}$$

$$\beta_7 = \frac{16}{\pi^3}\sum\sum\frac{(-1)^{(m-1)/2}}{m\,\Omega} \tag{9.175}$$

and where, in all cases, $m = 1, 3, 5,\ldots$ and $n = 1, 3, 5\ldots$ and

$$\Omega = \frac{m^2 b^2}{a^2} + n^2 \tag{9.176}$$

$$\rho = \frac{\pi^2}{2g}\,\frac{E}{G_{\text{eff}}}\,\frac{td}{b^2} \tag{9.177}$$

where

E = elastic modulus of faces
G_{eff} = shear modulus of core
ν = Poisson's ratio of face material
g = $1 - \nu^2$
t = thickness of each face
e = thickness of panel (centre to centre of faces)
D = bending stiffness per unit length = $Ete^2/2$

Obviously, when using the above equations, adequate accuracy can be obtained with a small number of terms in each series.

9.8 REFERENCES

9.1 Davies, J.M. (1986) The analysis of sandwich panels with profiled faces. In *8th International Speciality Conference on Cold-Formed Steel Structures*, St Louis, Missouri, November 11–12, pp. 351–369.

9.2 Davies, J.M. (1987) Axially loaded sandwich panels. *Journal of Structural Engineering*, **113** (11), ASCE, 2212–2230.

9.3 Allen, H.G. (1969) *Analysis and Design of Structural Sandwich Panels*, Pergamon Press, London.

9.4 Stamm, K. and Witte, H. (1974) *Sandwichkonstructionen – Berechnung, Fertigung, Ausführung* (*Sandwich Construction – Calculation, Manufacture and Use*), Springer-Verlag, Wien and New York.

9.5 Hartsock, J.A. and Chong, K.P. (1976) Analysis of sandwich panels with formed faces. In *Proceedings of ASCE, Journal of the Structural Division*, **102** (ST4).

9.6 Chong, K.P., Engen, K.O. and Hartsock, J.A. (1977) Thermal stresses and deflections of sandwich panels. *Proceedings of ASCE, Journal of the Structural Division*, **103** (ST1).

9.7 Berner, K. (1998) Praxisgerechte Nachweise zur Trag- und Gebrauchsfahigkeit von Sandwichelementen (Practical design checks for sandwich panels at the ultimate and serviceability limit states), *Stahlbau 67*, vol. 12, pp. 910–925.

Chapter 10

Load-bearing Capacity

10.1 PRINCIPLES OF THE EVALUATION OF RESISTANCE

In structural sandwich panels, there are several potential failure modes which may limit and determine the load-bearing capacity of the panel. The load-bearing capacity depends directly on the face and core materials and on the adhesion between the faces and the core. It also depends on the structural dimensions and structural system including the lengths of spans, the support widths and the fastening system, etc. This chapter considers the resistance of typical sandwich panels used in the building industry whose static system is either a single-span simply supported beam or a continuous multi-span beam with possible cantilevers at the ends. The faces of these panels are usually made of thin metal sheets and the core of polyurethane or polystyrene foam or of mineral wool.

In order to evaluate the structural performance of a sandwich structure, the strength of the sandwich panel with respect to the different possible failure modes has to be determined. Loading tests indicate directly the resistance of the panel. However, the test results are associated with a particular test specimen, span and loading arrangement and, therefore, very many tests would be required in order to cover all of the failure modes needed in the design for the entire range of different spans, static systems, panel depths and load combinations.

Alternatively, and preferably, the strength of the panel can be described by mathematical models based on the physical process of events which take place in each of the individual failure modes. In these mathematical models, data concerning the material properties are required in order to describe the stiffness and strength of the separate parts (e.g. layers of material) of the panel and, in most cases, these material properties have to be determined experimentally. In particular, the properties of the core materials may vary considerably inside the core of one sandwich panel product and there may also be variations between the panels produced during a certain time period. Therefore, statistically reliable characteristic values of the material properties incorporated in the mathematical models, taking into account the manufacturing process, have to be determined before the evaluation of the resistance values can be undertaken. It is also important that the material properties are controlled by regular testing in order to guarantee the level of the values used in the design calculations. Procedures for quality control are given in Chapter 13.

After a certain time in service, the strength of a panel in the wall or a roof of a building depends on the stress history, the time, temperature and relative humidity and other climatic loads at the location under the consideration. In other words, the strength is not constant but depends on the mechanical and climatic loading history of the faces, core and joints of the panel. The following observations, based on experiments and practical experience, have influenced the current design models for the resistance of sandwich panels.

- The shear stresses caused by typical permanent and short-term service loads are low and are usually of the order of 10–30% of the initial short-term shear strength of the core. The influence of the service stresses alone has not been found to have a significant effect on the long-term resistance of typical sandwich panels.
- The faces and also the joints between typical sandwich panels are water- and vapour-tight. The influence of humidity on the strength of the core and the adhesion between the core and faces can, therefore, usually be neglected. However, in some special constructions, for example in the walls and roofs of cold stores and freezing plants, the temperature and humidity gradients are more significant. In these cases, only a carefully detailed design together with high quality installation can guarantee the long-term, fault-free functioning of the structure.
- The strength and moduli of elasticity of plastic foams decrease with increasing temperature. The corresponding reduction in the strength and stiffness of the core has to be taken into account in the evaluation of those resistances which depend on the properties of the core at elevated temperature. The relevant failure modes are shear failure in the core and wrinkling failure of the outer face due to the stresses caused by the loads at high summer temperatures.
- The compressive strength of the face, the strength of the core and the adhesive bond between the core and faces all change with time alone or with time together with the climatic and mechanical loading history. For this reason, a reduced long-term shear strength for the core may be used in the design of roof panels against long-term loads such as snow and self-weight. An experimental method to determine the long-term shear strength is given in Chapter 6.
- With the chemical and physical changes in the core layer and its adhesive bonds due to the climatic loads, the resistances against the different failure modes may also change. The resistances at a certain time t_n may be compared with the stresses caused by the probable loads at that time, as shown in Fig. 10.1. In Chapter 6, accelerated ageing tests are presented with which to study the stability of the tensile strength of the core and its bond to the faces.

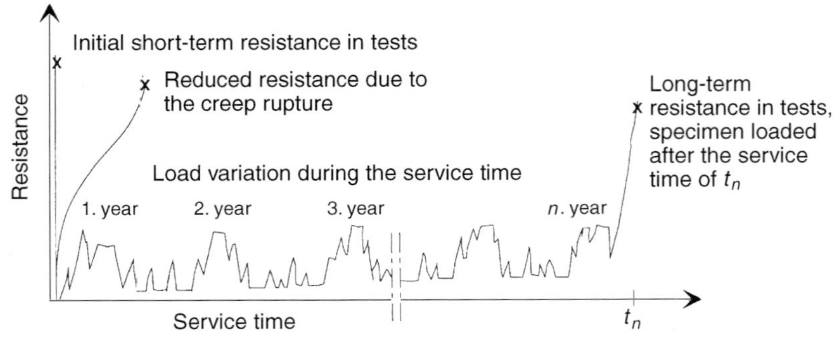

Fig. 10.1 Initial and long-term resistance of the face, core and joint.

The static behaviour and strength of sandwich panels is based on the composite action of the three structural layers, namely the two faces and the core. In order to carry the stresses arising from this composite action in the short and long terms, practical experience has shown that the core and its adhesion to the faces have to exceed a minimum strength level. A characteristic tensile strength value, which may be considered as a minimum requirement in order to guarantee the long-term composite action, is 0.075 N/mm^2.

In order to carry out design calculations, the geometrical dimensions of the sandwich panel are required. All of the dimensions which are of significance for the resistance, such as the panel depth and width and the depth and shape of the face profiling, have to correspond to the real dimensions of the sandwich product. Usually the nominal geometrical dimensions are used in the calculations. In this case the real dimensions have to agree with the dimensions used in calculations within close tolerances, because the static properties and resistances may be quite sensitive to changes in the dimensions. For example, typical allowable tolerances for the total depth are -1 mm and +2 mm.[10.4, 10.5] The dimensions of the panels should be regularly monitored during the manufacturing process. If the differences between the nominal and the actual dimensions in production are large and these differences have significant effects on the behaviour and resistances of the sandwich panel, the dimensional measurements resulting in the most unfavourable effects and lowest resistances should be used in the calculations.

The thicknesses of the sheet metal faces have a dominant influence on the bending resistance of a sandwich panel. If the steel sheet is manufactured using the 'special' thickness tolerances (S-tolerance according to EN 10143), the design thickness t_d to be used in design calculations is equal to the nominal core thickness of the hot-dip coated steel sheet. It is recommended that the nominal core thickness should be taken as:

$$t_d = t_{nom} - t_{zinc} - 0.5\ t_{tol}$$

where t_{nom} is the nominal thickness of the steel sheet, t_{zinc} is the total thickness of the two zinc layers and t_{tol} is the normal minus tolerance according to EN 10143. A zinc coating of 275 gram/m^2 results in a total zinc thickness of 0.04 mm. The calculation models in ENV 1993-1-3[10.6] are in accordance with this tight definition of the thickness value. It follows that the nominal thicknesses should be adjusted to maintain the equivalent reliability. In addition to this harmonised European method, different national systems are in use for the determination of the design thicknesses. The design thicknesses of other metal facing sheets, such as those made of aluminium, stainless steel and copper also have to be determined so that they represent statistically reliable minimum thickness values.

10.2 FAILURE MODES

The relevant failure modes of metal-faced sandwich panels with flat or slightly profiled faces are as follows:

- tensile failure of the faces
- wrinkling failure of the faces (due to compressive stress)

- shear failure of the core or the adhesion between the core and face
- crushing failure of the face and core at a support
- tensile or shear failure of fasteners

In sandwich panels with thin fully-profiled faces, two additional failure modes are introduced:

- shear strength of the webs of the profile
- support reaction capacity of the profile

These failure modes are illustrated in Fig. 10.2. Large deflections of wall and roof panels are also undesirable so that the maximum allowable deflection is also a criterion in the design. In sandwich panel roofs, the influence of long-term shear deformations on deflections is also significant as illustrated in Fig. 10.3.

10.2.1 Tensile and compressive failure of the faces

The yield stress and tensile strength of typical metal sheets are high. Therefore, the tensile failure mode only limits the load-bearing capacity of metal faced sandwich panels in rare cases. The tensile failure mode may be significant in an asymmetrically faced profile which is loaded in a manner which causes a high bending moment at an intermediate support. The tensile strength of timber and gypsum-based boards is considerably smaller than the tensile strength of metal sheets. Tensile stress, and especially the combined stress caused by tension and bending either in the span or at an intermediate support, may be critical for these materials.

The tensile strength of the face material is usually determined by tensile tests in accordance with the relevant ISO, European or national standards. The strength of non-metallic face materials may depend on the duration of the loading and on the relative humidity of the loading environment.

The strength of a compressed face may be limited by yielding of the face material or, more likely, by a short wave length buckling failure mode when the material reaches the 'wrinkling stress'. The compression strength, when limited by a wrinkling failure, depends on the thickness and, if relevant, the profile of the face together with the stiffness and the strength of the core material as shown in Fig. 10.4. The mathematical derivation of design expressions for the wrinkling stresses for flat, slightly profiled and for strongly profiled thin metal faces are given in more detail in Section 10.3.

A classical mathematical analysis results in expressions for the elastic buckling wave lengths, a, and the wrinkling stresses, σ_{cr}, of thin flat faces, i.e., for the eigenmodes and eigenvalues of the compressed face. Equations 10.1 and 10.2 are based on an elastic half-space foundation model in the formulation of which the core is assumed to consist of an ideal elastic isotropic material. Equations 10.3 and 10.4 are based on a simplified foundation model in which only the three most important stresses and strains in the core (σ_z, τ_{xz}, τ_{yz}, ε_z, γ_{xz} and γ_{yz}) are taken into account in the analysis. In the approximate expressions, the values of Poisson's ratio $\nu_F = 0.3$ and $\nu_C = 0.25$ for the face and core respectively have been used. In practice, the numerical results given by Equations 10.1 and 10.3 and also by Equations 10.2 and 10.4 respectively are rather close to each other:

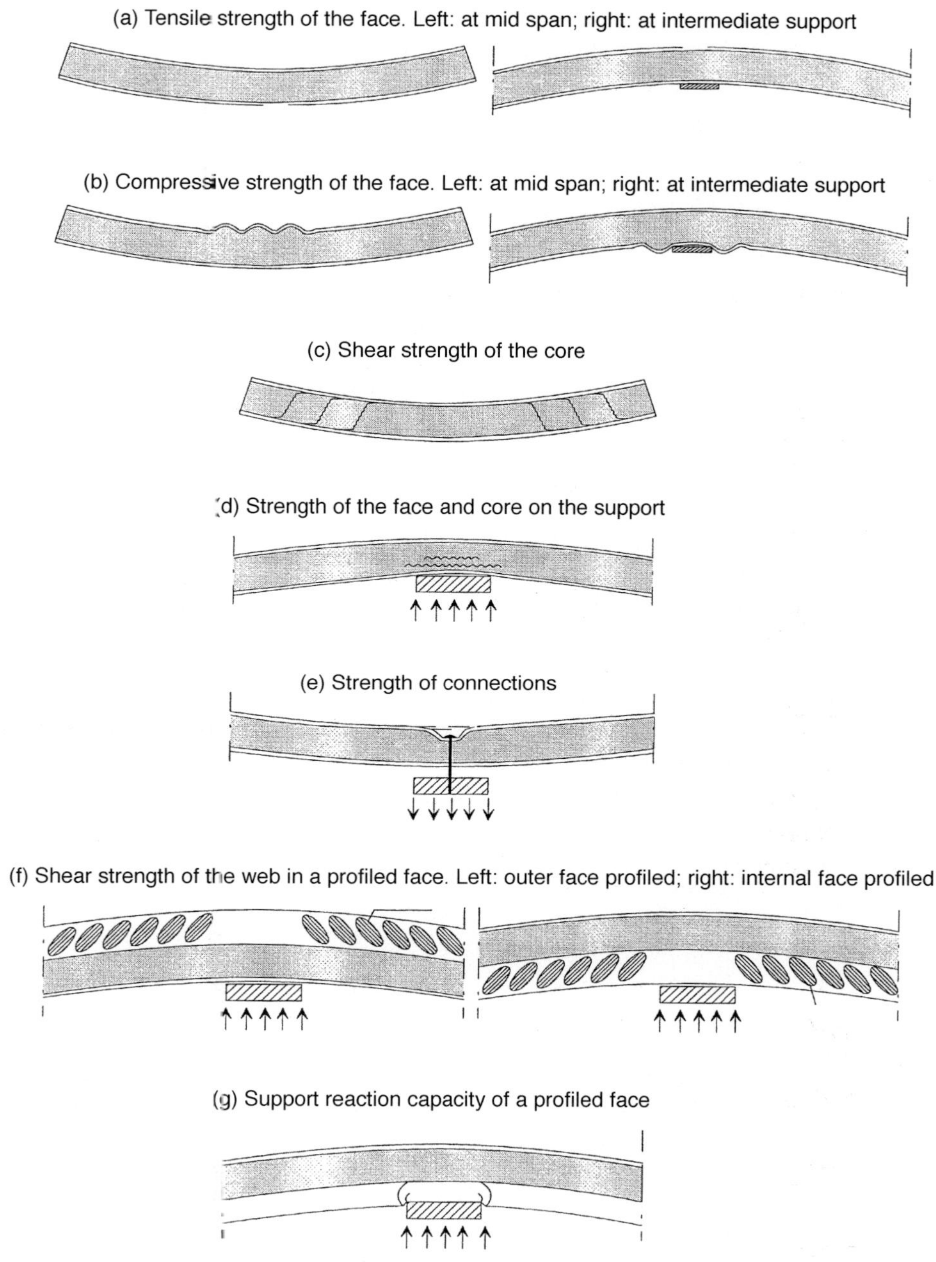

At the serviceability limit state, the failure modes at the intermediate supports generally have to be considered (failure modes (a (right)), (b (right)) and (e) to (g)).

At the ultimate limit state, the failure modes (a) to (e) for thin flat-faced sandwich panels and (a) to (g) (depending on the face geometry) for panels with profiled faces have to be considered.

Fig. 10.2 Failure modes of wall and roof panels.

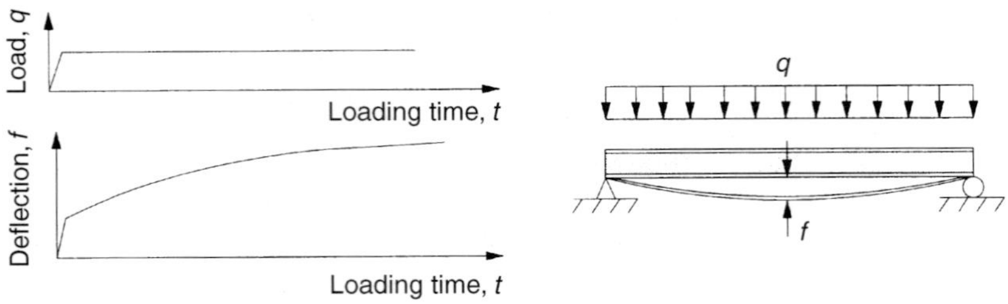

Fig. 10.3 Deflections: including the deflections caused by long-term shear deformations in the core.

$$a_1 = \pi t \left[\frac{(1+\nu_C)(3-4\nu_C)}{12(1-\nu_F^2)(1-\nu_C)}\right]^{1/3} \left(\frac{E_F}{E_C}\right)^{1/3} \approx 1.816t\left(\frac{E_F^2}{E_C G_C}\right)^{1/6} \tag{10.1}$$

$$\sigma_{cr,1} = \frac{3}{2}\left[\frac{2(1-\nu_C)^2}{3(1+\nu_C)^2(3-4\nu_C)^2(1-\nu_F^2)}\right]^{1/3} (E_C^2 E_F)^{1/3} \approx 0.823(E_C G_C E_F)^{1/3} \tag{10.2}$$

$$a_2 = \pi t \left[\frac{1}{6(1-\nu_F^2)}\right]^{1/3} \left(\frac{E_F^2}{E_C G_C}\right)^{1/6} \approx 1.784t\left(\frac{E_F^2}{E_C G_C}\right)^{1/6} \tag{10.3}$$

$$\sigma_{cr,2} = \frac{3}{2}\left[\frac{E_C G_C E_F}{6(1-\nu_F^2)}\right]^{1/3} \approx 0.852(E_C G_C E_F)^{1/3} \tag{10.4}$$

where E_F and E_C are the elastic moduli of the face and core materials, G_C is the shear modulus of the core and ν_F and ν_C are the values of Poisson's ratio for the face and core respectively.

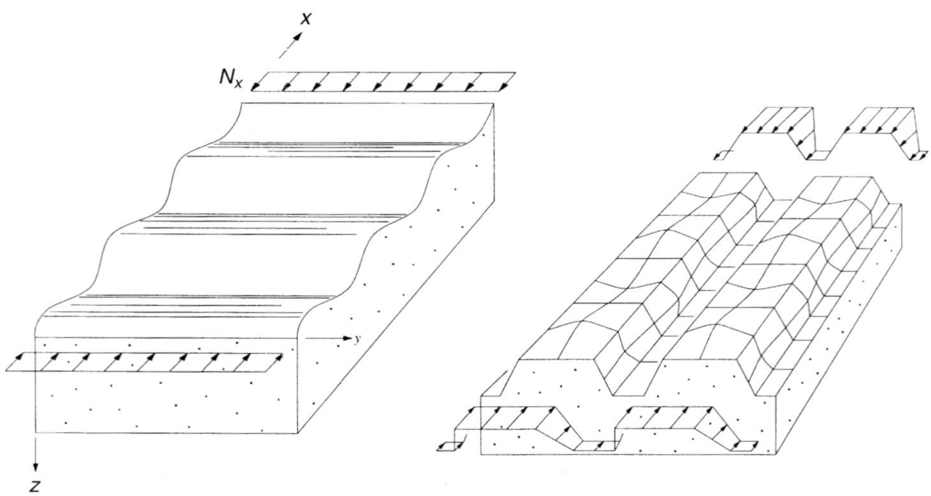

Fig. 10.4 Elastic buckling modes in a flat and in a profiled face made of thin metal sheet.

The above values of wrinkling stress are idealised values for a 'perfect' model and represent an upper bound to the compression strength of the thin plane faces of sandwich panels. A thin flat face fails by plastic buckling soon after the first elastic wrinkling waves have appeared. In the design equations used in practice, the compression strength of thin plane faces is evaluated on the basis of a reduced value of the wrinkling stress which takes account of initial imperfections such as the unevenness of the faces, the lack of homogeneity of the core material and any minor faults in the adhesion between the face and core. In the ECCS Recommendations,[10.4, 10.5] the expression for the compression strength is written as follows:

$$f_{\mathrm{Fc}} = k\sigma_{\mathrm{cr}} = 0.65(E_{\mathrm{F}}\, E_{\mathrm{C}}\, G_{\mathrm{C}})^{1/3} \tag{10.5}$$

where k is an empirical reduction factor ($k = 0.65/0.823 = 0.79$ or $0.65/0.852 = 0.76$).

If the compressive strength of the thin plane faces is to be evaluated for the stresses caused by the loads at high summer temperatures, the modulus of elasticity E_{C} and the shear modulus G_{C} of the core material in the above expression may be modified to take into account the influence of the higher temperatures, thus:

$$f_{\mathrm{Fc,T}} = k\sigma_{\mathrm{cr,T}} = 0.65(E_{\mathrm{F}}\, E_{\mathrm{C,T}}\, G_{\mathrm{C,T}})^{1/3} \tag{10.6}$$

In practice, the compressive strength of flat faces is often determined experimentally by means of full-scale panel tests which indicate the real resistance but also show some scattering of test results. This generally results in higher strengths than those given by Equation 10.5 which is rather conservative for rigid plastic foamed core materials. The strength at elevated temperature can then be estimated on the basis of the test results and the moduli of the core:

$$f_{\mathrm{Fc,Texp}} = \left(\frac{E_{\mathrm{C,T}}\, G_{\mathrm{C,T}}}{E_{\mathrm{C}}\, G_{\mathrm{C}}}\right)^{1/3} f_{\mathrm{Fc,exp}} \tag{10.7}$$

where $E_{\mathrm{C,T}}$ and $G_{\mathrm{C,T}}$ are the modulus of elasticity and the shear modulus at the elevated temperature and $f_{\mathrm{Fc,exp}}$ is the experimentally determined compressive strength at room temperature.

Profiling of the metal face increases its flexural stiffness and thus also the buckling stress and light profiling is often used for this purpose. The theoretical buckling wavelength and the wrinkling stress of slightly profiled faces ($d_{\mathrm{F}} < 10t$), based on the simplified elastic half-space foundation model, can be expressed as follows:

$$a_{\mathrm{cr,p}} = \pi\left(\frac{4\, B_{\mathrm{F}}^2}{E_{\mathrm{C}}\, G_{\mathrm{C}}}\right)^{1/6} \approx 3.958\left(\frac{B_{\mathrm{F}}^2}{E_{\mathrm{C}}\, G_{\mathrm{C}}}\right)^{1/6} \tag{10.8}$$

$$\sigma_{\mathrm{cr,p}} = \frac{3}{2\, A_{\mathrm{F}}}(2\, B_{\mathrm{F}}\, E_{\mathrm{C}}\, G_{\mathrm{C}})^{1/3} \approx \frac{1.89}{A_{\mathrm{F}}}(B_{\mathrm{F}}\, E_{\mathrm{C}}\, G_{\mathrm{C}})^{1/3} \tag{10.9}$$

In the above expressions, A_{F} and B_{F} are the cross-sectional area and flexural stiffness of the slightly profiled face per unit width and d_{F} is the depth of profiling.

At the present time, for the purposes of practical design, the compressive strength of slightly profiled faces is evaluated either experimentally or by using calculation models based on the theoretical buckling stress in Equation 10.9. As for the case of flat faces, for the analytical approach, this idealised buckling stress is reduced because of initial imperfections in the face, core and adhesive layer. There are, therefore, three alternative approaches to determine the compressive strength of slightly profiled faces which are in current use.

(1) The design value of compressive strength is the characteristic value of the test results obtained from wrinkling tests on full-scale sandwich panels ($f_{Fc,exp}$).

(2) The buckling stress given by Equation 10.9 is reduced in order to take into account imperfections and material non-linearities in the face, core and adhesive layer as well as deficiencies in the analysis. The resulting expression for the compressive strength according to the ECCS Recommendations[10.4] is:

$$f_{Fc} = k_p \, \sigma_{cr,p} = \frac{0.95}{A_F} (B_F \, E_C \, G_C)^{1/3} \tag{10.10}$$

where k_p is a reduction factor ($k_p = 0.95/1.89 = 0.50$).

(3) In order to take into account the increased bending stiffness of the slightly profiled face, the expression for the compression strength of a plane face (10.5) is modified by a factor α, which is greater than one. The factor α is based on the results of full-scale sandwich panel tests. This method has been used, notably in German type-approvals (i.e. in 'Zulassung' documents).

$$f_{Fc} = \alpha k \, \sigma_{cr} = 0.65\alpha(E_F \, E_C \, G_C)^{1/3} \tag{10.11}$$

It is recommended that the first approach, based on full-scale tests, should generally be used in the practical design of sandwich panels. The influence of higher (summer) temperatures on the compressive strength can then be estimated using Equation 10.7. When using the approaches based on calculation models (Equations 10.5 and 10.10), there are difficulties in the choice of the moduli of elasticity of the core and the flexural rigidity of the lightly profiled face layer. On the other hand, the calculation models given above are useful tools when new face and panel geometries are to be developed and optimised.

If the faces of a sandwich panel are more heavily profiled, local buckling of that plane part of the profile which is most highly stressed in compression generally dominates the behaviour. For the analysis, this most highly stressed part, usually the outer flange of the profile, is isolated from the rest of the profile and the compressive strength of this flange is evaluated by assuming it to be a long plate simply supported along its longitudinal edges and continuously supported by an elastic foundation consisting of the core layer as shown in Fig. 10.5. The buckling stress of a plane part of a profile, taking account of the restraint provided by the core material, can be estimated as follows:

$$\sigma_{cr} = \frac{N_{x,cr}}{t} = K_\sigma \frac{\pi^2 E_F}{12(1 - \nu_F^2)} \left(\frac{t}{b}\right)^2 \tag{10.12}$$

where

$$K_\sigma = (16 + 7R + 0.002\,R^2)^{1/2} \tag{10.13}$$

and

$$R = \frac{12(1 - \nu_F^2)\sqrt{E_C \, G_C}}{\pi^3 E_F} \left(\frac{b}{t}\right)^3 \approx 0.35 \frac{\sqrt{E_C \, G_C}}{E_F} \left(\frac{b}{t}\right)^3 \tag{10.14}$$

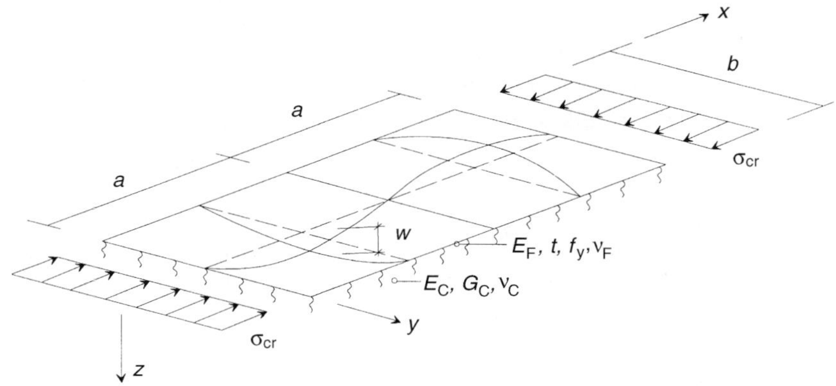

Fig. 10.5 Local buckling of the compressed flange of a profiled metal sheet face.

The dependence of the buckling coefficient K_σ on the parameter R is illustrated in Fig. 10.6. This value of the coefficient K_σ, which is recommended for use in practical design, is reduced from the theoretical value K_{theo} on the basis of test results.

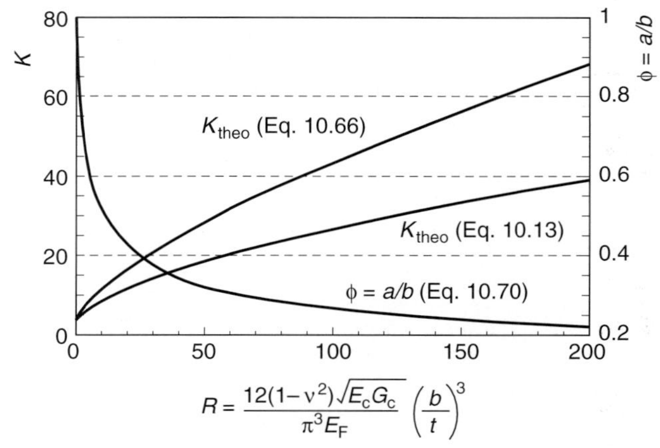

Fig. 10.6 Theoretical and design values of the buckling coefficient and the length of the critical buckling half wave of the flange of a profiled face layer.

This local buckling of the most highly stressed plane part of the face profile is then assumed to make the plane part partially ineffective in its ability to carry the compressive stresses. The evaluation of the compressive strength of the buckled plate is based on the well-known winter formula in which the buckled plate is replaced by effective widths carrying the uniformly distributed yield stress as shown in Fig. 10.7.

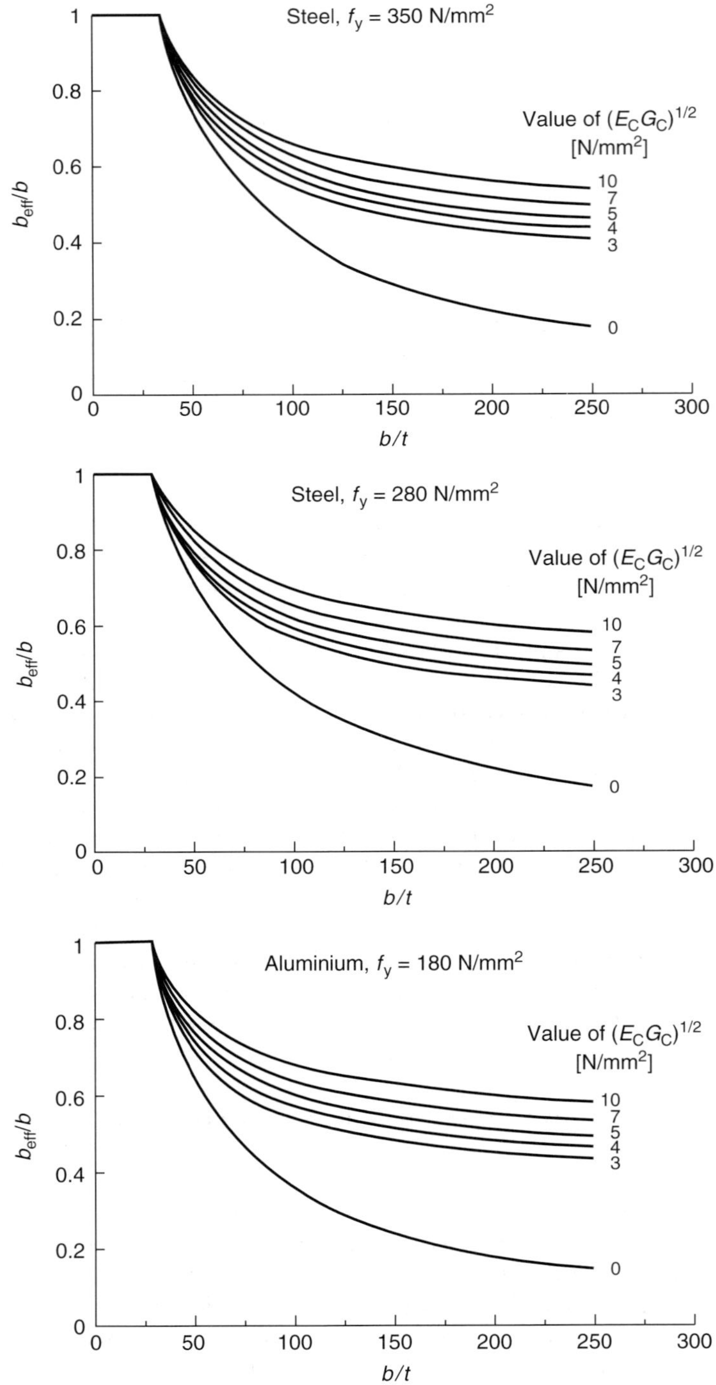

Fig. 10.7 Effective widths of the plane flanges of profiled face layers.

$$\frac{b_{\rm eff}}{b} = 1.0 \quad \text{for } \lambda \leq 0.673$$

and

$$\frac{b_{\rm eff}}{b} = \frac{1}{\lambda}\left(1 - \frac{0.22}{\lambda}\right) \quad \text{for } \lambda > 0.673 \tag{10.15}$$

where

$$\lambda = \left(\frac{f_{\rm y}}{\sigma_{\rm cr}}\right)^{1/2} \tag{10.16}$$

The buckling stress $\sigma_{\rm cr}$, taking account of the stiffening effect of the core, is calculated using Equations 10.12–10.14.

The compressive strength of the plane unstiffened flange of a face profile can now be expressed by

$$f_{\rm Fc} = \frac{b_{\rm eff}}{b} f_{\rm y} \tag{10.17}$$

In the above derivation, it is assumed that the ultimate limit state of a compressed plate element is defined by yielding of the most highly stressed part. The limit of application of this effective width approach may be taken to be $R \leq 200$ and $b/t \leq 250$.

The classification into slightly profiled faces (Equations 10.1–10.11) or strongly profiled faces (Equations 10.12–10.17) depends on the influence of the flexural rigidity of the faces on the normal stresses in a face layer (see Figs 9.3 and 9.9). In principle, if the normal stresses in a profiled face are calculated first using the theory of thin-faced sandwich beams and then using the theory of thick-faced sandwich beams, and the difference between the two calculated stresses is higher than about 10%, the theory and calculation models of thick-faced sandwich beams and strongly profiled faces should be used (see Section 9.3). However, the equations for thick-faced sandwich beams tend to become ill-conditioned at the interface between the two theories so that this criterion is of limited practical value. A more practical guideline is that the faces with a depth of profiling less than ten times the face thickness ($d_{\rm F} < 10t$) should be analysed as thin-faced sandwich panels and the others (with $d_{\rm F} > 10t$) as thick-faced sandwich panels.

10.2.2 Failure of the core and face at the support with positive support reaction

10.2.2.1 Definitions

The support reaction at an end or intermediate support may cause either compressive or tensile forces in the connection between the supporting structure and the sandwich panel. A support reaction resulting in compressive contact stresses is considered to be a positive support reaction and, correspondingly, a support reaction resulting in tensile forces in the fasteners between the sandwich panel and the supporting structure is considered to be a negative support reaction.

10.2.2.2 Evaluation of the stresses in the face and core

As shown in Fig. 10.8, a sandwich panel may fail at a support by either (a) a shear

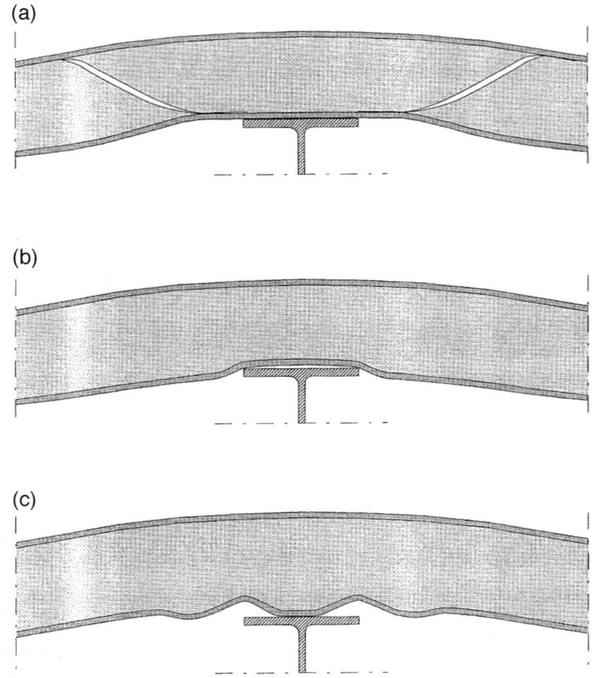

Fig 10.8 Three different failure modes may define the load-bearing capacity of a sandwich panel at a support. At an intermediate support, the failure may be a combination of modes (a), (b) and (c). At an end support, failure mods (a) and (b) may exist.

or (b) a crushing failure of the core or by (c) a buckling and bending (wrinkling) failure of the face. The ultimate load may also be determined by a combination of these failure modes. Calculation models for determining the resistance of the panels against the core crushing, face bending and buckling failure modes are given in Section 10.2.2 and against the core shear failure mode in Section 10.2.4.

The contact pressure distribution between the supporting structure and the sandwich panel depends on the flexibility of the upper flange of the supporting beam. Effectively rigid support beams, such as solid concrete or hot rolled steel sections, concentrate the support reaction at the edges of the upper flange of the supporting beam with the result that the support reaction may be described by two line loads. More flexible or slender supporting beams tend to concentrate the reaction force close to the web of the supporting profile with the result that there is only one line load at the support. These distributions of support forces are valid at the serviceability limit state. The contact pressure distribution changes after local failure in either the panel or in the support beam at the ultimate limit state.

The support reaction F causes a support pressure distribution q_s which, in turn, causes local compressive and shear stresses in the core and local shear forces and bending moments in the face which is placed against the support beam as shown in Fig. 10.9. If the interaction between the compressive axial force in the face layer and the transverse support reaction is neglected in the design calculations, four different assumptions are available for the distribution of support pressure in the

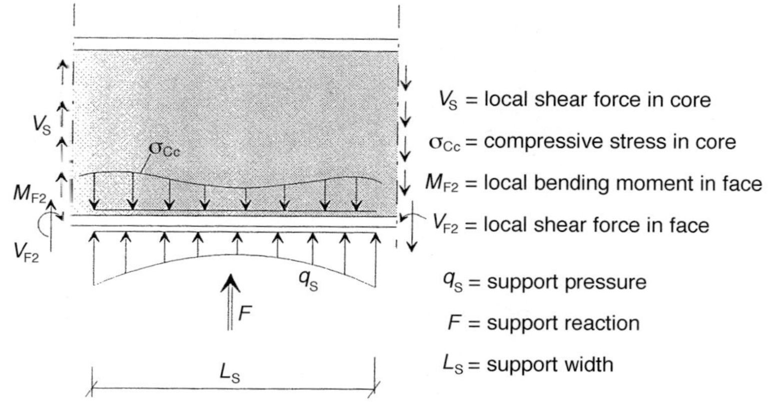

Fig. 10.9 Local stresses in the core and lower face caused by the support reaction F and the support pressure q_S.

core, as shown in Fig. 10.10(a)–(d). In the first model (Fig. 10.10(a)), the compressive stress in the core is assumed to be the same as the support pressure, which is assumed to be uniformly distributed over the width L_s of the support as in Equation 10.18:

$$\sigma_{Cc} = \frac{F}{L_s} \tag{10.18}$$

This model does not take into account the influence of the bending stiffness of the face, which causes an increase in the compressed area of the core. The influence of the bending stiffness and elasticity of the core is taken into account very approximately in the second model (Fig. 10.10(b)), in which the compressed area is increased by dispersion into the depth of the panel and it is assumed that the support reaction resistance of the core is described by the compressive stress at the mid-depth of the panel as in Equation 10.19. The angle of dispersion α, may be determined experimentally for different sandwich panel products. However, generally for rigid plastic foams, the angle α can be assumed to be given by $\alpha = \arctan\left(\frac{1}{2}\right) \sim 27°$. Thus, at the intermediate supports, the compressed width of the core is $L_{s\,C} = L_s + 2\,(e/2)\tan\alpha$ and, at an end support, assuming that the support structure is placed at the very end of the panel, $L_{s,C} = L_s + (e/2)\tan\alpha$.

$$\sigma_{Cc} = \frac{F}{L_{s,C}} = \frac{F}{L_s + ke} \quad \text{in which } k = 0.5 \tag{10.19}$$

The influence of the bending stiffness of the face layer and the flexibility of the core on the compressive stress distribution in the core can be taken into account more accurately by modelling the face by a beam which is supported by an elastic Winkler foundation which simulates the core layer. The foundation coefficient is derived simply as $k_w = E_{Cc}/d_C$ where E_{Cc} is the compressive modulus and d_C the depth of the core ($d_C = e - d_F$). The derivation of the expressions for the compressive stress in the core and the bending moment in the face is given in detail in Section 10.4. Two models are available for the support reaction pressure.

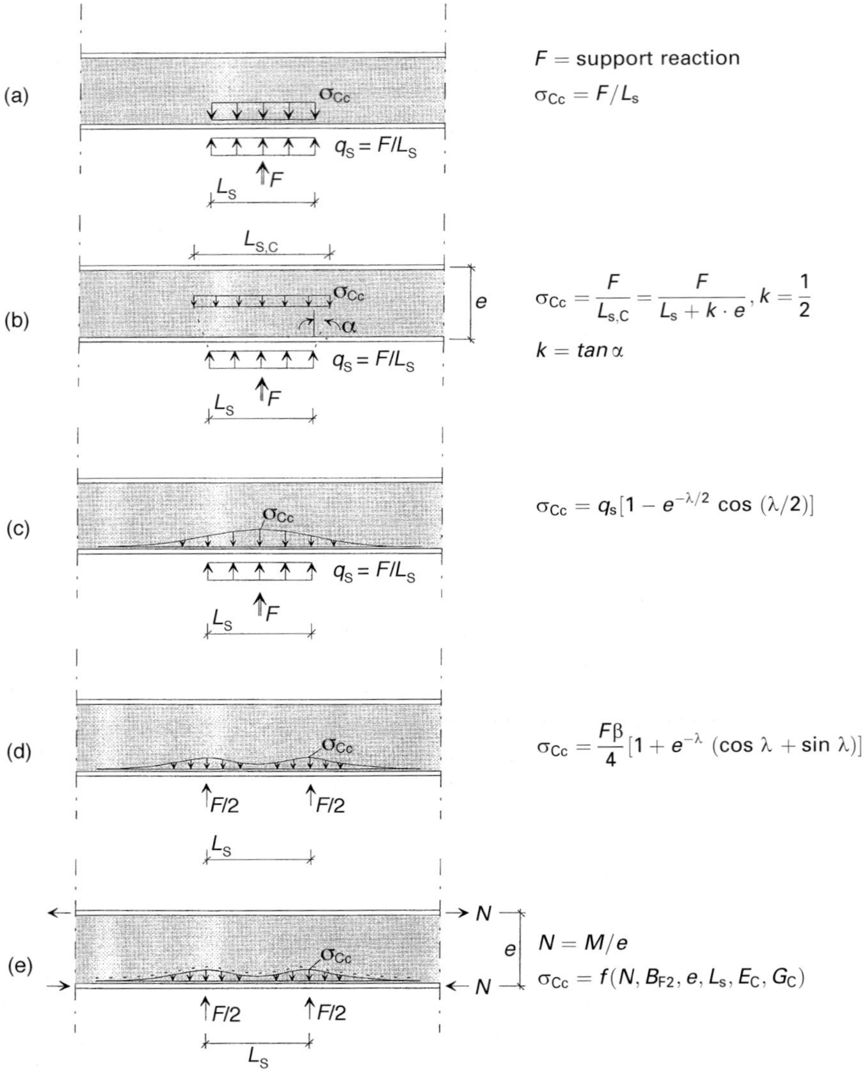

Fig. 10.10 Different assumptions for the compressive stress distribution in the core at a support. Model (e), where there is an interaction between the axial compressive force N and the support reaction F is closest to reality.

In the first, the support pressure is assumed to be uniformly distributed over the support width, as shown in Fig. 10.10(c), giving rise to Equations 10.20 and 10.21:

$$\sigma_{Cc} = q_s[1 - e^{-\lambda/2} \cos(\lambda/2)] \tag{10.20}$$

$$M_{F2} = \frac{q_s}{2\beta^2} e^{-\lambda/2} \sin(\lambda/2) \tag{10.21}$$

In the second model, the support reaction is assumed to consist of two line loads locating at the edges of the support beam, as shown in Fig. 10.10(d), giving rise to Equations 10.22 and 10.23:

$$\sigma_{Cc} = \frac{F\beta}{4}[1 + e^{-\lambda}\ (\cos\lambda + \sin\lambda)] \qquad (10.22)$$

$$M_{F2} = \frac{F}{8\beta}[1 + e^{-\lambda}\ (\cos\lambda - \sin\lambda)] \qquad (10.23)$$

where

$$\beta = \left(\frac{E_C B}{4 d_C B_{F2}}\right)^{1/4} \qquad (10.24)$$

and

$$\lambda = \beta L_s \qquad (10.25)$$

The bending stresses in the face are calculated using $\sigma_{F2,R} = M_{F2}/W_{F2}$, where W_{F2} is the elastic section modulus of the face.

10.2.2.3 Interaction between bending moment and support reaction

At an intermediate support, the face layer which is in contact with the supporting structure is loaded by an axial compressive force. This axial compressive force increases the local bending moment M_{F2} caused by the support reaction and so increases the stresses in the face and core as shown in Fig. 10.10(e). Three equations defining the limit states with respect to buckling and yielding of the face and crushing of the core can be written:

$$\sigma_{F2} \leq f_{F2} \qquad (10.26a)$$

$$\sigma_{F2} + \sigma_{F2,F} = \frac{N_S}{A_{F2}} + \frac{M_{F2} + \Delta M_{F2}}{W_{F2}} \leq f_y \qquad (10.26b)$$

$$\sigma_{Cc} = k_w w \leq f_{Cc} \qquad (10.26c)$$

The local deflection and local bending moment of the face can be described by the differential equations:

$$B_{F2} w'''' + (N_S - k_1) w'' + k_w w = q_s \qquad (10.27a)$$

$$M_{F2} + \Delta M_{F2} = -B_{F2} w'' \qquad (10.27b)$$

where k_w and k_1 are the Winkler and second foundation parameters respectively which take into account the compressive and shear flexibility of the core. The derivation of theoretical design equations based on Expressions 10.26 and 10.27 is given in Section 10.4. In practice, the definition of the foundation coefficients based on the elastic properties of the core results in difficulties because of the anisotropy and non-homogeneity of the typical core materials. This reduces the accuracy of the calculated stresses $\sigma_{F2,F}$ and σ_{Cc}.

In reality it is relatively difficult to evaluate this interaction. Factors to be mentioned in this connection are, in particular, face imperfections, core stiffness and strength, bonding to the core, irregular lightly profiled faces and, above all, the way in which the load is introduced into the internal support zone depending on the type and stiffness of the supporting beam or substructure.

The chances of being able to assess such a multiplicity of factors with sufficient accuracy by theoretical analysis would appear to be very small. This is indicated, above all, by the fact that efforts in this connection have failed, even for the relatively simple case of single-span panels with lightly profiled faces. For these components, it is as necessary now, as previously, to perform panel tests in order to determine the strength of the compressed face layer. In the case of multi-span panels, where the situation is made even more complicated by the problems of additional loading due to compressive stresses in the core over an internal support, the need for tests to analyse the failure criteria is indisputable.

The test arrangement for the interaction between bending moment and support reaction force is a single-span panel subject to a line load as shown in Chapter 13. This is often referred to as the *simulated central support test* because it simulates the conditions in the central support of a two-span beam. If, as is usually the case, the line load is applied vertically downwards, it is necessary to invert the panel in order to obtain the correct relationship between the reaction force and the contact face. Varying the span of this arrangement varies the relationship between the bending moment and the reaction force and the choice of the correct span is an important consideration when planning such tests. A great many tests of this type have been carried out and some representative results, for polyurethane and stone wool cored panels, have been described in two research reports.[10.1, 10.23] In the investigation of the load-bearing capacity at an internal support, the following parameters were varied:

- the producer and type and quality of the core material
- the panel type and the geometry of the lightly profiled faces
- the thickness of the panel
- the span of the simulated central support tests and
- the width of the support

A crucial point was the care taken in the tests to select the spans and the widths of the supporting beams, which introduce the loads, so that the compressive stresses in the core over the internal supports, which arise upon reaching the ultimate load, are of the same magnitude as the compressive strength of the core.

For a proper evaluation of the test results, it is very important to have also the results of single span beam tests with a uniformly distributed load which give the wrinkling stresses in the undisturbed zone and the results of the corresponding core material tests, i.e. the compressive, shear and tensile strengths and the corresponding moduli. It is then possible to relate the results of the simulated central support tests to the material parameters. A total of 92 simulated central support tests were evaluated by Berner[10.1] on the basis of results from the material tests and from single span panel tests on polyurethane foam sandwich panels.

In Fig. 10.11, the ratio of the failure stresses from the simulated central support tests, σ_u, to the compression strength from the single span panel tests, f_{Fc}, is plotted as the ordinate (σ_u/f_{Fc}). This figure summarises all of the results. In the diagram, σ_{Cc} is the compressive stress in the core directly above an internal support:

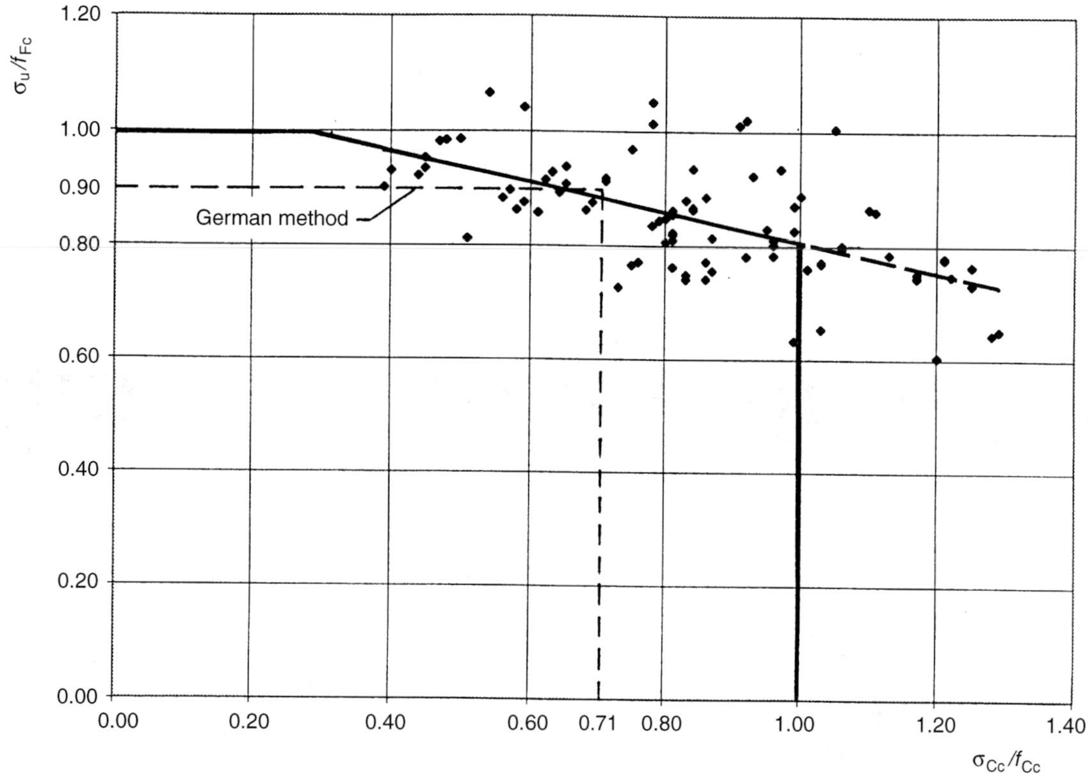

σ_u = wrinkling stress with interaction with the support force
f_{Fc} = wrinkling stress without interaction form single-span tests
σ_{Cc} = actual compressive stress
f_{Cc} = compressive strength of the core

Fig. 10.11 Interaction between bending moment and reaction force at intermediate supports.[10.1] Test results on polyurethane cored sandwich panels ◆, and German design approach - - - -.

$$\sigma_{Cc} = F/L_s \tag{10.28}$$

where F = support reaction per unit width
L_s = width of the support

The main advantage of this approach is that it is based on the measured values of the compressive strength of the face in the undisturbed zone (wrinkling stress in the middle of the span of the single span tests with uniformly distributed load). It is possible, therefore, to identify the effect of the additional compressive stresses directly because the results of the wrinkling strength tests are used as a basis. The above factors are then plotted on the abscissa in order to identify the behaviour over the support as a function of the key parameters.

A series of theoretical models of this type have been investigated in References 10.1 and 10.23. However, in order to obtain a practical design approach, it was found to be convenient to represent the summarised results in terms of an interaction between the ratio of the compressive face failure stress (wrinkling stresses) in the disturbed and undisturbed zone (σ_u/f_{Fc}) and the ratio of the existing compressive stress in the core over the support and the core compressive strength (σ_{Cc}/f_{Cc}).

As a first approximation in the interpretation of the results, the arithmetic mean of all the points was presented as a line having a distinctly falling trend. This was to be expected because the compression strength of the face decreases with the additional compressive stress in the core over an internal support (σ_{Cc}).

On the basis of the results given in Fig. 10.11, therefore, it is possible to read off the reduction in the wrinkling failure stress over an internal support as a function of the compressive strength of the face in the span in a very simple and practical manner on the basis of the compressive stress in the core over the support.

However, in Germany, for polyurethane panels, it has been found to be adequate to take a uniform global decrease of 10% of the wrinkling stress over the middle support. This is applied on the basis of the above evaluation together with the additional requirement that $1.4\ \sigma_{Cc} \leq f_{Cc}$ ($\gamma = 1.4$) and hence $\sigma_{Cc}/f_{Cc} = 1/1.4 = 0.71$. It is quite conceivable that this approach is also valid for stone wool panels.

Therefore, for practical design, the interaction between the bending moment M_s and the support reaction F can be taken into account, without any detailed interaction studies, by reducing the strength of the compressed face at an intermediate support by 10% and by evaluating the support reaction resistance using the conservative expression in Equation 10.18 with an additional reduction coefficient $1/1.4 = 0.71$, thus:

$$f_{Fc,support} = 0.9 f_{Fc} \tag{10.29a}$$

$$F_R = 0.71\, L_s f_{Cc} \tag{10.29b}$$

An alternative way to take into account the interaction between the bending moment and support reaction is illustrated by Fig. 10.12.[10.29]

Fig 10.12 shows the result of a comparison between test results for both PUR and stone wool cored panels and the calculation model:

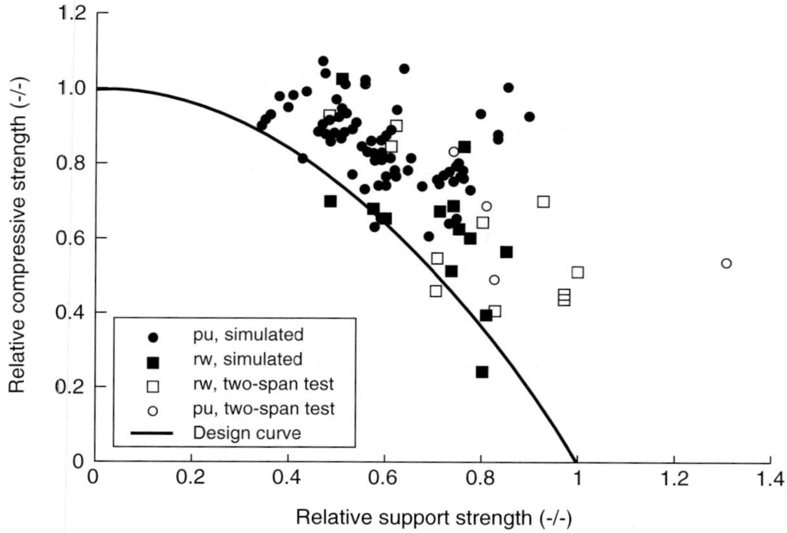

Fig. 10.12 Interaction between bending moment and support reaction force.

$$\frac{F}{F_R} - \sqrt{1 - \frac{\sigma_{F2}}{f_{F2c}}} = 0 \tag{10.30}$$

where F is the support reaction force at an intermediate support
F_R is the support reaction resistance:

$$F_R = f_{Cc} B (L_s + \eta e) \tag{10.31}$$

where σ_{F2} is the compressive axial stress in the lower face
f_{F2c} is the compressive strength of the lower face of the panel (in the span)
η is a distribution factor that takes the values:
0.5 for PUR cored panels
0.4 for stone wool cored panels

10.2.2.4 Post-buckling bending resistance at intermediate supports

A thin flat face does not lose all of its compressive strength immediately after first failure at an intermediate support. As shown in Fig. 10.13 for some stone wool cored specimens of 100-mm depth and 1200-mm width with thin steel faces and with an intermediate support width of 200 mm, the reducing 'plastic' compressive strength after the maximum bending moment has been attained depends on the initial failure mode. If the initial failure mode is a bending and buckling (wrinkling) failure, the face may have a significant residual plastic compressive strength which is nearly independent of the deflection and thus of the plastic rotation at the support as shown in Fig. 10.13(c). However, it is recommended that this residual plastic compressive strength is not taken into account in design because of the practical difficulty of sorting out the shear and crushing failures of the core and the bending and buckling failures of the face.

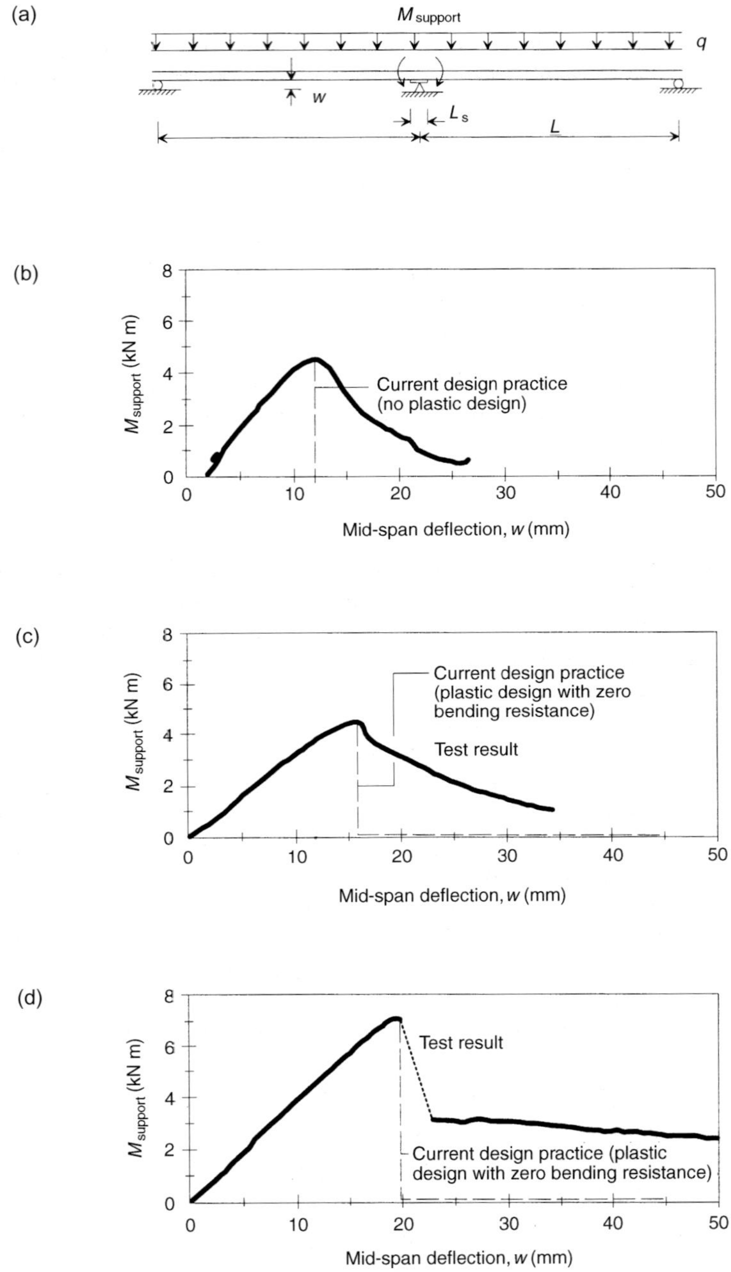

Fig. 10.13 Experimental curves showing the plastic bending resistance at the intermediate support in sandwich panels with a stone wool core.[10.23] (b) Shear failure ($L = 2500 + 2500\,\text{mm}$); (c) core crushing failure ($L = 3250 + 3250\,\text{mm}$) and (d) face bending and buckling failure ($L = 4500 + 4500\,\text{mm}$).

10.2.2.5 *Strongly profiled faces*

If the face in contact with the supporting structure is strongly profiled, its characteristic support reaction resistance can be evaluated using the semi-empirical expression derived for trapezoidal sheets.[10.6]

$$F_{R2} = 0.15\,n\,t^2(E_F f_y)^{1/2}\left[1 - 0.1\left(\frac{r}{t}\right)^{1/2}\right]\left[0.5 + \left(\frac{L_s}{50t}\right)^{1/2}\left[2.4 + \left(\frac{\theta}{90}\right)^2\right]\right] \qquad (10.32)$$

where n is the number of webs per unit width
r is the radius of the bend between the lower flange and the web
L_s is the support width
θ is the angle between the support beam and the web of the profile

Expression 10.32 is valid for $45° \leq \theta \leq 90°$. In addition, it is assumed that the relative difference between the transverse shear forces on each side of the support is small, i.e.:

$$\frac{|V_{Sd,1}| - |V_{Sd,2}|}{|V_{Sd,1}| - |V_{Sd,2}|} \leq 0.2 \qquad (10.33)$$

where $|V_{Sd,1}|$ and $|V_{Sd,2}|$ are the absolute values of the transverse shear forces on each side of the intermediate support.[10.6] If the difference between these values is large, the contact pressure between the panel and the supporting structure is no longer symmetrical but concentrates towards a small area at the edge of the supporting beam. In such cases, Equation 10.33 is no longer valid and the value of the support width L_s in Equation 10.32 has to be reduced.

Because the face has a high bending stiffness and distributes the support reaction effectively, the compressed area in the core is large. Therefore, the checks of the compression strength of the core or the interaction between the support reaction force F and the axial compressive force N_S are no longer required.

When strongly profiled faces are placed against the supporting structure, the dominant stress resultants are the bending moment M_{F2} and the support reaction F. If the lateral support provided by the core against local buckling of the flanges and the web of the profile is neglected, the resistance of the face profile against the combined failure mode can be evaluated using the semi-empirical equations derived for the trapezoidal steel sheets[10.6] which is shown in Fig. 10.14:

$$\frac{M_{F2}}{M_{2R}/\gamma_M} \leq 1 \text{ when } \frac{F}{F_{R2}/\gamma_M} \leq 0.25 \qquad (10.34a)$$

$$\frac{M_{F2}}{M_{2R}/\gamma_M} + \frac{F}{F_{R2}/\gamma_M} \leq 1.25 \text{ when } 0.25 \leq \frac{F}{F_{R2}/\gamma_M} \leq 1 \qquad (10.34b)$$

$$\frac{F}{F_{R2}/\gamma_M} \leq 1 \text{ when } \frac{M}{M_{2R}/\gamma_M} \leq 0.25 \qquad (10.34c)$$

where M_{2R} and F_{R2} are the bending resistance and support reaction resistance of the strongly profiled face profile without the influence of the core. Obviously, this approach is very conservative.

If the depth of the face profile in contact with the supporting structure is between about 10 and 30 mm, the face profile and the core both contribute to the resistance against the support reaction force F. There are no design expressions available for

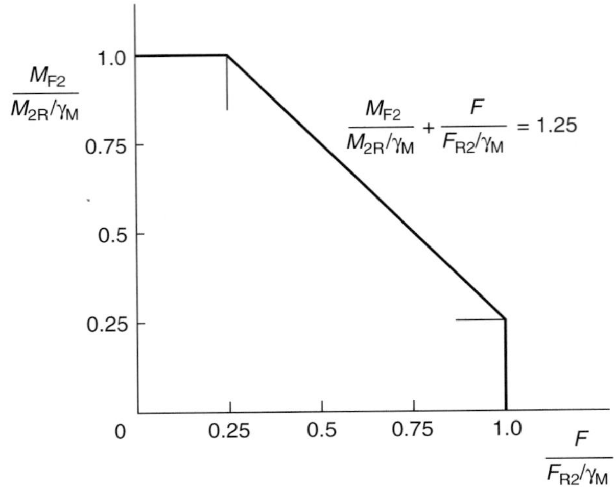

Fig. 10.14 Interaction diagram for support reaction and bending moment in a strongly profiled metal sheet face layer.

this case. Thus, the evaluation of the strength of profiled faces with moderate depth, d_2, at intermediate supports has to be based on tests.

10.2.3 Failure of the core and face at a support subject to negative support reaction

The ultimate failure mode of the core and face layers at a support in a sandwich panel which is subject to an uplift load caused by wind suction and/or a temperature difference between the faces depends completely on the connection system of the panel. If the panel is supported by effectively rigid continuous covering plates which are fixed to the building frame by screws or bolts, the analysis of the resistance of the panel against the negative support reactions can be based on similar assumptions and calculation models to those proposed for the analysis for positive support reactions. However, if the panel is fixed to the frame directly with screws, the flexibility and resistance of the panel against the negative support reactions is quite different. When loaded by a tensile force, a discrete screw connector causes a significant imperfection in the face and, therefore, reduces the compression strength of the face at the support. The reduction in strength depends on the thickness and geometry of the face and on the location of the screw in the face.

The head of the screw causes compressive stresses in the core and results in local crushing in the core. The mode of ultimate failure is typically a pull-through type, the resistance of which is affected by both the face and the core. Up to now, there are no calculation models available for the evaluation of the combined failure consisting of face buckling and pull-through of the fastener. The compression strength of the face and the tensile strength of the fasteners have been studied separately based on experimental observations as summarised in Table 10.1. The tensile and shear resistance of the screws alone is studied in Section 10.2.6.

Table 10.1 Design of screw fasteners and special fastening systems against negative support reactions

	Screw fasteners	Special fastening systems in the longitudinal joints of the panels
Compression strength of the face	Compressive strength of the face is strongly reduced.	Compressive strength of the face is reduced. The reduction has to be determined experimentally.
Tensile strength of the fasteners	See Section 10.2.6.	Tensile strength of special fastening system has to be determined experimentally.
Flexibility (see Chapter 12)	Flexibility of the screw connection is moderate.	Flexibility of the fastening is large and has effects on the bending moment and shear force distributions and on the deflections.

Berner[10.1] has also attempted to investigate the various factors experimentally with a series of simulated central support tests in which the support reaction forces were introduced through the screws. When designing such a test series, it is necessary to consider the following parameters in particular:

- the number of screws
- the screw forces
- the geometry of the face
- the thickness of the panel
- the span of the substitute support tests

As in the case of the simulated central support tests under downward load, it is important for the interpretation that the corresponding results from single-span tests under uniformly distributed load and the core material tests are also available for these support tests. It is, therefore, again possible to relate the test results from the simulated central support tests to the wrinkling stresses in the panel and the corresponding material parameters. Berner[10.1] evaluated thirty tests on polyurethane panels in this way. Corresponding test results on stone wool cored sandwich panels have been reported by Martikainen and Hassinen.[10.23]

After investigating a series of theoretical approaches for analysing the individual parameters, a trial evaluation was undertaken by drawing up interaction diagrams. These were similar to Fig. 10.11, which was used in the corresponding evaluation for compressive loads, so that the ratio of the failure stresses from the simulated central support tests (σ_u) to the wrinkling stresses from the single span beam tests (f_{Fc}) were again plotted as the ordinate (σ_u/f_{Fc}). The advantage of this approach is that it is based once again on reliable values in the undisturbed zone (wrinkling stresses from the single-span panel tests). If the ratio σ_u/f_{Fc} is smaller than 1, it is possible to read the decrease of compressive strength in the internal support zone off the ordinate. After investigating various alternatives, the following reference values were plotted on the abscissa:

$$\frac{\sigma_{Cc}}{f_{Cc}}$$

Knowing from the evaluation of the interaction relationships for compressive loads that there is a practically linear relationship between the reducing failure stresses in the faces and the compressive stresses in the core, a corresponding compressive stress (σ_{Cc}) was also calculated for the introduction of tensile forces. An area available for the introduction of the tensile force was determined in the form of a type of co-bearing width. This area depends greatly on the number of screws. The compressive stress in the core can then be calculated using the following simple formula:

$$\sigma_{Cc} = \frac{\text{screw force}}{\left(\dfrac{\text{panel width}}{\text{number of screws} + 1}\right)^2} \tag{10.35}$$

The resulting theoretical model is shown in Fig. 10.15 together with some test results. When the distribution of screws is determined, at least to a first approximation, by considering the compressive stress, it is possible to identify a relatively good and sensible distribution. In Germany this procedure is recommended as offering a practical approach to the design of polyurethane panels. A similar approach for mineral wool panels is also conceivable.

The interaction between the transverse forces caused by the screws and the axial compressive load in the face can be taken into account by reducing the compression strength of the face at the intermediate supports. In Germany the compression strength is reduced by 20% if the number of screws at the support is less than or equal to 3 in a panel width of 1 metre. If the number of screws is larger than three, the compression strength of the face is reduced according to Equation 10.36b.

$$f_{Fc,\text{support}} = 0.8 f_{Fc} \quad \text{if } n \leq 3 \tag{10.36a}$$

$$f_{Fc,\text{support}} = 0.10\,(11 - n) f_{Fc} \quad \text{if } n > 3 \tag{10.36b}$$

where n is the number of screws in a panel width of 1 metre.

Special fasteners, often called concealed fasteners, may be placed in the longitudinal joints between two sandwich panels (see Chapter 12). When loaded by the tensile forces caused by negative support reactions, these fasteners stress the 'beam' consisting of the joint details and the face located close to the longitudinal joint and also the region of the core near the fastener. Local tensile forces in the longitudinal joints also cause transverse curvature of the panel and a transverse tensile stress field in the compressed face, both of which tend to stabilise the face against buckling. At the ultimate limit state, the fastener screws or the steel strip in the fastening may fail or the plastic bending resistance of the joint 'beam' or the crushing failure mode of the core may determine the ultimate load. No calculation models are available with which to evaluate the tensile resistance of these special fastening systems against the negative support reaction forces. The design of special fastening systems has, therefore, to be based on experimental observations.

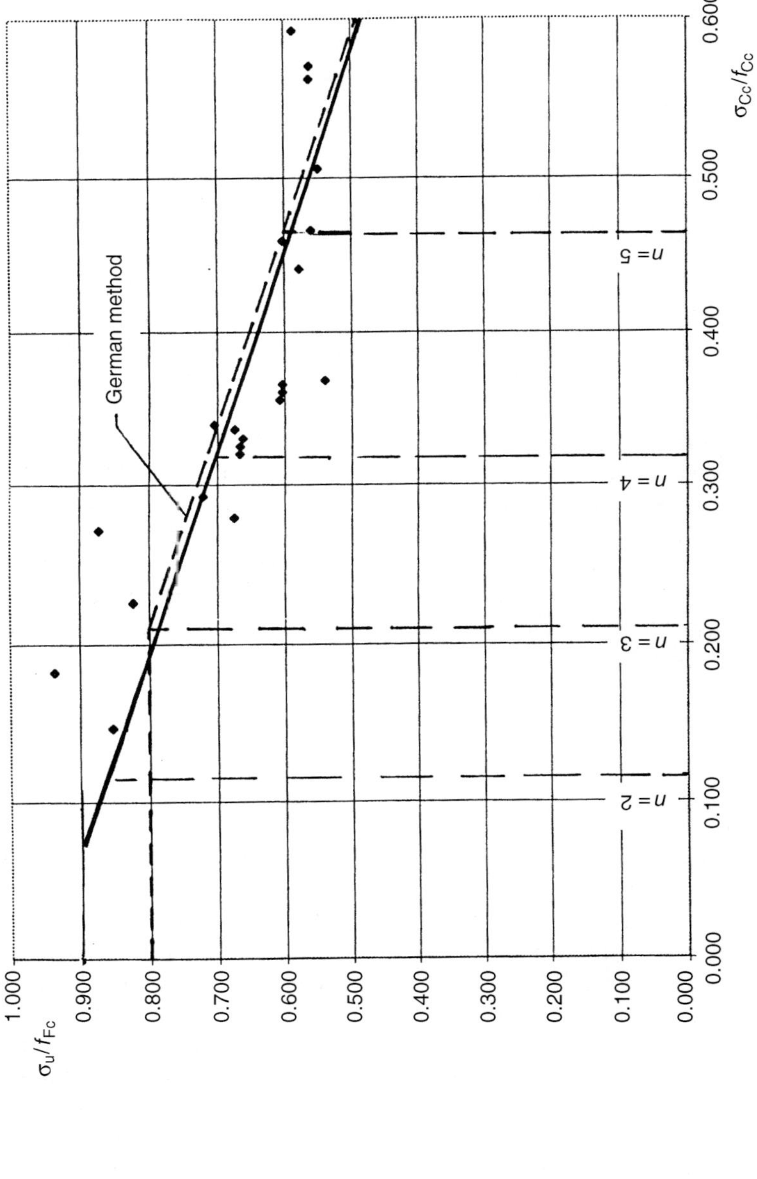

Fig. 10.15 Interaction of bending moment and negative support reaction at an intermediate support.[10.1] Test results on polyurethane cored sandwich panels showing the influence of screw fasteners on the compressive strength of the face ◆; and German design approach - - - -.

10.2.4 Shear failure of the core

The theoretical expression for the maximum shear stress in the cross-section of a symmetric thin-faced three-layer sandwich beam is:

$$\tau_C = \frac{V(\sum E_i S_i)_{max}}{(\sum E_i I_i)B} = \frac{V\left(\frac{E_F te}{2} + \frac{E_C d_C^2}{8}\right)}{\left(\frac{E_F te^2}{2} + \frac{E_C d_C^3}{12}\right)B} \approx \frac{V\left(1 + \frac{E_C e}{4E_F t}\right)}{\left(1 + \frac{E_C e}{6E_F t}\right)eB} \tag{10.37}$$

in which a core depth $d_C \approx e$ has been assumed. V is the shear force in the cross-section and S_i and I_i are the first and second moments of inertia of part i of the cross-section respectively.

As shown in Fig. 10.16, for typical sandwich panels faced with steel or aluminium sheets, because the term $(E_C e/4E_F t)$ is less than 0.03, Equation 10.37 can be simplified to:

$$\tau_C \approx \frac{V}{eB} \tag{10.38}$$

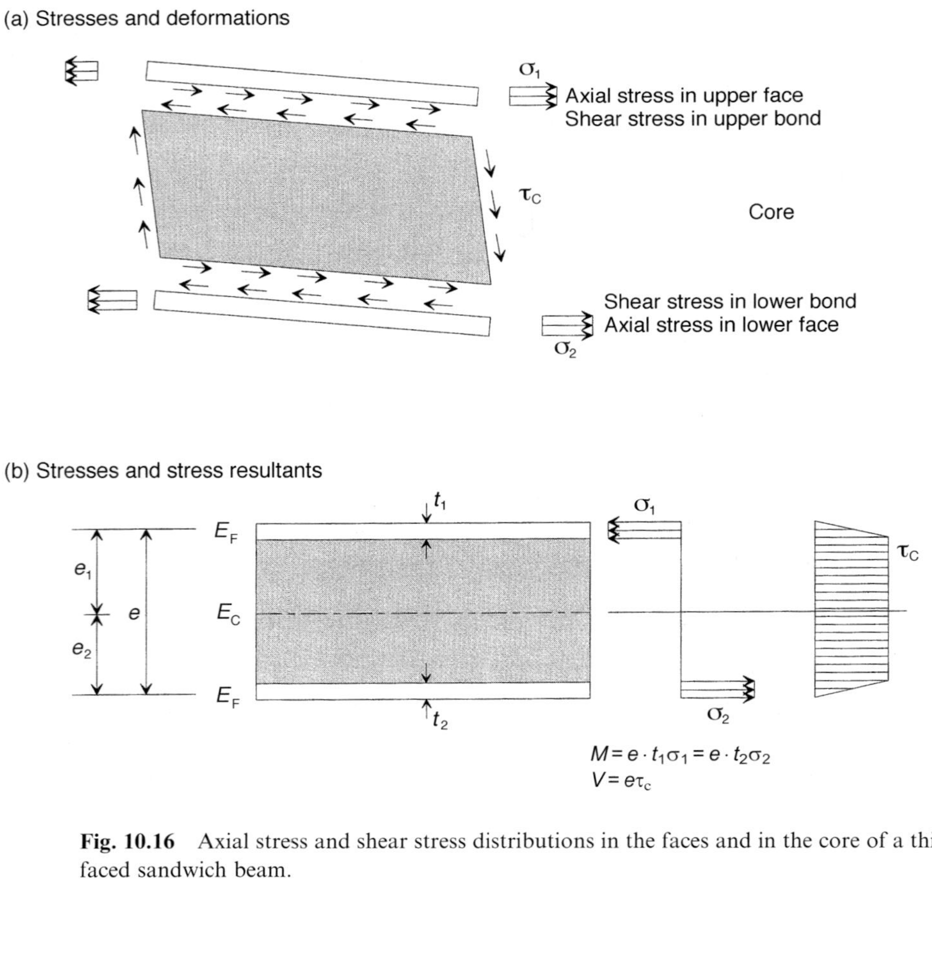

Fig. 10.16 Axial stress and shear stress distributions in the faces and in the core of a thin-faced sandwich beam.

It follows that the shear resistance of a thin-faced sandwich beam can be evaluated using:

$$V_{CR} = eBf_{Cv} \tag{10.39}$$

The critical section of the core with regard to shear strength is either the mid-depth of the core, where the density of some foams has its minimum value, or in the bond between the core and the faces. The stress–strain curves of typical core materials usually have an approximately linear and a non-linear part. A core may fail like a brittle material or it may show plastic shear resistance by reaching a maximum value where $f_{Cv} = \tau_C$ $(\partial\tau_C/\partial\gamma = 0)$. In design calculations, the characteristic shear strength values based on the highest shear stresses measured in experiments are used in the expression (10.39). If the core material shows a very non-linear behaviour in a test, or if the ultimate shear deformation is very high, additional criteria may be defined in order to establish the shear strength. A criterion of ten percentage deformation $f_{Cv} = \tau_C$ $(\gamma = 10\%)$ may be used, for instance. Methods to determine the shear strength experimentally are given in Chapter 13.

If the faces of a sandwich beam are flat and thick, the faces themselves will carry a part of the shear force V. However, in this case also, the shear resistance of the panel can be evaluated conservatively using Equation 10.39 because the core will carry the major part of the shear force and because the shear strength of the core or of the bond between the core and the faces always determines the shear resistance of the panel. For practical purposes, the non-linear stress–strain behaviour of the core materials at the ultimate limit state calls into question the precision of the exact analysis for the shear stresses so that a reasonable approximation is sufficient.

If the faces are profiled, the shear stress distribution is no longer constant through the width of the panel (in the y-direction) but the shear stresses tend to concentrate in the regions of the core where the distances between the upper and lower face profiles are at their smallest. In the cross-section of a sandwich panel with profiled faces, numerical methods are most effective tools with which to study the shear stress distributions in the depth (z-direction) and width (y-direction). The analysis becomes more complicated if the variations of the density and the stiffness of the foam core material in the depth and width directions of the panel are taken into account. When the highest shear stress in a cross-section has been determined, the shear resistance can be evaluated by comparing the highest shear stress with the shear strength f_{Cv}:

$$\tau_{max}(V) = f_{Cv} \Rightarrow V_{CR} = V(\tau_{max}, f_{Cv}) \tag{10.40}$$

In practice, the shear resistance of a panel with profiled faces is often evaluated using Expression 10.39, where e is the distance between the centroids of the faces. Because of the complicated nature of the problem, if this gives rise to a critical design limitation, it is recommended that the validity of Expression 10.39 for profiled faced sandwich panels should be verified experimentally.

10.2.5 Shear failure of a profiled face

Profiled faces carry a part of the total shear force in the panel because of their finite bending stiffness. Shear stresses in the webs of the face profiles are usually assumed to be uniformly distributed over the depth of the web.

$$\tau_{Fi} = \frac{V_{Fi}}{n s_w t} \qquad i = 1, 2 \tag{10.41}$$

where n is the number of webs per unit width, s_w is the length of the web measured along the slope and $i = 1, 2$ denotes the upper or lower face.

The shear strength of a thin face profile is determined by either the yield stress or by the buckling strength of the web of the profile. If the lateral support provided by the core against local shear buckling is neglected, in the absence of any information on this subject, the shear resistance of a profiled face with unstiffened webs can be evaluated on the basis of the design expressions given for trapezoidal metal sheeting in ENV 1993-1-3.[10.6]

$$V_{FR} = n f_{Fv} s_w t \tag{10.42}$$

where

$$f_{Fv} = \frac{f_y}{\sqrt{3}} \quad \text{if } \bar{\lambda}_w \leq 0.83 \tag{10.43a}$$

$$f_{Fv} = \frac{0.48 f_y}{\bar{\lambda}_w} \quad \text{if } 0.83 < \bar{\lambda}_w \leq 1.40 \tag{10.43b}$$

$$f_{Fv} = \frac{0.67 f_y}{\bar{\lambda}_w^2} \quad \text{if } 1.40 < \bar{\lambda}_w \tag{10.43c}$$

and

$$\bar{\lambda}_w = \sqrt{\frac{f_{yv}}{\tau_{cr}}} = \frac{1}{\pi}\sqrt{\frac{12(1-\nu^2)}{k_\tau \sqrt{3}}}\frac{s_w}{t}\sqrt{\frac{f_y}{E_F}} = 0.346\frac{s_w}{t}\sqrt{\frac{f_y}{E_F}} \tag{10.44}$$

10.2.6 Connections

Sandwich panels are usually fastened to the building frame with long screws going through the panel or with special connectors placed in the longitudinal joints between the panels (see Chapter 12). Fasteners between the panel and the supporting structure are loaded by tensile and shear forces. Tensile forces are caused by wind suction and temperature loads and shear forces by temperature expansion of the faces, by diaphragm (stressed skin) action and sometimes also by the self-weight of the panel. The characteristic strength of connections loaded in tension and shear by either static or repeated loads may be determined by testing according to the procedures given in Chapter 13.

The failure modes and the flexibility of screw connections in sandwich panels differ from those in trapezoidal sheeting as illustrated in Fig. 10.17. In tension, the screws which pass right through the sandwich panel fix the external face to the supporting structure. The presence of the elastic core between the head of the screw

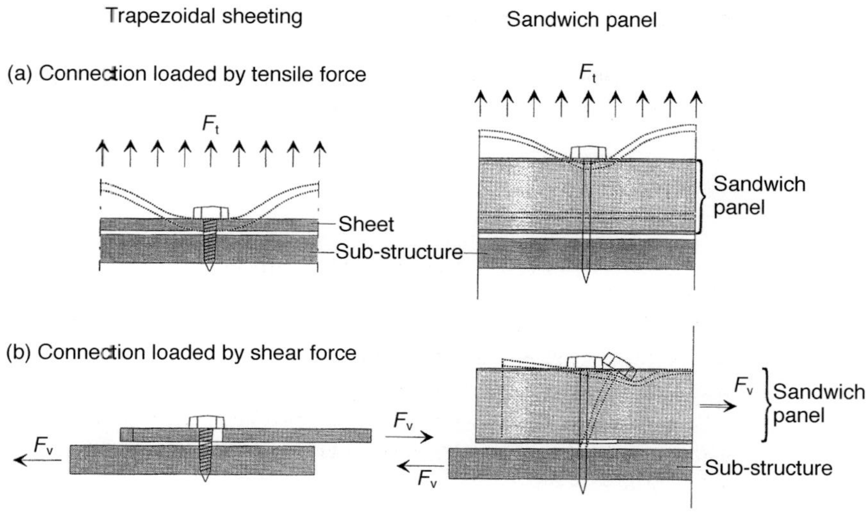

Fig. 10.17 Local deformations of screw connections in trapezoidal sheeting and sandwich panels when loaded by (a) tensile and (b) shear forces.

and the supporting structure increases the flexibility and changes the failure mode. Conversely, in shear, the through fixing screws effectively only transfer the shear force between the inner face of the panel and the supporting structure.

There are many kinds of special connection systems which have been developed for sandwich panels. These special connectors are placed in the longitudinal joints between the panels in order to conceal the connections in walls and in order to guarantee the water-tightness of the connections in roofs. Typical special connection systems usually comprise a steel sheet strip, which is fixed to the supporting structure with screws and which is locked into the outer face of the sandwich panel with cold-formed folds. The steel sheet strip increases the tensile and possibly also the shear flexibility of the connection. The tension and shear strength of the special connection systems have to be defined experimentally.

The following design expressions, which were derived for screw connections between two metal sheets, provide very approximate expressions for the common screw connections used in the erection of sandwich panels. These expressions should only be used in the preliminary phases of a design.

The shear resistance of a screw connection according to ENV 1993-1-3[10.6] is:

$$F_{\mathrm{Rvd}} = \min\begin{Bmatrix} \alpha f_u\, d_n\, t/\gamma_M \\ A_n f_u/\gamma_M \end{Bmatrix} < \frac{F_{v,\,\mathrm{Rd}}}{1.2} \qquad (10.45, 10.46)$$

where

$$\alpha = 3.2\sqrt{\frac{t}{d_n}} \leq 2.1 \text{ for } t = t_1 \qquad (10.47a)$$

$$\alpha = 2.1 \text{ for } t_1 \geq 2.5t \qquad (10.47b)$$

where d_n is the nominal diameter of the fastener

t is the thickness of the face which is against the supporting structure

A_n is the net cross-sectioned area of the connected part
f_u is the tensile strength of the sheet material
t_1 is the thickness of the thickest sheet in the connection
$F_{v,Rd}$ is the shear resistance of the shaft of the fastener itself

The shear resistance for thickness relationships $1 \leq t_1/t \leq 2.5$ should be determined by linear interpolation between the above equations.

10.2.7 Deflections

The deflections of sandwich beams are due mainly to the axial and bending deformations of the faces and to the shear deformations in the core. Local compressive deformations of the core and deformations in the connections at the supports may further increase the deflections. However, deflections caused by the local deformations at the supports are small and are usually ignored in the design calculations (see Section 12.4).

The total deflection of a sandwich beam with *thin plane faces* can be expressed as the sum of the partial deflections caused by the bending deformations and shear deformations. Table 10.2 summarises the expressions for the end and intermediate support reactions, for the bending moments in end spans and at the intermediate supports and for the largest deflections of one-, two- and three-span thin-faced sandwich beams with equal spans. The exact location of the largest deflection in a multi-span sandwich beam depends on the relative magnitudes of the bending and shear deformations and it is located between $x_{max} = 3/8L$ and $1/2L$ from the end support. The exact expressions for deflection are complex, so approximate expressions are given in Table 10.2.

The static behaviour of *thick-faced* sandwich beams, i.e. sandwich beams with profiled faces, is influenced also by the individual bending stiffnesses of the outer and inner face layers. The equations for the deflection of thick-faced sandwich beams are even more complicated than those for thin-faced beams and they can no longer be divided into two separate parts containing the bending or shear deformation terms, as is the case with thin-faced sandwich beams. Analytical expressions and graphs for the evaluation of the deflections of thick-faced sandwich beams are given in Chapter 9.

Among the typical core materials, the plastic foams are viscoelastic materials whose deformations increase with the time of loading. It follows that the data on the elastic material properties alone are not sufficient and the viscoelastic nature of the core has also to be modelled in the design calculations. The shear creep under low shear stress levels can be described by linear creep models. A common creep model in structural design calculations is the Findlay model:

$$J(t) = \frac{\gamma(t)}{\tau_0} = c_0(1 + c_1 t^{c_2}) = \frac{1}{G_{0.1}}(1 + \phi(t)) \tag{10.48}$$

In the above expression, c_0, c_1 and c_2 are experimental parameters and $G_{0.1}$ is the initial shear modulus corresponding to the loading time $t = 0.1$ hours.

For sandwich panels, instead of using the theory of viscoelasticity, the calculations carried out in practice are based on the linear theory of elasticity and the shear creep is taken into account by reducing the initial shear modulus by a creep coefficient $\phi(t)$:

Table 10.2 Summary of solutions for one-, two- and three-span panels with flat or lightly profiled faces loaded by a uniformly distributed load q and by a temperature difference between the outer (T_1) and inner (T_2) face. ($k = 3B_s/A_C GL^2$; $\theta = \alpha_T (T_2 - T_1)/e$)

	End support reaction, R_0	Intermediate support reaction, R	Bending moment in end span, M	Bending moment at internal support, M	Maximum deflection in span, w_{max}
Single span of L Uniform load, q	$\frac{qL}{2}$		$\frac{qL^2}{8}$		$\frac{5qL^4}{384B_S}(1+3.2k)$
Temperature difference, $(T_2 - T_1)$					$\frac{\theta L^2}{8}$
Two spans of L Uniform load, q	$\frac{qL}{2}\left[1-\frac{1}{4(1+k)}\right]$	$qL\left[\frac{1}{(1+k)}\right]$	$\frac{qL^2}{8}\left[1-\frac{1}{4(1+k)}\right]$	$-\frac{qL^2}{8}\left[\frac{1}{(1+k)}\right]$	$\frac{qL^4}{48B_S}\frac{(0.26+2.625k+2k^2)}{(1+k)}$
Temperature difference, $(T_2 - T_1)$	$-\frac{3B_S\theta}{2L}\frac{1}{(1+k)}$	$\frac{3B_S\theta}{L}\frac{1}{(1+k)}$	$-\frac{3B_S\theta}{4}\frac{1}{(1+k)}$	$-\frac{3B_S\theta}{2}\frac{1}{(1+k)}$	$\frac{\theta L^2}{32}\frac{(1.089+3.96k)}{(1+k)}$
Three spans of L Uniform load, q	$\frac{qL}{2}\left[1-\frac{1}{(5+2k)}\right]$	$qL\left[1+\frac{1}{2(5+2k)}\right]$	$\frac{qL^2}{8}\left[1-\frac{1}{(5+2k)}\right]$	$-\frac{qL^2}{(10+4k)}$	$\frac{qL^4}{24B_S}\frac{(0.826+5.625k+2k^2)}{(5+2k)}$
Temperature difference, $(T_2 - T_1)$	$-\frac{6B_S\theta}{L}\frac{1}{(5+2k)}$	$\frac{6B_S\theta}{L}\frac{1}{(5+2k)}$	$-3B_S\theta\frac{1}{(5+2k)}$	$-6B_S\theta\frac{1}{(5+2k)}$	$\frac{\theta L^2}{4}\frac{(1.056+k)}{(5+2k)}$

$$G(t) = \frac{G_{0.1}}{1 + \phi(t)} \tag{10.49}$$

Furthermore, the initial shear modulus $G_{0.1}$ may be replaced by the shear modulus G_0 determined in the short-term beam tests. The deflections of simply supported single-span thin-faced sandwich beams can now be evaluated from

$$w(t) = w_B + w_{S0}(1 + \phi(t)) \tag{10.50}$$

where w_B is the deflection caused by axial deformations in the faces and w_{S0} is the initial deflection caused by shear deformation of the core.

The shear creep of the core not only increases the deflections of a sandwich beam, it also changes the bending moment and shear force distributions of statically indeterminate multi-span beams. For this reason, the static calculations have to be repeated with either two or three different values of the shear modulus corresponding to the significant time-related loading cases. In practice, the stress resultants and deflections corresponding to the following loading cases are considered[10.4]:

- short-term loads such as wind loads and daily temperature changes: $G(t = 0)$
- snow load, in many parts of Europe: $G(t = 2000$ hours) and
- self-weight and other permanent loads: $G(t) = 100\,000$ hours

Evidently, the second of these may vary depending on the likelihood of snow lying for a significant length of time. $G(t = 2000$ hours) is used in Germany and Northern Europe. In Southern Europe (excluding the mountainous regions) and in the maritime climates of England and Wales that are influenced by the Gulf Stream, $G(t = 0)$ may be adequate for snow load calculations.

The initial short-term shear modulus and the shear creep coefficient of typical core materials are also influenced by the temperature. At high temperature the creep coefficient increases and the shear modulus decreases, an effect which should be taken into account if the panels are loaded by long-term loads at a higher temperature than the common test temperature of +20°C.

10.3 DERIVATION OF THE THEORETICAL EXPRESSIONS FOR BUCKLING STRESS

The total potential energy of an uniaxially compressed rectangular plate (a × 6) which is supported by an elastic foundation and undergoing buckling may be expressed as:

$$U = U_B + U_C - V \tag{10.51}$$

where the strain energy of the bending in the plate is:

$$U_B = \frac{D}{2} \int_0^a \int_0^b (\Delta w)^2 \, dx \, dy \tag{10.52}$$

the strain energy of a foundation which is assumed to consist of an elastic half space is:

$$U_C = \frac{1}{2} \int_0^a \int_0^b \int_0^\infty (\sigma_x \varepsilon_x + \sigma_y \varepsilon_y + \sigma_z \varepsilon_z + \tau_{xy} \gamma_{xy} + \tau_{yx} \gamma_{yz} + \tau_{zx} \gamma_{zx}) \, dx \, dy \, dz \qquad (10.53)$$

or, alternatively, of a foundation with negligible horizontal displacements and stresses in the xy-plane:

$$U_C = \frac{1}{2} \int_0^a \int_0^b \int_0^\infty (\sigma_z \varepsilon_z + \tau_{yz} \gamma_{yz} + \tau_{zx} \gamma_{zx}) \, dx \, dy \, dz \qquad (10.54)$$

and the work done by the axial force $N_x(y)$ is:

$$V = \int_0^b N_x(y) \, u(x = a, y) \, dy = \frac{1}{2} \int_0^a \int_0^b N_x(y) \left(\frac{\partial w}{\partial x} \right)^2 dx \, dy \qquad (10.55)$$

When evaluating the linear buckling stresses of a long plate which is continuously supported by an elastic half space and simply supported along its longitudinal edges (x, $y = 0$, a) the following expressions can be assumed for the displacements in the three directions (see Fig. 10.5):

$$u_n(x, y, z) = \sum h_{1n}(z) \cos \alpha_1 x \sin \alpha_{2n} y \qquad (10.56a)$$

$$v_n(x, y, z) = \sum h_{2n}(z) \sin \alpha_1 x \cos \alpha_{2n} y \qquad (10.56a)$$

$$w_n(x, y, z) = \sum h_{3n}(z) \sin \alpha_1 x \sin \alpha_{2n} y \qquad (10.56c)$$

where

$$\alpha_1 = \frac{\pi}{a}; \qquad \alpha_{2n} = \frac{n\pi}{b} \qquad (10.57a, b)$$

The total potential energy can now be written:

$$U = \frac{\pi^4 abD}{8} \sum h_{3n,0}^2 \left(\frac{1}{a^2} + \frac{n^2}{b^2} \right)^2 + U_C - \frac{\pi^2 N_{x0}}{8} \left(\frac{b}{a} \right) \sum h_{3n,0}^2 \qquad (10.58)$$

where $h_{3n,0} = h_{3n}(z = 0)$.

The functions $h_i(z)$ are found on the basis of Navier's equations for a homogeneous isotropic material. Complicated mathematical calculations produce the following expression for the strain energy of a thick core layer with vanishing displacements at the bottom of the core ($h_{1n}(z = \infty) = h_{2n}(z = \infty) = h_{3n}(z = \infty) = 0$) and vanishing horizontal displacements at the top of the foundation ($h_{1n}(z = 0) = h_{2n}(z = 0) = 0$). A more detailed derivation can be found in Reference 10.24:

$$U_C = \frac{\pi ab(1 - \nu_C)E_C}{4(1 + \nu_C)(3 - 4\nu_C)} \sum h_{3n,0}^2 \left(\frac{1}{a^2} + \frac{n^2}{b^2} \right)^{\frac{1}{2}} \qquad (10.59)$$

The condition that $C = \sigma_z(z = 0)/w(z = 0)$ results in an expression for the foundation coefficient which is dependent on the material properties of the core, the width of the plate and the buckled shape:

$$C_1 = \frac{2(1-\nu_C)\pi E_C}{(1+\nu_C)(3-4\nu_C)}\sqrt{\frac{1}{a^2}+\frac{n^2}{b^2}} \tag{10.60}$$

In the second foundation model, Equation 10.54, only the most important stress components are included in the analysis. The displacement functions, $h_{1n}(z)$ and $h_{2n}(z)$ are assumed to vanish and the function $h_{3n}(z)$ to have the expression:

$$h_{3n}(z) = h_n e^{-kz} \tag{10.61}$$

where k is a decay factor which is found from the condition $\partial U_C/\partial k = 0$. Assuming the core material to be isotropic ($G_{xz} = G_{yz} = G_C$), the analysis results in the following expressions for the strain energy of the core, the decay factor and the foundation coefficient:

$$U_C = \frac{\pi ab\sqrt{E_C G_C}}{8}\sum\left(\frac{1}{a^2}+\frac{n^2}{b^2}\right)^{\frac{1}{2}} \tag{10.62}$$

$$k = \pi\sqrt{\frac{G_C}{E_C}}\sqrt{\frac{1}{a^2}+\frac{n^2}{b^2}} \tag{10.63}$$

and

$$C_2 = \pi\sqrt{E_C G_C}\sqrt{\frac{1}{a^2}+\frac{n^2}{b^2}} \tag{10.64}$$

Minimising the total potential energy expression, 10.58, with respect to $h_{3n,0}$ gives the following equation for the buckling stress:

$$\sigma_{cri} = K_i \frac{\pi^2 E_F}{12(1-\nu_F^2)}\left(\frac{t}{b}\right)^2 \quad i = 1, 2 \tag{10.65}$$

where

$$K_i = \left(\frac{1}{\phi}+n^2\phi\right)^2 + R_i\phi(1+n^2\phi^2)^{\frac{1}{2}} \tag{10.66}$$

$$\phi = a/b \tag{10.67}$$

and

$$R_1 = \frac{E_C}{E_F}\frac{2(1-\nu_C)}{(1+\nu_C)(3-4\nu_C)}\frac{12(1-\nu_F^2)}{\pi^3}\left(\frac{b}{t}\right)^3 \tag{10.68}$$

and

$$R_2 = \sqrt{\frac{E_C G_C}{E_F}}\frac{12(1-\nu_F^2)}{\pi^3}\left(\frac{b}{t}\right)^3 \tag{10.69}$$

The critical buckling stress can be found by minimising the buckling coefficient K_i with respect to the wavelength ϕ. The condition $\partial K_i/\partial\phi = 0$ gives:

$$2n^4\phi - \frac{2}{\phi^3} + R_i(2n^2\phi^2+1)(n^2\phi^2+1)^{-\frac{1}{2}} = 0 \tag{10.70}$$

If the numerical values of the parameters R_1 or R_2 are known, the critical buckling stress σ_{cr} is found by solving Equation 10.70 for ϕ using a suitable numerical method and, hence, by evaluating the buckling coefficient K in Equation 10.66. The first eigenvalue ($n = 1$) is found to be critical in the cases $0 < R_2 < 200$.

In the above expression, the buckling coefficient depends on the width-to-thickness ratio of the plane element and the moduli of elasticity and Poisson's ratios of the face and core. Several approximate formulae have been found for the critical buckling coefficient K as shown in Table 10.3. The first two expressions in the table are exactly valid for steel plates ($\nu_F = 0.3$) and for foundations with a Poisson's ratio of $\nu_C = 0.25$. The third expression in the table is also a best-fit curve for the calculated buckling stress values but based on a different formulation from the second expression. The fourth expression is based on experimental results found for small-scale polyurethane foam sandwich panels and this is recommended for use in practical design calculations.

Table 10.3 Approximate buckling coefficients for uniaxially compressed long plates simply supported along their longitudinal edges and continuously supported by the core

Foundation	Buckling coefficient	Parameter
Elastic half space, E_C, ν_C	$K = 0.703R^2 - 0.415R + 4.00$	$R = \frac{b}{t}\left(\frac{E_C}{E_F}\right)^{\frac{1}{3}}$
Simplified elastic half space, E_C, G_C	$K = 0.985R^2 - 0.474R + 4.00$	$R = \frac{b}{t}\left(\frac{E_C G_C}{E_F{}^2}\right)^{\frac{1}{6}}$
Simplified elastic half space, E_C, G_C	$K = (16 + 11.8R + 0.055R^2)^{\frac{1}{2}}$	$R = \frac{12(1-\nu_F{}^2)(E_C G_C)^{\frac{1}{2}}}{\pi^3 E_F}\left(\frac{b}{t}\right)^3$
Design elastic half space, E_C G_C	$K = (16 + 7R + 0.002R^2)^{\frac{1}{2}}$	$R = 0.35\left(\frac{b}{t}\right)^3 \frac{(E_C G_C)^{\frac{1}{2}}}{E_F}$

If, in the analysis, the width of the plate b increases to infinity, the important special case of a wide flat face is obtained. On the basis of Equation 10.65 and $b \to \infty$, the buckling stress of a flat face is:

$$\sigma_{cri} = \frac{\pi^2 D}{a^2 t} + R'_i \frac{a}{\pi t} \tag{10.71}$$

where

$$R'_1 = \frac{2(1-\nu_C)E_C}{(1+\nu_C)(3-4\nu_C)} \tag{10.72}$$

$$R'_2 = \sqrt{E_C G_C} \tag{10.73}$$

Minimising the critical stress with respect to the wavelength a gives expressions for the critical half wavelength and the buckling stress:

$$a_{cri} = \pi\left(\frac{2D}{R'_i}\right)^{\frac{1}{3}} = \pi t\left(\frac{E_F}{6(1-\nu_F^2)R'_i}\right)^{\frac{1}{3}} \tag{10.74}$$

$$\sigma_{cri} = \frac{3}{2}\frac{(2R'_i D)^{\frac{1}{3}}}{t} = \frac{3}{2}\left(\frac{R'_i E_F}{6(1-\nu_F^2)}\right)^{\frac{1}{3}} \tag{10.75}$$

10.4 DERIVATION OF EXPRESSIONS FOR THE INTERACTION AT INTERMEDIATE SUPPORTS

The governing differential equation for a beam-column which is supported by a Winkler foundation can be written as:

$$B_{F2}\frac{d^2w}{dx^4} + N_s\frac{d^2w}{dx^2} + k_w w = q(x) \tag{10.76}$$

In this equation, B_{F2} and $N_s = M_s/e$ are the flexural ridigity and the axial compressive force of the beam per unit width and k_w is the foundation parameter, $k_w = E_{Cc}/d_C$. Explicit solutions for a great many beam on elastic foundation problems have been given by Hetenyi.[10.16] A general (exact) finite-element solution has been given by Davies.[10.7] For the case:

$$N_s < 2\sqrt{k_w B_{F2}}$$

the general solution of the differential Equation 10.76 is:

$$\begin{aligned} w(x) &= (C_1 e^{\beta_0 x} + C_2 e^{-\beta_0 x}) \cos \alpha_0 x \\ &+(C_3 e^{\beta_0 x} + C_4 e^{-\beta_0 x}) \sin \alpha_0 x + w_0(x) \end{aligned} \tag{10.77}$$

where

$$\alpha_0 = \sqrt{\sqrt{\frac{k_w}{4B_{F2}}} + \frac{N_s}{4B_{F2}}} \tag{10.78a}$$

and

$$\beta_0 = \sqrt{\sqrt{\frac{k_w}{4B_{F2}}} - \frac{N_s}{4B_{F2}}} \tag{10.78b}$$

and $w_0(x)$ is the particular integral of the equation.

For a two-parameter foundation (k_w, k_1; see Equation 10.27a), the solutions of the equations are the same except that the term N_s is replaced by $(N_s - k_1)$.

In the first loading case, the beam is loaded by a point load F at $x = 0$ as shown in Fig. 10.18(a). Because the local deflection $w(x)$ and the local bending moment $M_{F2}(x)$ have to vanish at some distance from the origin, the constants C_1 and C_3 have to equal zero. The constants C_2 and C_4 can be determined from the condition of equilibrium and from the symmetry of the deflected shape at $x = 0$:

$$-B_{F2}\frac{d^3w}{dx^3} = -\frac{F}{2} \quad \text{and} \quad \frac{dw}{dx} = 0$$

The solution for the local deflection $w(x)$ and the local bending moment $M_{F2}(x)$ of the lower face and for the local compressive stress $\sigma_{Cc}(x)$ in the core for $x \geq 0$ can be written as:

$$w(x) = \frac{F}{4\sqrt{k_w B_{F2}}}\frac{e^{-\beta_0 x}}{\alpha_0 \beta_0}(\alpha_0 \cos \alpha_0 x + \beta_0 \sin \alpha_0 x) \tag{10.79a}$$

$$M_{F2}(x) = \frac{F}{4}\frac{e^{-\beta_0 x}}{\alpha_0\,\beta_0}(\alpha_0 \cos \alpha_0 x - \beta_0 \sin \alpha_0 x) \tag{10.79b}$$

$$\sigma_{Cc}(x) = k_w w(x) = \frac{F}{4}\sqrt{\frac{k_w}{B_{F2}}}\frac{e^{-\beta_0 x}}{\alpha_0\,\beta_0}(\alpha_0 \cos \alpha_0 x + \beta_0 \sin \alpha_0 x) \tag{10.79c}$$

If $N_s = 0$, Expressions 10.79a, b and c can be simplified to $(x \geq 0)$

$$w(x) = \frac{F}{4\sqrt{k_w\,B_{F2}}}\frac{e^{-\alpha x}}{\alpha}(\cos \alpha x + \sin \alpha x) \tag{10.80a}$$

$$M_{F2}(x) = \frac{F}{4}\frac{e^{-\alpha x}}{\alpha}(\cos \alpha x - \sin \alpha x) \tag{10.80b}$$

$$\sigma_{Cc}(x) = k_w w(x) = \frac{F}{4}\sqrt{\frac{k_w}{B_{F2}}}\frac{e^{-\alpha x}}{\alpha}(\cos \alpha x + \sin \alpha x) \tag{10.80c}$$

where

$$\alpha = 4\sqrt{\frac{k_w}{4\,B_{F2}}} \tag{10.81}$$

In the second loading case, the beam is loaded by two point loads $F/2$ at $x = -L_s/2$ and $x = +L_s/2$ (see Fig. 10.18(b)). The solutions for the local deflection and local bending moment of the lower face and for the local compressive stress in the core can be derived from the equations for the first loading case:

$$w(x) = \begin{cases} \dfrac{F}{8\sqrt{k_w\,B_{F2}}}\dfrac{e^{-\beta_0 L_s/2}}{\alpha_0\beta_0}[\alpha_0 f_1(x) + \beta_0 f_2(x)],\ 0 \leq x \leq L_s/2 & (10.82a)\\[2ex] \dfrac{F}{8\sqrt{k_w\,B_{F2}}}\dfrac{e^{-\beta_0 x}}{\alpha_0\,\beta_0}[\alpha_0 f_3(x) + \beta_0 f_4(x)],\ L_s/2 \leq x & (10.82b)\end{cases}$$

$$M_{F2}(x) = \begin{cases} \dfrac{F}{8}\dfrac{e^{-\beta_0 L_s/2}}{\alpha_0\,\beta_0}[\alpha_0 f_1(x) - \beta_0 f_2(x)],\ 0 \leq x \leq L_s/2 & (10.82c)\\[2ex] \dfrac{F}{8}\dfrac{e^{-\beta_0 x}}{\alpha_0\,\beta_0}[\alpha_0 f_3(x) - \beta_0 f_4(x)],\ L_s/2 \leq x & (10.82d)\end{cases}$$

$$\sigma_{Cc}(x) = \begin{cases} \dfrac{F}{8}\sqrt{\dfrac{k_w}{B_{F2}}}\dfrac{e^{-\beta_0 L_s/2}}{\alpha_0\,\beta_0}[\alpha_0 f_1(x) + \beta_0 f_2(x)], 0 \leq x \leq L_s/2 & (10.82e)\\[2ex] \dfrac{F}{8}\sqrt{\dfrac{k_w}{B_{F2}}}\dfrac{e^{-\beta_0 x}}{\alpha_0\,\beta_0}[\alpha_0 f_3(x) + \beta_0 f_4(x)],\ L_s/2 \leq x & (10.82f)\end{cases}$$

where

$$f_1(x) = e^{-\beta_0 x} \cos \alpha_0(x + L_s/2) + e^{\beta_0 x} \cos \alpha_0(x - L_s/2) \tag{10.83a}$$

$$f_2(x) = e^{-\beta_0 x} \sin \alpha_0(x + L_s/2) - e^{\beta_0 x} \sin \alpha_0(x - L_s/2) \tag{10.83b}$$

$$f_3(x) = e^{-\beta_0 L_s/2} \cos \alpha_0(x + L_s/2) + e^{\beta_0 L_s/2} \cos \alpha_0(x - L_s/2) \tag{10.83c}$$

$$f_4(x) = \mathrm{e}^{-\beta_0 L_s/2} \sin \alpha_0(x + L_s/2) + \mathrm{e}^{\beta_0 L_s/2} \sin \alpha_0(x - L_s/2) \tag{10.83d}$$

If $N_s = 0$, Expressions 10.82a–f can be simplified to:

$$w(x) = \begin{cases} \dfrac{F}{8\sqrt{k_w B_{F2}}} \dfrac{\mathrm{e}^{-\alpha L_s/2}}{\alpha} [f_1(x) + f_2(x)], \ 0 \le x \le L_s/2 & (10.84a) \\ \dfrac{F}{8\sqrt{k_w B_{F2}}} \dfrac{\mathrm{e}^{-\alpha x}}{\alpha} [f_3(x) + f_4(x)], \ L_s/2 \le x & (10.84b) \end{cases}$$

$$M_{F2}(x) = \begin{cases} \dfrac{F}{8} \dfrac{\mathrm{e}^{-\alpha L_s/2}}{\alpha} [f_1(x) - f_2(x)], \ 0 \le x \le L_s/2 & (10.84c) \\ \dfrac{F}{8} \dfrac{\mathrm{e}^{-\alpha x}}{\alpha} [f_3(x) - f_4(x)], \ L_s/2 \le x & (10.84d) \end{cases}$$

$$\sigma_{Cc}(x) = \begin{cases} \dfrac{F}{8} \sqrt{\dfrac{k_w}{B_{F2}}} \dfrac{\mathrm{e}^{-\alpha L_s/2}}{\alpha} [f_1(x) + f_2(x)], \ 0 \le x \le L_s/2 & (10.84e) \\ \dfrac{F}{8} \sqrt{\dfrac{k_w}{B_{F2}}} \dfrac{\mathrm{e}^{-\beta_0 x}}{\alpha} [f_3(x) + f_4(x)], \ L_s/2 \le x & (10.84f) \end{cases}$$

In the third loading case, Fig. 10.18(c), the beam is loaded by a uniformly distributed support pressure $q_s = F/L_s$ at $-L_s/2 \le x \le +L_s/2$. Solutions for the local deflection and local bending moment of the lower face and for the local compressive stress in the core can be integrated from the equations for the first loading case:

$$w(x) = \begin{cases} \dfrac{F}{2L_s k_w} \left[2 - \mathrm{e}^{-\beta_0 L_s/2} f_1(x) + \dfrac{N_s}{\sqrt{4k_w B_{F2} - N_s^2}} \mathrm{e}^{-\beta_0 L_s/2} f_2(x) \right], \ L_s/2 \le x \le L_s/2 & (10.85a) \\ \dfrac{F}{2L_s k_w} \mathrm{e}^{-\beta_0 x} \left[\dfrac{N_s}{\sqrt{4k_w B_{F2} - N_s^2}} f_6(x) - f_5(x) \right], \ L_s/2 \le x & (10.85b) \end{cases}$$

$$M_{F2}(x) = \begin{cases} \dfrac{F}{L_s} \dfrac{B_{F2}}{\sqrt{4k_w B_{F2} - N_s^2}} \mathrm{e}^{-\beta_0 L_s/2} f_2(x), \ -L_s/2 \le x \le L_s/2 & (10.85c) \\ \dfrac{F}{L_s} \dfrac{B_{F2}}{\sqrt{4k_w B_{F2} - N_s^2}} \mathrm{e}^{-\beta_0 x} f_6(x), \ L_s/2 \le x & (10.85d) \end{cases}$$

$$\sigma_{Cc}(x) = \begin{cases} \dfrac{F}{2L_s} \left[2 - \mathrm{e}^{-\beta_0 L_s/2} f_1(x) + \dfrac{N_s}{\sqrt{4k_w B_{F2} - N_s^2}} \mathrm{e}^{-\beta_0 L_s/2} f_2(x) \right], \ -L_s/2 \le x \le L_s/2 & (10.85e) \\ \dfrac{F}{2L_s} \mathrm{e}^{-\beta_0 x} \left[\dfrac{N_s}{\sqrt{4k_w B_{F2} - N_s^2}} f_6(x) - f_5(x) \right], \ L_s/2 \le x & (10.85f) \end{cases}$$

where

$$f_5(x) = \mathrm{e}^{-\beta_0 L_s/2} \cos \alpha_0(x + L_s/2) - \mathrm{e}^{\beta_0 L_s/2} \cos \alpha_0(x - L_s/2) \tag{10.86a}$$

$$f_6(x) = \mathrm{e}^{-\beta_0 L_s/2} \sin \alpha_0(x + L_s/2) - \mathrm{e}^{\beta_0 L_s/2} \sin \alpha_0(x - L_s/2) \tag{10.86b}$$

If $N_s = 0$, Expressions 10.85 a–f can be simplified to:

$$w(x) = \begin{cases} \dfrac{F}{2L_s k_w}\left\{2 - e^{-\alpha L_s/2}[e^{-\alpha x}\cos\alpha(x + L_s/2) + e^{\alpha x}\cos\alpha(x - L_s/2)]\right\}, \; -L_s/2 \le x \le L_s/2 & (10.87a) \\ \dfrac{F}{2L_s k_w} e^{-\alpha x}\left[-e^{-\alpha L_s/2}\cos\alpha(x + L_s/2) + e^{\alpha L_s/2}\cos\alpha(x - L_s/2)\right], \; L_s/2 \le x & (10.87b) \end{cases}$$

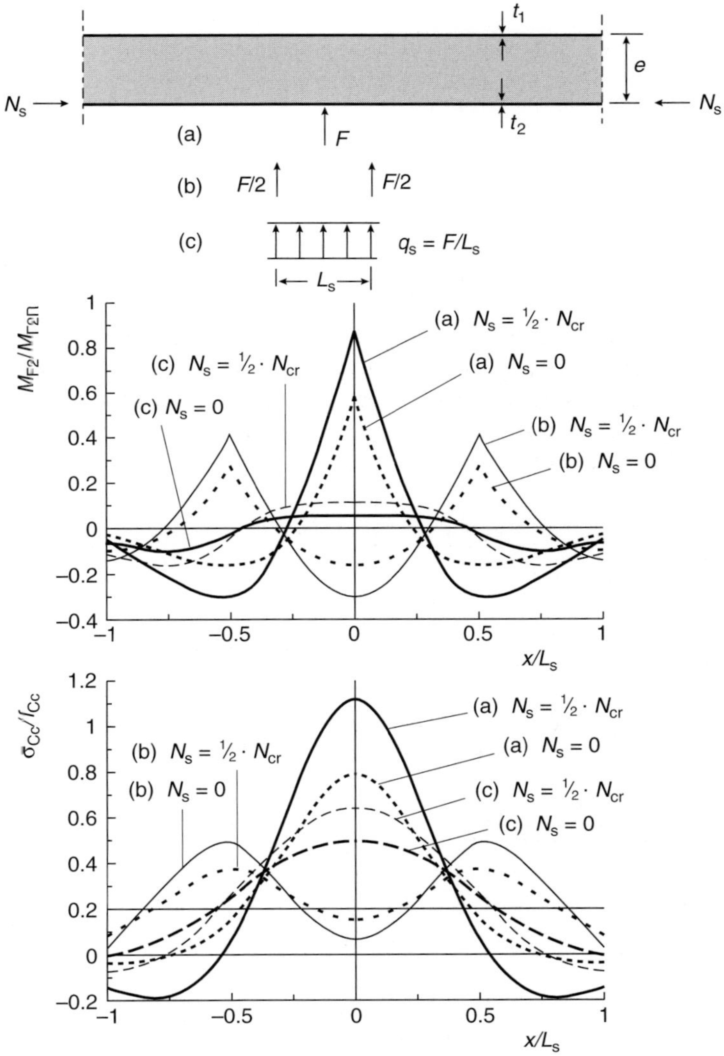

Fig. 10.18 Bending moment and foundation pressure distributions for three loading cases for an axially loaded beam supported by a Winkler foundation. (a) Point load F at the origin $x = 0$; (b) point loads $F/2$ at $x = -L_s/2$ and $x = +L_s/2$ and (c) uniformly distributed load $q_s = F/L_s$ at $-L_s/2 \le x \le +L_s/2$.

$$M_{F2}(x) = \begin{cases} \dfrac{F}{2L_s} \dfrac{B_{F2}}{\sqrt{k_w B_{F2}}} e^{-\alpha L_s/2} [e^{-\alpha x} \sin \alpha(x + L_s/2) - e^{\alpha x} \sin \alpha(x - L_s/2)], \ -L_s/2 \leq x \leq L_s/2 & (10.87c) \\ \dfrac{F}{2L_s} \dfrac{B_{F2}}{\sqrt{k_w B_{F2}}} e^{-\alpha x} \left[e^{-\alpha L_s/2} \sin \alpha(x + L_s/2) - e^{\alpha L_s/2} \sin \alpha(x - L_s/2)\right], \ L_s/2 \leq x & (10.87d) \end{cases}$$

$$\sigma_{Cc}(x) = \begin{cases} \dfrac{F}{2L_s} \left\{2 - e^{-\alpha L_s/2} [e^{-\alpha x} \cos \alpha(x + L_s/2) + e^{\alpha x} \cos \alpha(x - L_s/2)]\right\}, \ -L_s/2 \leq x \leq L_s/2 & (10.87e) \\ \dfrac{F}{2L_s} e^{-\alpha x} \left[-e^{\alpha L_s/2} \cos \alpha(x + L_s/2) + e^{\alpha L_s/2} \cos \alpha(x - L_s/2)\right], \ L_s/2 \leq x & (10.87f) \end{cases}$$

The beam-on-elastic-foundation model in Fig. 10.18 describes the behaviour of the lower face and the core of a sandwich panel with lightly profiled faces at an intermediate support. The flexural rigidity of the beam is $B_{F2} = 2.432 \times 10^5$ N mm^2/mm, the foundation coefficient $k_w = 0.06$ N/mm^3, the support width $L_s = 200$ mm. The elastic bending resistance of the face is $M_{F2R} = 247$ N mm/mm and compressive strength of the core is $f_{Cc} = 0.10$ N/mm^2.

10.5 CONCLUSIONS

In this chapter, methods were presented with which to evaluate the resistance of sandwich panels with respect to different failure modes. The calculation models given in the chapter are partially based on theoretical analyses and theoretical models but also on wide experience of the known behaviour of sandwich panels products with either a plastic foam or stone wool core and metal sheet faces when used as simply supported and continuous multi-span beam structures.

At the serviceability limit state, the first design criterion is the deflection w. Deflections are evaluated by calculations that are based on the data for the short-term and long-term shear modulus of the core G_0, $G(t)$. The second important criterion is the initial failure mode at the intermediate supports of multi-span continuous sandwich panels. The initial failure may be a core shear failure or a crushing failure in the core or a bending and buckling failure of the compressed face. Evaluation of the resistances of multi-span sandwich panels with respect to these failure modes is based on experimental data for the strength and moduli of the core material and the compressive strength of the face layer and, furthermore, on calculation models with varying degrees of sophistication.

At the ultimate limit state, the important criteria are the compression resistance of the face layer and the shear resistance and the support reaction resistance of the sandwich panel. It is recommended that the compressive strength of the face, f_{Fc}, the shear strength of the core, f_{Cv}, and the tensile resistance of the fasteners, F_{Rt}, should be evaluated experimentally. The resistance to core crushing failure at the support, f_{Cc}, can then be defined by calculations based on the compressive strength and modulus of the core layer.

Profiling of one or both of the metal sheet faces changes the compressive stress distribution in the face and the shear stress distribution in the core. Because of the more complicated stress distributions and the geometry of the cross-section, some additional failure modes exist at the ultimate limit state in panels with strongly profiled faces. In the case of very strongly profiled thick faces, the influence of the core can be neglected in connection with some of the failure modes and the analysis of the resistances can conservatively be based on the design codes used for trapezoidal sheeting. Such failure modes are the compression failure of the face and the crushing failure of the webs of the profile at a support.

10.6 REFERENCES

10.1 Berner, K. (1995) Erarbeitung vollständiger Bemessungsgrundlagen im Rahmen bautechnischer Zulassungen für Sandwichbauteile. *Forschungsbericht*, **IV**, 1-5-618/90, DIBt, Berlin. 10 pp. [in German].

10.2 Berner, K., Hassinen, P. and Heselius, L. (1995) Present design methods for lightweight sandwich panels. In *Proceedings of Nordic Steel Construction Conference '95*, Malmö, Sweden, 19–21 June 1995, pp. 637–644.

10.3 European Convention for Constructional Steelwork (ECCS) (1983) *European Recommendations for Steel Structures: The Design of Profiled Sheeting*, ECCS Publication 40, 111 pp.

10.4 European Convention for Constructional Steelwork (ECCS) (1991) *Preliminary European Recommendations for Sandwich Panels: Part I, Design*, ECCS Publication 66, 131 pp.

10.5 International Council for Building Research, Studies and Documentation (CIB) *European Recommendations for Sandwich Panels (2000) Part 1: Design*, CIB Publication 147, 161 pp.

10.6 European Committee for Standardization (1996) Document ENV 1993-1-3. *Design of steel structures: Part 1.3: Supplementary rules for cold formed thin gauge members and sheeting*, 128 pp.

10.7 Davies, J.M. (1986) An exact finite element for beam on elastic foundation problems. *Journal of Structural Mechanics*, **14**(4), 489–499.

10.8 Davies, J.M. (1987) Design criteria for structural sandwich panels. *The Structural Engineer*, **65A** (12), 435–441.

10.9 Davies, J.M. and Hakmi, M.R. (1990) Local buckling of sandwich plates. *IABSE Report*, **60**, 533–538.

10.10 Davies, J.M., Hassinen, P. and Hakmi, R. (1991) Face buckling stresses in sandwich panels. In *Proceedings of the Nordic Steel Colloquium*, Odense, Denmark, September 9–11 1991, pp. 99–110.

10.11 Frostig, Y. and Baruch, M. (1990) Bending of sandwich beams with transversely flexible core. *AIAA Journal*, **28**(3), 523–531.

10.12 Hakmi, R. (1988) Local buckling of sandwich panels. Ph.D. Thesis, University of Salford, 236 pp.

10.13 Hassinen, P. and Martikainen, L. (1994) Analysis and design of continuous sandwich beams. In *Proceedings of the 12th International Speciality Conference on Cold-Formed Steel Structures*, St. Louis, Missouri, U.S.A., 18–19 October 1994, pp. 523–538.

10.14 Hassinen, P. and Martikainen, L. (1996) Serviceability and ultimate limit states of continuous sandwich panels. In *Sandwich Construction 3* (ed. H.G. Allen), Vol. I, Engineering Materials Advisory Services Ltd, pp. 195–204.

10.15 Hassinen, P. and Martikainen, L. (1996) Design models of continuous sandwich panels. In *Proceedings of the 13th International Specialty Con-*

ference on Cold-Formed Steel Structures, St. Louis, Missouri, USA, 17–18 October 1996, pp. 293–307.
10.16 Hetenyi, M. (1974) *Beams on Elastic Foundations*. Ann Arbor, The University of Michigan Press, 255 pp.
10.17 Jungbluth, O. and Berner, K. (1986) *Verbund- und Sandwichtragwerke* (*Composite and Sandwich Structures*). Springer-Verlag. pp. 336–559 [in German].
10.18 Just, M. (1992) Zum Langzeitverhalten von Sandwichplatten mit einem Polyurethan-Hartschaumstoffkern (The long-term behaviour of sandwich panels with PUR rigid foam core). *Bauingenieur*, **67**, 83–89 [in German].
10.19 Kech, J. (1991) Druckbeanspruchbarkeit der biegeweichen Deckschicht eines Sandwichelementes (Compression load bearing capacity of the flexible facing of a sandwich panel). *Stahlbau*, **60** (7), 203–210 [in German].
10.20 Kilpeläinen, T. and Hassinen, P. (1995) Long-term behaviour and strength of EPS foam sandwich panels. In *Proceedings of Nordic Steel Construction Conference '95*, Malmö, Sweden, June 19–21 1995, pp. 645–652.
10.21 Kilpeläinen, T. and Hassinen, P. (1996) Long-term and fire behaviour of an eps-foam sandwich panel product. *Sandwich Construction*, **3** (ed. H.G. Allen), Vol. II, Engineering Material Advisory Services Ltd, pp. 911–920.
10.22 Linke, K.-P. (1978) Zum Tragverhalten von Profilsandwichplatten mit Stahldeckschichten und einem Polyurethan-Hartschaum-Kern bei kurz- und langzeitiger Belastung (Load bearing capacity of sandwich panels with steel sheet profiled faces and a polyurethane foam core loaded by short-term and long-term loads). Technische Hochschule Darmstadt, Dissertation D17, 258 pp. [in German].
10.23 Martikainen, L. and Hassinen, P. (1996) Load-bearing capacity of continuous sandwich panels. Helsinki University of Technology, Department of Structural Engineering, Publication No 135, 222 pp.
10.24 Stamm, K. and Witte, H. (1974) *Sandwichkonstruktionen* (*Sandwich Structures*). Springer-Verlag, Vienna, 337 pp. [in German].
10.25 Stamm, K. (1984) Sandwichelemente mit metallischen Deckschichten als Wandbauplatten im Bauwesen (Metal sheet faced sandwich panels in wall structures). *Stahlbau*, **53** (5), 135–143 [in German].
10.26 Stamm, K. (1984) Sandwichelemente mit metallischen Deckschichten als Dachbautafeln im Bauwesen (Metal sheet faced sandwich panels in roof structures). *Stahlbau*, **53** (8), 231–236 [in German].
10.27 Vlasov, V.Z. and Leont'ev, U.N. (1966) *Beams, Plates and Shells on Elastic Foundations*. Israel Program for Scientific Translations Ltd, Jerusalem, 357 pp.
10.28 Zenkert, D. (1995) *An Introduction to Sandwich Construction*. Engineering Materials Advisory Services Ltd, 277 pp.
10.29 Hassinen, P. (1999) Modelling of continuous sandwich panels. In *Proceedings of 4th International Conference on Lightweight Steel and Aluminium Structures*, Espoo, Finland, 20–23 June 1999, Elsevier Applied Science Ltd, pp. 189–196.

Chapter 11

Loads, Load and Material Factors and Design Procedure

11.1 DESIGN EQUATION

The safety philosophy embraced by modern limit state codes was defined in the Introduction to Chapter 9 and, in particular, in Fig. 9.1. The basic equation of limit state design is Equation 11.1. Load effects caused by the most severe combination of loads in a structure have to be limited so that they have an adequate reserve of safety when compared with the corresponding resistance of the structure. Neither the loads nor the resistances of real structures can be defined or calculated exactly because there are stochastic variations in natural loads, scattering in material properties and deviations in the geometrical dimensions of the structure. Furthermore, there are inaccuracies in the structural models that form the basis of design calculations and tests can never truly replicate the behaviour of a real structure. It follows that a structure has to be designed to carry 'characteristic' loads corresponding to a specified probability that they will not be exceeded during the life of the structure. In the case of wind and snow loads, these characteristic values of load effects, S_{ki}, are considerably higher than the load effects that exist every year in reality. The stresses caused by the characteristic loads then have to be multiplied by load factors γ_{Fi} in order to define the design stresses. The consideration that it is unlikely that all of the possible loads which act on a structure will reach their maximum values simultaneously is taken into account by multiplying the loads other than one principal non-permanent load by combination factors $\psi_i \leq 1.0$. In such combinations, only the loads giving unfavourable effects are included.

In the evaluation of the resistances, the possible changes in the material properties due to the loading history, loading time, temperature and other factors that may have an influence on the resistances have to be taken into account. In order to do this, it is necessary to define the expected service life. Having taken all of the necessary effects into consideration, the calculated or experimental characteristic resistance values R_k are divided by material safety factors γ_M corresponding to the specific failure modes. Thus, where subscript d denotes 'design' values and k 'characteristic' values:

$$S_d \leq R_d \tag{11.1}$$

where

$$S_d = \sum \gamma_{Fi} \psi_i S_{ki} \tag{11.2}$$

$$R_d = \frac{R_k}{\gamma_M} \tag{11.3}$$

In this chapter, the application of the above equation in order to carry out the structural design of sandwich panels with a specified level of safety is considered.

However, it should be borne in mind that the same principles may be applicable to design with respect to other requirements such as thermal, moisture and sound insulation. Of course, the safety factors vary according to the physical requirement being considered.

In the future, the basic values of the loads will be defined for the whole of Europe in Eurocode 1, ENV 1991-1.[11.3] At the present time, they are to be found in national codes and standards. The characteristic values of resistance corresponding to the different failure modes of a structure have to be defined by calculations and/or tests keeping in mind the expected service life and the loading and environmental conditions of the structure. In elements such as sandwich panels, where the materials are rather variable and their properties are defined on the basis of tests, the initial level of resistances assumed in the design have to be controlled continuously using appropriate methods of quality control.

Two limit states are defined in the structural design of sandwich panels used to cover the walls and roofs of buildings. At the *serviceability* limit state, the structure has to maintain the functional, comfort and aesthetic requirements which means that high deflections, inconvenient vibrations and permanent deformations are not allowed. At the *ultimate* limit state, the structure has to carry the most severe load combination without reaching a state of final collapse.

At the intermediate supports of multi-span sandwich panels, the first failure modes cannot always be classified clearly into one of the two limit states given above. From the technical point of view, local buckling, bending and crushing failures at an intermediate support or local failure at the point of attachment of a fastener define a serviceability limit state. However, from the aesthetic point of view, someone may classify these failures to be more serious than is warranted by the typical classification of a serviceability limit state failure. Therefore, the material safety factors corresponding to the buckling, bending and crushing failure modes at the intermediate supports of continuous sandwich panels are increased at the serviceability limit state from $\gamma_M = 1.0$ to $\gamma_M = 1.1$.

In Eurocode 1,[11.3] several alternative load combinations have been defined. At the ultimate limit state, the combinations of persistent and transient situations are used in the design of wall and roof panels and these constitute the fundamental combinations. In addition, if the risks are deemed to be real and probable, the combinations of the accidental or seismic situations, including unforeseen load cases, have to be included. At the serviceability limit state, three combinations are given, namely; the characteristic or rare combination, the frequent combination and quasi-permanent combination. The design guidelines given in this chapter follow the definition of the rare combination given in Eurocode 1.

Wall and roof panels are typical beam structures and generally have no load-bearing function in the axial direction or in the plane of the wall or roof. Thus, the loads, load combinations and design equations in this chapter are given to cover the design of beam-type sandwich structures. If the design requires consideration of diaphragm action, plate or shell action or the axial capacity of a sandwich panel in a wall or roof, additional investigations are required in order to verify the strength of the sandwich panels, joints and fastening systems.

11.2 LOADS ON SANDWICH PANELS

The self-weight of lightweight sandwich panels is relatively low and is typically in the range 10–25 kg/m^2. Self-weight is a permanent load, together with the weights of other structural components such as coverings that are supported by the panels. The intensities of wind pressure and suction on the walls and roofs of buildings typically vary from 0.3 to 1.0 kN/m^2. At the edges and corners of buildings, the wind suction loads may be much higher than this. The durations of the characteristic wind load levels given in the design codes are some few minutes. Thus, the wind loads can be classified as short-term loads. The characteristic snow load varies in Nordic countries from 1.0 to 2.5 kN/m^2 but, in mountainous areas, even in Central Europe, this can increase up to 5 kN/m^2. Snow load is a semi-permanent load with a strongly varying annual intensity and duration. In some areas its annual duration can be as much as 5 months.

An important load case for sandwich panels with a well-insulating core layer is caused by the difference of temperature between the external and internal faces. The temperature of the internal face is usually relatively constant and is around 20°C. The temperature of the external face has large daily and annual variations. Typical extreme values in Central Europe in winter and summer are −20 and +80°C respectively. In Northern Europe, and especially in the arctic regions, the temperature may decrease down to −40°C and even to −60°C. In tropical climates, the temperatures of dark-coloured outside faces may be considerably higher than those found in Central Europe. There are also special applications, such as cold stores and deep freeze rooms in which low inside temperatures, in the range of about +5°C to −40°C, increase the temperature difference between the outer and inner faces in summertime. Relevant design temperatures which follow the guidance given in the ECCS and CIB Recommendations[11.1, 11.2] are given in Table 11.1. Because of possible changes in the colours of the outer coatings and climatic conditions, for the ultimate limit state design, a temperature of +80°C in summer is generally chosen in order to ensure adequate long-term security and reliability.

For aesthetic or technical reasons, sandwich panels are sometimes covered by partially transparent glass or plastic plates or by non-transparent steel or aluminium sheeting, mounted directly on the external face of a sandwich panel wall or roof. There may also be an air space between this additional sheeting and the external face. Additional sheeting of this type may change the temperature of the outside face considerably. Provided that there is no direct contact between the sheeting and external the face of the panel, a ventilated light-coloured profiled sheeting, which is mounted in the vertical direction, may considerably decrease the temperature of the outer face. Another extreme case arises when an unventilated transparent curtain wall is fixed in front of a sandwich panel. This may increase considerably the temperature of the outer face when compared with the corresponding uncovered face. Unexpectedly high temperatures may also be caused when a dark-coloured profiled sheeting is mounted in the horizontal direction with its flanges in direct contact with the outer face of the sandwich panels. Depending on the nature of the fasteners between the sheeting and the sandwich panel, additional stresses may be caused in the sandwich structure because of composite action between the covering and the sandwich panel. When the covering and face

Table 11.1 Design temperatures (°C) for sandwich panels

	Serviceability limit state	Ultimate limit state
Summer (European conditions)		
• inside	+25	+25
• inside, cold stores, etc.	Ambient operating temperature	Ambient operating temperature
• outside		
○ very light colour[a]	+55	+80
○ light colour[a]	+65	+80
○ dark colour[a]	+80	+80
Winter (European conditions)		
• inside	+20	+20
• inside, cold stores, etc.	Ambient operating temperature	Ambient operating temperature
• outside[b]		
○ maritime climate (UK)	−10	−10
○ Central Europe	−20	−20
○ Nordic countries	−30	−30
• outside, simultaneous snow load on the roof	0	0

[a] Colour of the coating of the outer face (References 11.1 and 11.2).
[b] The outside design temperature is very much dependent on the latitude and on the climatic conditions.

are made of materials having different coefficients of thermal expansion, these stresses may be considerable. Because of the lack of detailed knowledge, no general guidance regarding design temperature values or additional stresses can be given for sandwich panels covered by additional sheeting.

However, Fig. 11.1 shows a schematic figure depicting the factors influencing the design temperatures of the outer face of sandwich panels covered by partially transparent glass or a non-transparent metal sheeting. Depending on the transparency of the covering structure, and the air flow between the external face and the covering structure, it is shown that the temperature of the external face may either increase or decrease in comparison with the temperature of the similar face without any covering.

Internal partition walls may be loaded by wind pressure and suction loads or internal air pressure differences, especially if there are large openings in the walls of the building. Internal walls may also be loaded by local impact loads. Internal ceiling structures are, in theory, only exposed to the self-weight of the structure. However, in practice, ceilings are sometimes used like walkways and goods may be stored on them. Loads and loading cases for internal walls and ceilings made of sandwich panels vary considerably and they have to be defined case by case on the basis of risk analyses.

The mechanical and thermal loads applied to wall and roof panels do not usually cause any significant accelerations or additional dynamic stresses in the panels. Thus, the common loads on sandwich panels are considered to be static actions. Accidental impact actions may cause dynamic stresses which have an influence on the structural behaviour and failure modes of sandwich panels and their fastenings

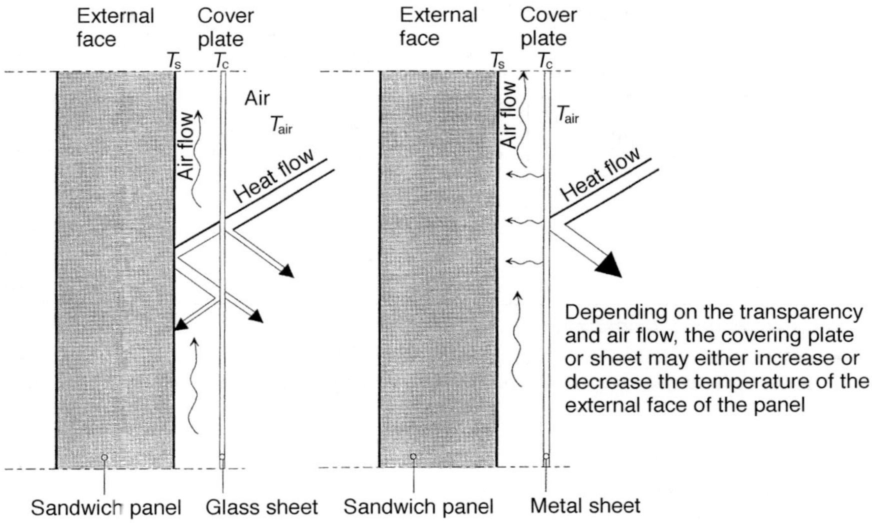

Fig. 11.1 The factors influencing the design temperature of the outer face of a sandwich panel with a secondary covering.

and these, therefore, may need to be taken into account in design. Local impact stresses which are repeated a number of times, such as the stresses caused by walking on a roof or ceiling panel, may also have an influence on the resistance of the panel. The resistance of sandwich panels against local repeated loads has to be determined experimentally or else the sandwich panel structure has to be protected or strengthened in order to exclude the additional local failure modes.

11.3 CHOICE OF THE METHOD OF ANALYSIS

Because of the conventional requirement for reversible deformations in the structure, the evaluation of the deflections and stress resultants at the serviceability limit state is based on the linear theory of elasticity. In this elastic analysis, both the short-term and long-term shear flexibility of the core of the sandwich structure have to be taken into account. This shear flexibility increases the deflections and changes the bending moment and shear force diagrams in comparison with the conventional 'engineers' theory of bending. A range of analytical and numerical methods of elastic analysis are explained and relevant design expressions derived in Chapter 9.

Experimental studies on full-scale multi-span sandwich panels have shown that some of the commonly-encountered failure modes at intermediate supports do not result in final collapse of the panel structure but effectively turn the sandwich panels into a series of single-span panels with a rather small or even negligible bending resistance in the failed cross-section at the support. They are, therefore, able to continue to carry more load. The failure modes where this happens are crushing failure of the core and buckling and bending failure of the compressed face and, in some cases also, shear failure of the core. Therefore, if the bending resistance is critical, the design of multi-span sandwich panels at the ultimate limit state may be based on the static model of a series of single-span sandwich beams. Because of the use of this pseudo-plastic type of static model, the thermal stresses at the ultimate limit state disappear in the case of thin-faced sandwich panels and are considerably reduced in the case of thick-faced sandwich panels.

The shear failure mode of rigid plastic foams is typically a brittle type of failure. If the shear strength is critical in the design of multi-span sandwich panels, the first failure, which is the shear failure, generally leads to the ultimate limit state of the structure without any plastic deformations. It follows, therefore, that no plastic hinges can be assumed at the intermediate supports and that the design calculations for both the serviceability and ultimate limit states have to be based on an elastic analysis which includes all of the mechanical and thermal loads. However, some new experimental observations have shown that core layers made of structural stone wool material do not fail in a totally brittle way in shear but have some plastic-type shear resistance.[11.6] Utilisation of the full or partial plastic shear resistance of stone wool cores changes one of the basic assumptions of the design of the shear-critical sandwich panels. However, this observation may not be valid for all types of wool-cored sandwich panels so that the problem has to be studied carefully case by case.

Instead of assuming zero plastic bending resistance, the non-vanishing resistance at intermediate supports could be utilised in the ultimate limit state design of continuous sandwich panels. Experiments and numerical analyses have shown that, if the first failure at the intermediate support is determined by the buckling and bending of the compressed face layer, the remaining bending resistance may have a significant non-vanishing value which is retained up to large rotations. If the first failure is caused by the crushing of the core over a narrow supporting beam, the remaining bending resistance tends to zero though still allowing large plastic rotations at the support. This remaining bending resistance changes the static behaviour and the ultimate load capacity of real sandwich panels but, in the present state of the art, it is not practical to make use of it in design calculations. Indeed, if, as is often the case, the design is limited by the behaviour at the serviceability limit state, it is of no benefit to do so.

11.4 SAFETY FACTORS

The stiffness and strength of typical core materials are likely to show considerably more variability than the stiffness and strength of the metal sheet faces. In the case of stiffness, the safest design calculations are not necessarily given by using the minimum value. Therefore, two different stiffness values are used in design calculations. The analysis of deflections and stress resultants is based on the mean shear modulus, which is determined experimentally at room temperature. However, in the evaluations of all of the resistance values which depend on the core properties, characteristic values are used, as discussed in Chapter 10.

In Table 11.2, the material safety factors against the failure modes in which the metal face sheet or the fasteners are critical are similar to the factors given in Eurocode 3.[11.4] The material safety factor against the failures at an intermediate support in a continuous sandwich panel is increased from 1.0 to 1.1 because this

Table 11.2 Material property values, material safety factors and creep coefficients used in the design of sandwich panels

		Serviceability limit state	Ultimate limit state
Material property values of the core	Analysis of sandwich panels		
	shear modulus, G_C	Average value ($T = 20°C$)	Average value ($T = 20°C$)
	Evaluation of resistances		
	moduli, E_{Cc}, E_{Ct}, G_C	Characteristic values at a design temperature	Characteristic values at design temperatures
	strength, f_{Cc}, f_{Cv}	Characteristic values at design temperatures	Characteristic values at design temperatures
Material safety factors, γ_M	Failure mode		
	crushing of the core	1.0	1.25
	shear of the core	1.0	1.25
	yielding of the face	1.0	1.1
	wrinkling of the face	1.0	1.25
	failure of the face at intermediate supports	1.1[a]	
	fastenings	1.0	1.33
Creep coefficient of typical core materials, ϕ_t	All rigid plastic foams		
	snow load	2.4	2.4
	permanent load	7.0	7.0
	Mineral wool[b]		
	snow load	1.0	1.0
	permanent load	2.0	2.0

[a] $\gamma_M = 1.0$ is increased by 10% in order to avoid visible defects at intermediate supports.
[b] Structural wools typically have low creep coefficients which depend very much on the structure of the wool.

failure mode is sometimes classified as being more serious than a typical serviceability limit state failure.

The partial load factors can be determined on the basis of the general directions given in Eurocode 1[11.3] with the exception of the load case creep, for which the load factor $\gamma c_c = 1.00$ both for the serviceability and the ultimate limit state designs is given in the ECCS and CIB Recommendations[11.1, 11.2] (see Tables 11.3 and 11.4).

11.5 LOAD COMBINATIONS

The design loads of typical sandwich wall panels are usually composed of wind pressure and suction loads and of temperature differences between the external and internal faces. Load combinations covering all of the failure modes of thin- and thick-faced single-span and multi-span wall panels are given in Table 11.3 which is an interpretation of the ECCS and CIB Recommendations and Eurocode 1. In general, all of the load cases from W1 to W8 have to be considered in order to determine the critical stresses of sandwich panels and fasteners at the serviceability and at the ultimate limit states, because there are structures, such as multi-span wall panels with unequal spans and wall panels with cantilevers, for which the dominant load case cannot be determined without calculations.

Design of single-span thin-faced wall panels at the serviceability and ultimate limit states can be made on the basis of load cases W3 & W4 and W5 & W6. Similarly, when a thin-faced sandwich panel spans continuously over three or more supports, the serviceability limit state design, where the dominant failure modes are at the intermediate supports, may be made on the basis of load cases W3 & W4 and W5 & W6. However, if the shear strength of the sandwich panels is critical and, therefore, the ultimate limit state design is based on the elastic method of analysis, load cases W1 & W2 and W7 & W8 may also be important because of the necessity to consider the shear stresses and support reactions at the end supports.

The primary loads on roof panels are usually the self-weight and snow load, but the wind suction load and the temperature difference between the faces may also be important loads in the design. In addition, the shear creep of the core causes changes in the stress distributions which have to be taken into account, especially in the calculations of the stress combinations for the serviceability limit state. Because the effect of the self-weight is in the opposite sense to wind suction and some temperature loads, one or other of two partial load factors (1.35 or 1.00) has to be applied to the self-weight in order to determine the most unfavourable load combinations. Table 11.4 is also an interpretation of the ECCS and CIB Recommendations and Eurocode 1 which gives the load cases for roof panels in general. In the case of a thin-faced single span roof panel, load cases R6, R10 and R11 give the dominant stresses and deflections at the serviceability and ultimate limit states.

Table 11.3 Load and combination factors and load cases for the design of wall panels and wall panel connections (This table is a modification of Eurocode 1 proposed by a working group of CIB commission W56. The load cases which are relevant for a given design depend on the loads, the face geometry, the static system and the method of analysis used in the calculations. When calculating the deflections, the characteristic wind loads can be reduced by 25% in combinations of wind and temperature loads.)

				Serviceability limit state	Ultimate limit state
Load factors		Wind	γ_W	1.0	1.5
		Temperature	γ_T	1.0	1.5
		Combination factor	ψ	0.6	0.6
Load cases					
	W1	Wind pressure + Summer temperature		1.0 (Wind + 0.6 Temp.) or	1.5 (Wind + 0.6 Temp.) or
	W2	Wind pressure + Summer temperature		1.0 (0.6 Wind + Temp.)	1.5 (0.6 Wind + Temp.)
	W3	Wind suction + Summer temperature		1.0 (Wind + 0.6 Temp.) or	1.5 (Wind + 0.6 Temp.) or
	W4	Wind suction + Summer temperature		1.0 (0.6 Wind + Temp.)	1.5 (0.6 Wind + Temp.)
	W5	Wind pressure + Winter temperature		1.0 (Wind + 0.6 Temp.) or	1.5 (Wind + 0.6 Temp.) or
	W6	Wind pressure + Winter temperature		1.0 (0.6 Wind + Temp.)	1.5 (0.6 Wind + Temp.)
	W7	Wind suction + Winter temperature		1.0 (Wind + 0.6 Temp.) or	1.5 (Wind + 0.6 Temp.) or
	W8	Wind suction + Winter temperature		1.0 (0.6 Wind + Temp.)	1.5 (0.6 Wind + Temp.)

Table 11.4 Load and combination factors and load cases for the design of roof panels and roof panel connections (This table is a modification of Eurocode 1. The load cases which are relevant for a given design depend on the loads, the face geometry, the static system and the method of analysis used in the calculations. When calculating the deflections, the characteristic wind and snow loads may be reduced by 25% in combinations of snow, wind and temperature loads.)

			Serviceability limit state	Ultimate limit state
Load factors	Self weight	γ_G	1.0	1.35 or 1.00[a]
	Snow	γ_s	1.0	1.5
	Wind	γ_w	1.0	1.5
	Temperature	γ_T	1.0	1.5
	Creep	γ_c	1.0	1.0
	Combination factor	ψ	0.6	0.6
Load cases				
R1	Self-weight + Snow		1.0 (Self-weight + Snow) or	1.35 Self-weight + 1.5 Snow or
R2	Self-weight + Snow		1.0 (Self-weight + Snow + Creep)	1.35 Self-weight + 1.5 Snow + 1.0 Creep
R3	Self-weight + Winter temperature ($T_0 = -10°C/-20°C/-30°C$)		1.0 (Self-weight + Temp.) or	1.35 Self-weight + 1.5 Temp. or
R4	Self-weight + Winter temperature ($T_0 = -10°C/-20°C/-30°C$)		1.0 (Self-weight + Temp. + Creep)	1.35 Self-weight + 1.5 Temp. + 1.0 Creep
R5[b]	Self-weight + Snow + Winter temperature ($T_0 = 0°C$)		1.0 (Self-weight + Snow + Temp.) or	1.35 Self-weight + 1.5 (Snow + Temp.) or
R6[b]	Self-weight + Snow + Winter temperature ($T_o = 0°C$)		1.0 (Self-weight + Snow + Temp. + Creep)	1.35 Self-weight + 1.5 (Snow + Temp.) + 1.0 Creep
R7	Self-weight + Wind suction		1.0 (Self-weight + Wind)	1.0 Self-weight + 1.5 Wind
R8	Self-weight + Summer temperature		1.0 (Self-weight + Temp.) or	1.35 Self-weight + 15 Temp. or
R9	Self-weight + Summer temperature		1.0 (Self-weight + Temp.)	1.0 Self-weight + 1.5 Temp.

	R10	Self-weight + Wind suction + Summer temperature	1.0 (Self-weight + Wind + 0.6 Temp.) or	1.0 Self-weight + 1.5 (Wind + 0.6 Temp.) or
	R11	Self-weight + Wind suction + Summer temperature	1.0 (Self-weight + 0.6 Wind + Temp.)	1.0 Self-weight + 1.5 (0.6 Wind + Temp.)
	R12	Self-weight + Wind suction + Winter temperature ($T_0 = -10°C/-20°C/-30°C$)	1.0 (Self-weight + Wind + 0.6 Temp.) or	1.0 Self-weight + 1.5 (Wind + 0.6 Temp.) or
	R13	Self-weight + Wind suction + Winter temperature ($T_0 = -10°C/-20°C/-30°C$)	1.0 (Self-weight + 0.6 Wind + Temp.)	1.0 Self-weight + 1.5 (0.6 Wind + Temp.)

[a] If the influence of self-weight is favourable, the load factor $\gamma_G = 1.00$ has to be used.
[b] In some cases the wind pressure load has to be taken into account simultaneously with snow and wind by the use of a combination factor of $\psi = 0.6$.

11.6 PERMISSIBLE SHORT-TERM AND LONG-TERM DEFLECTIONS

The limitations to be applied to the deflections at the serviceability limit state have to be decided after bearing in mind the technical and functional requirements of the sandwich panel walls and roofs. Large deflections cause large deformations and potential damage in the joints at the end supports of sandwich panels and in the joints between the panels and other structural elements. Large deflections in roof panels may also cause large permanent, non-recoverable deflections with increased water and ice load phenomena. Very large deflections may even change the static behaviour of the structure. Furthermore, large deflections may cause stresses in other parts of the structure which are in contact with the sandwich panels. Finally, large deflections may give the false impression that a structure is close to collapse.

The limits given in published guidelines and recommendations regarding the highest allowable deflections vary in general from span/100 to span/200. In the ECCS Recommendations, the deflection limit of span/100 is recommended, in general, for both wall and roof panels[11.1] although it is stated that more stringent limits may be applicable where there are higher technical and functional requirements. The deflections of sandwich wall panels, especially single-span panels, are increased significantly by the thermal deformations. Long-term creep deformations increase the deflections of roof panels.

Table 11.5 Design equations for single-span simply supported sandwich wall panels with flat or slightly profiled faces

Design criterion	Design expression	Remarks
(1) Wrinkling	$\sigma_{Fd} = \dfrac{\gamma_F q_k L^2}{8eBt} \le \dfrac{f_{Fck}}{\gamma_M}$	f_{Fck} defined experimentally (recommendation)
(2) Shear	$\tau_{Cd} = \dfrac{\gamma_F q_k L}{2eB} \le \dfrac{f_{Cvk}}{\gamma_M}$	f_{Cvk} defined experimentally
(3) Support reaction	$\sigma_{Cd} = \dfrac{\gamma_F q_k L}{2(L_s + e/4)B} \le \dfrac{f_{Cck}}{\gamma_M}$ [a] or $\sigma_{Cd} = \dfrac{\gamma_F q_k L}{2L_s B} \le \dfrac{f_{Cck}}{\gamma_M}$ [a]	f_{Cck} defined experimentally
(4) Tension resistance of fasteners	$F_d = \dfrac{\gamma_F q_{sk} L}{2} \le \dfrac{F_{Rtk}}{\gamma_M}$	F_{Rtk} defined experimentally
(5) Deflection	$w = \dfrac{5}{384}\dfrac{q_k L^4}{B} + \dfrac{1}{8}\dfrac{q_k L^2}{G_C A_C} + \dfrac{\alpha_T (T_2 - T_1) L^2}{8e} \le w_{allowable}$	G_C defined experimentally

[a] Depending on the core material and the bending stiffness of the face.
Different design codes may have different versions of this equation.
The subscript k in the above expressions denotes a characteristic value; s denotes a wind suction load.

11.7 CONCLUSIONS

The well-known general equations for limit state design can be applied to sandwich panels in order to compare the effects of loads and the corresponding resistances at both the serviceability and the ultimate limit states. The partial load factors can be chosen on the basis of the general guidelines given for thin-walled metal structures, but the choice of material factors has to be based on the failure modes which are typical for lightweight sandwich panels and the associated variability of the failure loads. It is important that the material and load safety factors given in different guidelines should not be mixed because these will have been chosen to give an appropriate global level of safety.

Table 11.6 Design equations for multi-span sandwich wall panels with flat or slightly profiled faces

Design criterion	Design expression	Remarks
Serviceability limit state		
(1) Deflection	$w_{max} \leq w_{allowance}$	G_c defined experimentally
(2) Intermediate support	$\begin{cases} \sigma_{Fd} \leq 0.9 f_{Fck}/\gamma_M \\ F_d \leq 0.71\, L_s f_{Cck}/\gamma_M \end{cases}$	$\gamma_M = 1.1$ $\gamma_M = 1.0$
	$\begin{cases} \sigma_{Fd} \leq 0.8 f_{Fck}/\gamma_M \\ F_d \leq F_{Rtk}/\gamma_M \end{cases}$	$\gamma_M = 1.1$ ($n \leq 3$) $\gamma_M = 1.0$
Ultimate limit state, elastic analysis (σ_{Fck}, τ_{Ck}, σ_{Cck} and F_k) are maximum values of the stresses and forces caused by the characteristic loads based on elastic analysis of continuous multi-span sandwich beams.		
(3a) Wrinkling	$\sum \gamma_{Fi}\psi_i\sigma_{Fcki} \leq f_{Fck}/\gamma_M$	In a span
(4a) Shear	$\sum \gamma_{Fi}\psi_i\tau_{Cki} \leq f_{Cvk}/\gamma_M$	At a support
(5a) Support reaction	$\sum \gamma_{Fi}\psi_i\sigma_{Ccki} \leq f_{Cck}/\gamma_M$	At a support
(6a) Tension resistance of fasteners	$\sum \gamma_{Fi}\psi_i F_{ki} \leq F_{Rtk}/\gamma_M$	At a support
Ultimate limit state, plastic analysis for single span, simply-supported sandwich beams.		
(3b) Wrinkling	$\sigma_{Fd} = \dfrac{\gamma_F q_k L^2}{8eBt} \leq \dfrac{f_{Fck}}{\gamma_M}$	In a span
(4b) Shear	$\tau_{Cd} = \dfrac{\gamma_F q_k L}{2eB} \leq \dfrac{f_{Cvk}}{\gamma_M}$	At a support
(5b) Support reaction	$\sigma_{Cd} = \dfrac{\gamma_F q_k L}{2(L_s + e/4)B} \leq \dfrac{f_{Cck}}{\gamma_M}$ [a]	At end support
	or $\sigma_{Cd} = \dfrac{\gamma_F q_k L}{2L_s B} \leq \dfrac{f_{Cck}}{\gamma_M}$ [a]	
(6b) Tension resistance of fasteners	$F_d = \dfrac{\gamma_F q_{sk} L}{2} \leq \dfrac{F_{Rtk}}{\gamma_M}$	At end support

[a] Depending on the core material and the bending stiffness of the face.
Different design codes may have different versions of this equation.
The subscript k in the above expressions denotes a characteristic value; s denotes a wind suction load.

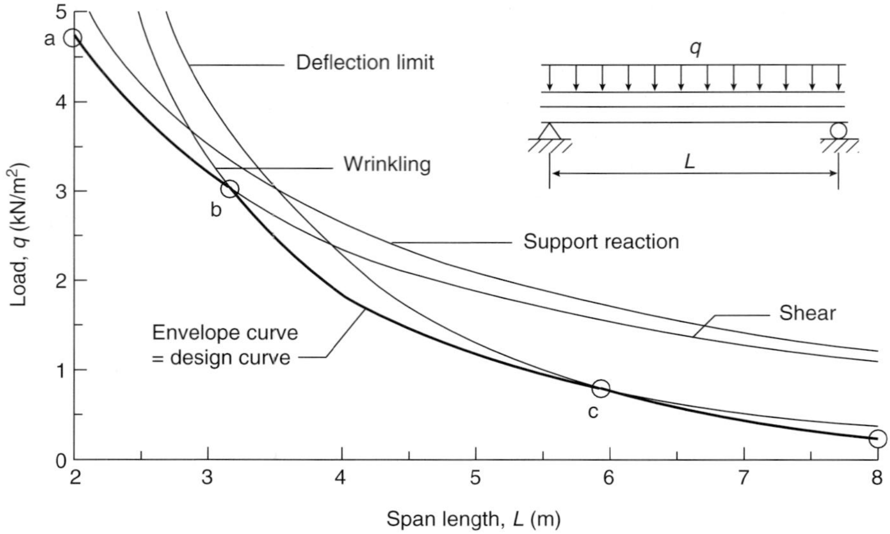

Fig. 11.2 Example of typical load–span curves for a single-span simply supported wall panel with thin flat faces.

A large number of load combinations may have to be studied during the design process because of the necessity of considering the effects associated with the shear flexibility of the core and because of the many potential failure modes. Changes in the outside temperature result in thermal loads and it is essential that these are considered in the design of sandwich panels. Tables 11.5 and 11.6 summarise the equations needed in the design of single- and multi-span wall panels with a uniformly distributed load and a temperature difference between the faces. Figure 11.2 shows typical characteristic load–span curves for a simply supported single-span sandwich panel resulting from the application of the design equations given in Table 11.5. The design curve is the lower bound curve to the four alternative load-span curves determined by the four design criteria given in Table 11.5. In this example, between a and b, the shear strength of the core is dominant, between b and c, the wrinkling strength of the face is critical and, between c and d, the deflection limits the design.

11.8 REFERENCES

11.1 European Convention for Constructional Steelwork (ECCS) (1991) *Preliminary European Recommendations for Sandwich Panels; Part I, Design*, ECCS Publication 66, 131 pp.

11.2 International Council for Building Research, Studies and Documentation (CIB) *European Recommendations for Sandwich Panels (2000) Part 1: Design*, CIB Publication 257, 161 pp.

11.3 European Committee for Standardisation (1996) *Document ENV 1991-1. Basis of Design and Actions on Structures: Part 1: Basis of Design* (Eurocode 1), 85 pp.

11.4 European Committee for Standardisation (1996) *Document ENV 1993-1-3. Design of Steel Structures: Part 1.3: Supplementary Rules for Cold Formed Thin Gauge Members and Sheeting* (Eurocode 3: Part 1.3), 128 pp.

11.5 Institut für Bautechnik (1996) *Design Loads and Statical Analysis for Sandwich Panels – Core of Polyurethane Rigid Foam with Metal Faces*, Version 3.93, 7 pp.

11.6 Martikainen, L. and Hassinen, P. (1996) *Load-Bearing Capacity of Continuous Sandwich Panels*. Helsinki University of Technology, Department of Structural Engineering, Publication 135, 222 pp.

Chapter 12
Additional Structural Considerations

12.1 INTRODUCTION

Chapter 12 complements Chapters 9, 10 and 11 which considered formal structural design. Chapters 9, 10 and 11 present the current design practice for load-bearing sandwich panels and give the background to the existing calculation models, whereas this chapter introduces new ideas and new areas of interest including those where the favourable properties of sandwich structures may be utilised to a greater extent than is general in current practice.

The first part of this chapter gives some examples of good details for the joints between adjacent sandwich panels and the connections with other load-bearing and non-load-bearing structural elements of a building. Section 12.3 describes some typical fasteners which may be used to connect the sandwich panels to the supporting structures. Section 12.4 then gives an overview of the influence of small and large openings on the static behaviour and resistance of the panels.

The following sections in Chapter 12 introduce some potential possibilities which may be utilised to take advantage of the properties of sandwich panels in the design of building structures. These sections present methods which may be used to analyse the influence of the transverse flexibility of the fastenings on the stress distributions in sandwich panels and the influence of the lateral flexibility of the fastenings on the stability of the supporting purlin profiles. These sections also demonstrate the possible utilisation of the axial loading capacity of a sandwich panel to carry vertical loads alone or as a part of the structural framing and the in-plane shear capacity, i.e. the diaphragm action, of sandwich panels in stiffening the complete frame of a building.

12.2 DETAILS FOR SANDWICH STRUCTURES

Construction details are an important factor in building design. This is not just a matter of selecting the right materials, it is equally important that they are connected together properly. When drawing up a detail solution, there are three requirements to be met:

(1) *Function* This includes properties such as air- and water-tightness, fire resistance, thermal insulation, sound insulation and mechanical strength.
(2) *Appearance* As well as material selection, the appearance is influenced by the fixing method and the window framing details, flashings, etc.
(3) *Workmanship* Much depends on the workmanship (execution) and the details should be chosen for their 'buildability'. This is particularly important on the facades of tall buildings. It may be necessary to execute the detail solution in bad weather conditions on a 30-metre high scaffold.

There are a great many viable details which may be used in sandwich panel construction and it is not possible to describe all of the feasible arrangements here. Many manufacturers have their own proprietary details which may also include provision for fixing so that details may often be specific to a particular manufacturer. However, all details have similar performance requirements which are generally a matter of building physics rather than structural strength. These are perhaps best discussed in terms of the four generic uses to which sandwich panels may be put.

12.2.1 Details for external walls

External wall panels may be installed as single-span or multi-span structures. In the case of multi-span panels, particular attention should be paid to the fixings at the intermediate supports where there is a risk of indentation of the panel surface as a consequence of high wind suction and thermal loads. Visible fixings should be made with corrosion-resistant materials in order to avoid rust streaks on the facade of the structure.

The primary performance requirements for joints in external walls are:

- strength of the connection to the supporting structure (this is dealt with in Section 12.3 and will not be considered in detail here)
- reliable thermal performance (avoidance of cold bridges)
- avoidance of air leakage
- avoidance of moisture penetration
- performance in fire (details have been given in Chapter 5 which are specific to fire performance requirements and these will not be repeated here)
- appearance

The flashings play an important role and these should be folded for rigidity and in order to avoid any edge corrosion. Flashings around windows and penetrations should be made in such a way that no water can flow along the facade and result in streaks of dirt.

Sealants are important in external walls for both water- and air-tightness, particularly in tall buildings where the facades are subject to substantial water and wind pressure. Sealants are also important behind flashings. Age-resistant materials should be used.

Some typical details for external wall panels are shown in Fig. 12.1.

12.2.2 Details for external roofs

The primary performance requirements for joints in external roofs are similar to those for walls:

- strength of the connection to the supporting structure
- reliable thermal performance (avoidance of cold bridges)
- avoidance of air leakage
- avoidance of moisture penetration
- performance in fire

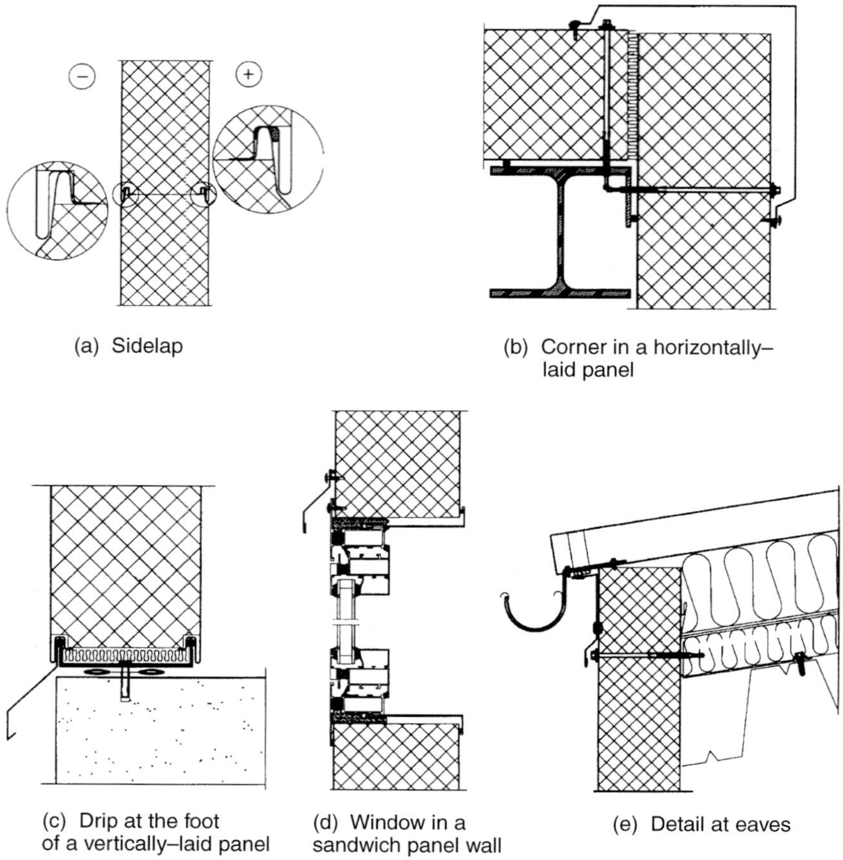

Fig. 12.1 Typical details for external wall panels.

It is important to note that the edges of roofs can be subject to very high wind suction forces and that these make major demands on the fastening systems.

Roof systems should be sealed from the inside in order to avoid moisture penetration to the outer face where it can condense.

Some typical details for external roof panels are shown in Fig. 12.2.

12.2.3 Details for internal walls

The primary performance requirements for joints in internal walls are concerned with sound insulation. For this reason, they need to be carefully sealed, though with rather different criteria than for joints in external walls. Figure 12.3(a) shows a typical corner detail. Internal walls also accommodate doors and a typical detail at a door jam is shown in Fig. 12.3(b).

Internal walls may also have an important role to play with regard to the fire safety of the building as a whole by dividing it up into fire compartments. This aspect of the design is considered in more detail in Chapter 5.

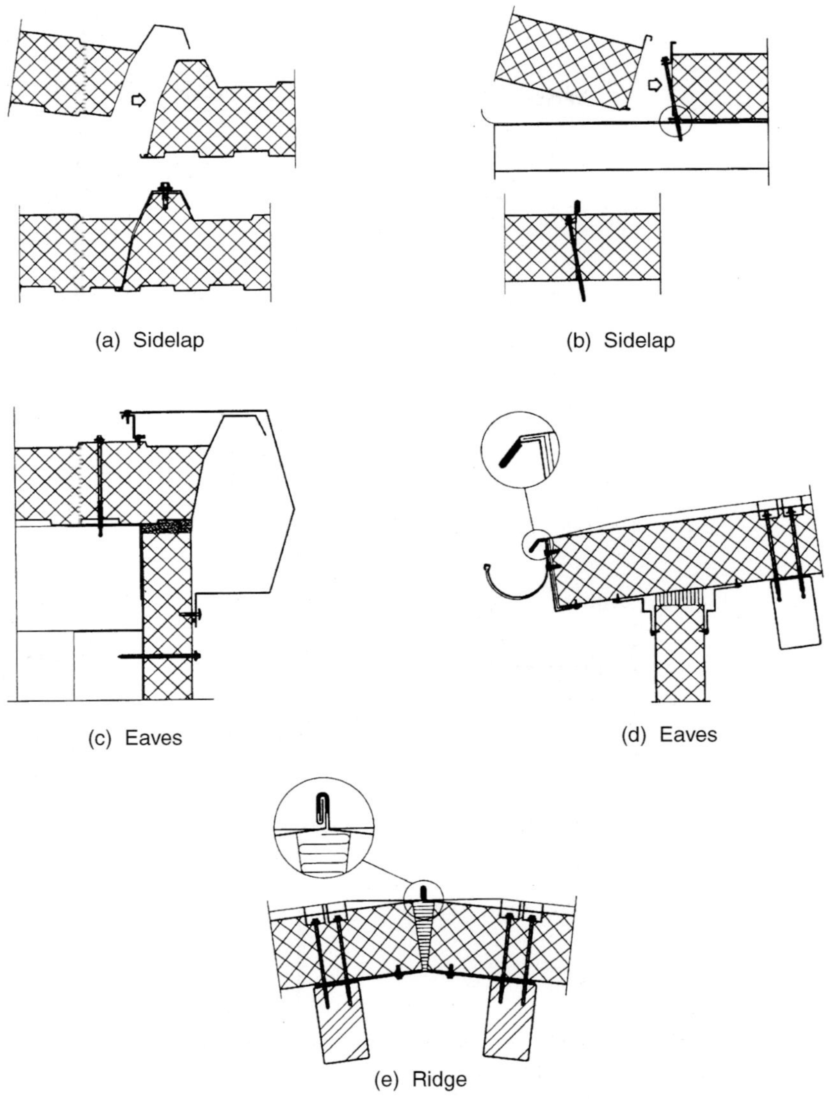

Fig. 12.2 Typical details for roof panels.

It should also be borne in mind that internal walls may be connected to other parts of the construction that are subject to movement during the life of the building. For example, they may be connected to floors that deflect or external walls that are subject to temperature movements. Flashing details need to be designed to accommodate the predicted movements.

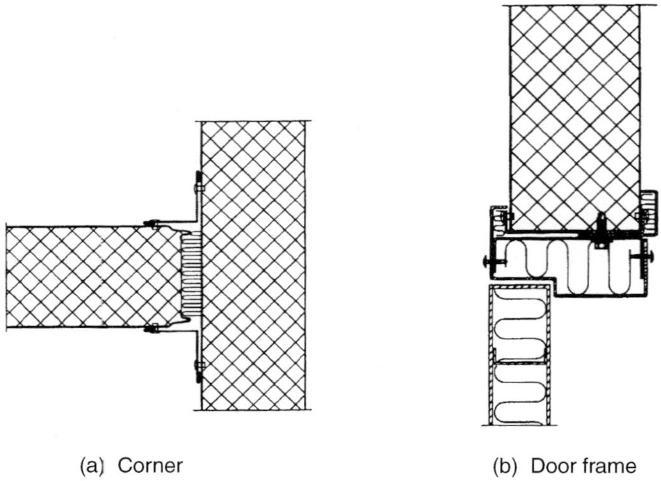

Fig. 12.3 Details for internal walls.

12.2.4 Details for ceilings

Ceilings may also be subject to sound insulation requirements although their primary performance criteria are more likely to be concerned with fire; this topic was considered in detail in Chapter 5. Ceilings are generally suspended from the structure above and a typical detail is shown in Fig. 12.4.

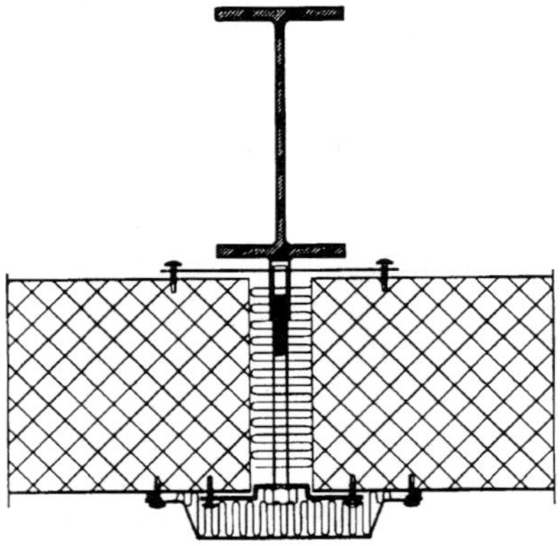

Fig. 12.4 Ceiling detail at a suspension point.

12.3 CONNECTIONS BETWEEN A SANDWICH PANEL AND THE SUPPORTING STRUCTURE

Typically, wall panels are fastened to the supporting structure with screws drilled through the panel. However, many manufacturers have their own proprietary systems, notably those which conceal the screw heads in the longitudinal joints.

Figure 12.5(a) shows such a system. Fastening systems of this type are often used for wall panels and have the advantages of being invisible as well as minimising the risks of water and air leakages. Sometimes also, fastening systems in which only the inside face is fixed to the support structure are met in the market, as shown in Fig. 12.5(b). This type of system is not generally recommended because the tensile force induced by wind suction is transferred through the core material and this may result in a sudden brittle failure as a consequence of tensile failure in the core and joint. Another variant is to use a clip fixing system similar to that used in a conventional standing seam roof.

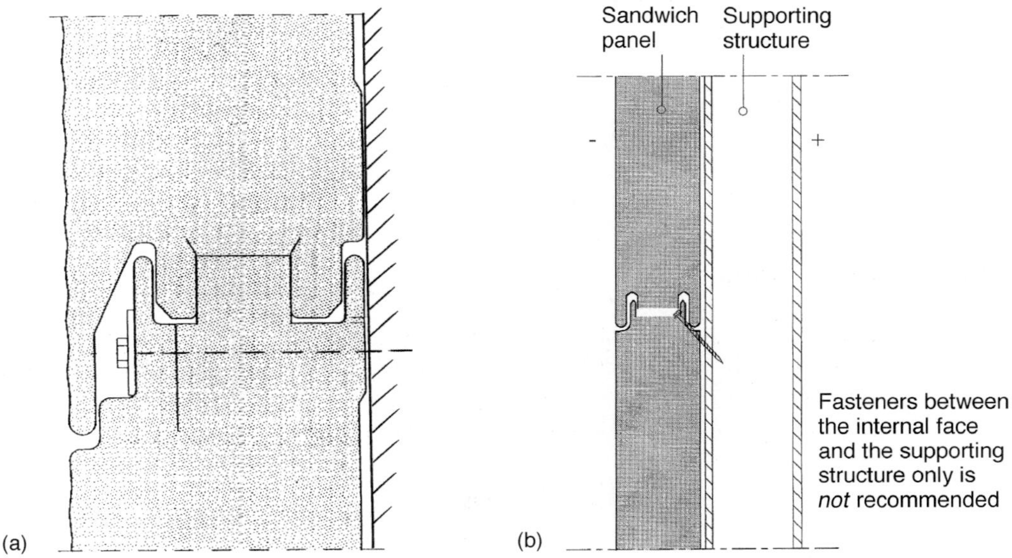

Fig. 12.5 Fastening systems connecting wall panels to the supporting structure.

With a steel supporting structure, self-drilling screws are the most common form of fixing but self-tapping screws are also used, especially in cases when the material thickness of steel supporting structure is large. Pre-drilled holes are needed for self-tapping screws. In screws which connect sandwich panels to timber supporting structures, the threads are wider than in those designed for steel structures. For concrete structures and other similar support structures special anchor fasteners may be used which are driven into pre-drilled holes. However, these anchor fasteners are not allowed in all countries.

Structural screws may be made of either hardened carbon steel or stainless steel. Stainless steel is preferred because of its longer life in a corrosive environment. In some countries, stainless steel may be the only material allowed for structural fasteners.

Fastening screws are loaded by the tensile and shear forces caused by wind suction loads, temperature differences and the self-weight of the panel. In through-panel connections, tensile forces in the screws cause bending and tensile stresses in the outer face of a sandwich panel and compressive and shear stresses in the core layer of the panel. The tensile resistance of a screw fastener depends very much on the diameter of the washer (see Section 10.2.6). Shear forces result primarily in shear stresses in the internal face of the panel that is placed against the supporting structure. The shear resistance of a screw fastener is mainly dependent on the diameter of the screw and on the thickness and strength of the face sheet (Section 10.2.6).

Typical nominal diameters of fixing screws measured from the outer surface of the thread are 5.5 and 6.3 mm. Typical diameters of the washers are 16, 19, 22 and 29 mm. The length of the screw fasteners is chosen to correspond to the depth of the panel. Some typical fixing screws are shown in Fig. 12.6. The detailed design of the screws themselves depends on the material of the supporting structure and on the general arrangement of the connection. A typical through-panel connection using self-drilling screws is shown in Fig. 12.7(a).

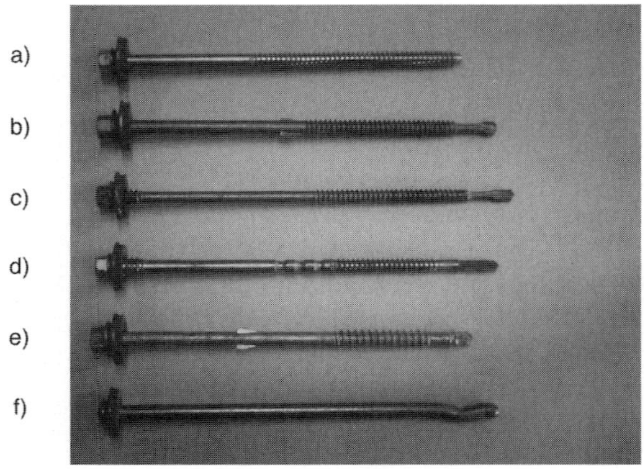

Fig. 12.6 Examples of screws for fastening sandwich panels to the supporting structure: (a) self-tapping screw for a steel supporting structure; (b) self-drilling screw for a steel supporting structure – the small blades in the shaft enlarge the hole in the external face by 1.0–1.5 mm in order to reduce the stresses caused by thermal movements; (c) self-drilling screw for a steel supporting structure – the additional threads near the head tighten the external face sheet against the washer in order to minimise water leakages; (d) self-drilling stainless steel screw for a steel supporting structure – the three thinner regions in the shaft increase the bending flexibility of the screw and so reduce the stresses caused by thermal movements; (e) self-drilling screw for a timber supporting structure; and (f) anchor fastener for a concrete supporting structure.

External wall panels may also be fastened to a steel supporting structure using bolts and a typical system is shown in Fig. 12.7(b). Here, the sandwich panels are supported by small steel plates which are fixed to the supporting structure with bolts and pre-welded nuts. Separate cover plates are needed to protect and to cover the connection. However, this system is not recommended because of the risk of unfastening in a fire and because of the large self-weight of sandwich panels in tall buildings.

The joints and connections of roof panels have to fulfil stringent long-term requirements for water- and air-tightness. Roof panels are typically fastened to the

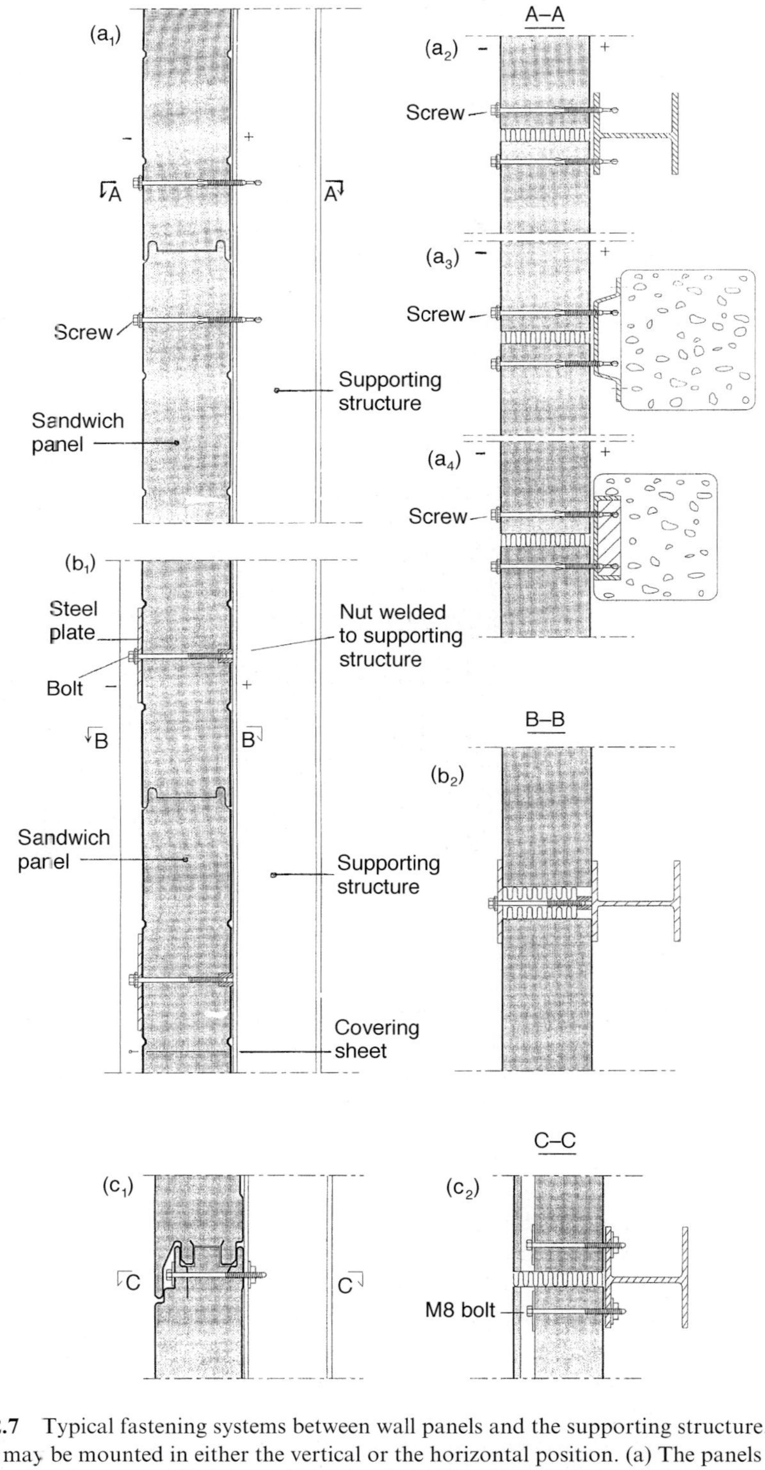

Fig. 12.7 Typical fastening systems between wall panels and the supporting structure. The panels may be mounted in either the vertical or the horizontal position. (a) The panels have been fixed with screws drilled through the panel. (b) The fastening bolts have been placed in the transverse joints of the panels and protected with covering plates. (c) The bolts have been placed in longitudinal joints where they remain invisible.

purlins with screws drilled through the panels or with special hidden fastening systems which are placed in the longitudinal joints of the panels as illustrated by Fig. 12.8. In roof panels which are fastened using screws drilled directly through the panel into the supporting structure, the screw heads and washers are usually placed against the upper flange of the external face profile. The screws may alternatively be placed against the lower flange of the external profile. However, this is not generally recommended because of the greater risk of water penetration. Special screws, in which the face sheet is pressed against the washer by additional threads near the head of the screw, have been developed in order to guarantee the long-term water-tightness of the roof. The middle two screws in Figs 12.6(c) and (d) are of this type. Some of the hidden fastening systems are based on an additional cold-formed steel strip that is fitted into the longitudinal joint between two panels and fastened to the purlin with screws. The tensile resistance is then dependent on the width of the steel strip and on the number of screws.

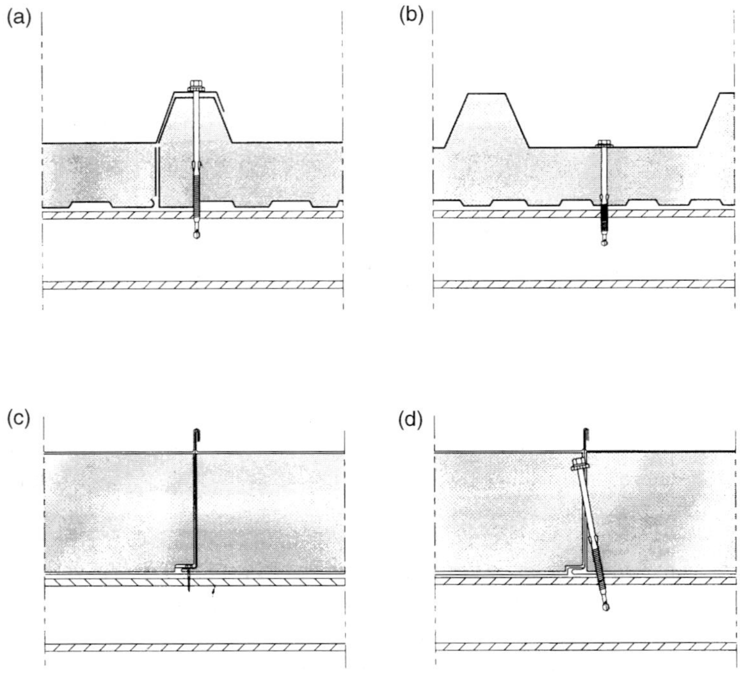

Fig. 12.8 Typical fastening systems for roof panels. (a) and (b) Typical screw fastening systems drilled through the sandwich panels; (c) and (d) special fastening systems placed in the longitudinal joints between the sandwich panels.

In addition to the load-bearing fasteners described above, different kinds of screws may be used to fasten the additional covering plates needed in the joints between sandwich panels and to connect sandwich panels to other parts of the construction such as foundations, window and door frames and openings. The diameters and lengths of the screws for details of this type are usually smaller than those of the structural screws described above.

12.4 EFFECT OF OPENINGS

Very often it is necessary to make openings in a sandwich panel wall or roof. Openings may be cut for windows, doors, ventilation pipes and for different small and large fittings and devices. As illustrated in Fig. 12.9, openings in sandwich panels may extend over the whole panel width or they may cut only a part of the panel symmetrically or asymmetrically in relation to the mid-span and to the mid-line of the panel. Openings are typically either rectangular or circular and rectangular openings may have sharp corners or may have varying amounts of curvature at the corners. Structurally optimal shapes of openings allow a smooth flow of the stresses in the faces and core without causing serious stress concentrations at the corners of the openings. From the practical point of view, optimal shapes such as ellipses are seldom used.

Fig. 12.9 Typical openings in sandwich panels.

Large openings and cut-outs reduce the flexural, shear and torsional rigidity and the bending and shear resistance of sandwich panels. Because of the reduced rigidity, a part of the load is transferred through the longitudinal joints to the adjacent complete sandwich panels. If a sandwich panel is completely cut, the whole load has to be carried by the adjacent panels. The precise way that the loads are transferred from the panel with openings as line loads along the longitudinal joints of one or more adjacent panels depends on the flexural, shear and torsional rigidity of the sandwich panels. Methods to evaluate the torsional rigidity of complete sandwich panels have been given by Stamm and Witte.[12.12] The torsional rigidity of a sandwich panel with thin flat faces can be evaluated using the expression:

$$V = GI_{\mathrm{T}} = \frac{4e^2 BG_{\mathrm{C}}}{\lambda^2 d_{\mathrm{C}}}\left(1 - \frac{\tanh\,(0.5\,\lambda B)}{0.5\,\lambda B}\right) \tag{12.1}$$

where

$$\lambda = \sqrt{\frac{G_{\mathrm{C}}}{G_{\mathrm{F}}}\frac{(t_1 + t_2)}{d_{\mathrm{C}}\,t_1\,t_2}} \tag{12.2}$$

Höglund[12.7] has derived, for the torsional rigidity of sandwich panels with thin flat or lightly profiled faces, the following expression based on Bredt's formula for closed thin-walled sections. This expression has been verified experimentally:

$$V = G_F I_T = G_F \frac{4A_0^2}{\oint_s \frac{ds}{t}} = \frac{8e^2 B^3 t G_F G_C}{6B^2 G_C + 27etG_F} \tag{12.3}$$

In Expressions 12.1, 12.2 and 12.3, the shear modulus for the face is defined in the xy-plane by $G_F = G_{xy}$ and the shear modulus for the core in xz-plane by $G_C = G_{Cxz}$, where x, y and z denote the longitudinal, width and depth directions, respectively.

Expressions 12.1 and 12.3 are different. They also result in different numerical values for the torsional rigidity of sandwich panels in such a way that Expression 12.3 gives lower values for panel widths $B \leq 1200$ mm, which are the most interesting from the practical point of view.

Openings reduce the torsional rigidity of sandwich panels given by Expressions 12.1 and 12.3. The influence of openings on the torsional rigidity needs to be studied experimentally or by using advanced methods of analysis such as the finite-element method.

In order to define the stress distribution in a sandwich panel wall which includes openings, Höglund[12.7] has proposed a method based on the compatibility of deflections at the mid-span of the shear stiff longitudinal joints of simply supported sandwich panels. The deflected shape caused by bending on the one hand and, on the other hand, by the shear and torsional deformations in the longitudinal joints of the sandwich panels are not affine and, in reality, the shear loads q_i in the longitudinal joints are not uniformly distributed. Thus, the method is approximate. The method is illustrated by the example shown in Fig. 12.10.

The bending, shear and torsional rigidities of sandwich panels may vary because of the different face thicknesses and panel depths and finally because of the openings. These quantities are denoted by B_i, S_i and V_i respectively calculated for

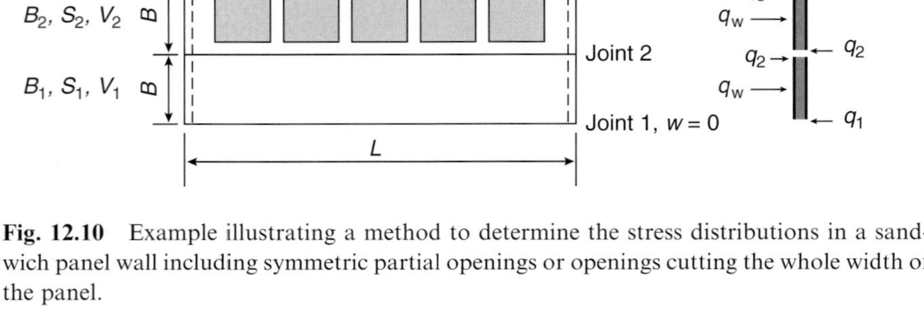

Fig. 12.10 Example illustrating a method to determine the stress distributions in a sandwich panel wall including symmetric partial openings or openings cutting the whole width of the panel.

the full width B of the sandwich panel. The lower end of the sandwich panel wall in the example is fastened to the foundations. The deflections in the upper end of the wall are free. The temperature difference $(T_1 - T_2)$ causes stresses because of the restraint of the deflections by the fastenings at the lower end and because of the different depths e_i and different face materials, α_{F1i}, α_{F2i}, not because of the openings in the second sandwich panel.

The compatibility equation for the first joint is (Fig. 12.10):

$$(q_w - q_1 + q_2)(w_{B1} + w_{S1}) - (q_1 + q_2)w_{V1} + w_{T1} = 0 \tag{12.4}$$

for the second joint:

$$(q_w - q_1 + q_2)(w_{B1} + w_{S1}) + (q_1 + q_2)w_{V1} + w_{T1} = (q_w - q_2 + q_3)(w_{B2} + w_{S2}) - (q_2 + q_3)w_{V2} + w_{T2} \tag{12.5}$$

for the third joint:

$$(q_w - q_2 + q_3)(w_{B2} + w_{S2}) + (q_2 + q_3)w_{V2} + w_{T2} = (q_w - q_3 + q_4)(w_{B3} + w_{S3}) - (q_3 + q_4)w_{V3} + w_{T3} \tag{12.6}$$

for the fourth joint:

$$(q_w - q_3 + q_4)(w_{B3} + w_{S3}) + (q_3 + q_4)w_{V3} + w_{T3} = (q_w - q_4 + q_5)(w_{B4} + w_{S4}) - (q_4 + q_5)w_{V4} + w_{T4} \tag{12.7}$$

and, finally, for the fifth joint:

$$(q_w - q_4 + q_5)(w_{B4} + w_{S4}) + (q_4 + q_5)w_{V4} + w_{T4} = (q_w - q_5)(w_{B5} + w_{S5}) - q_5 w_V 5 + w_{T5} \tag{12.8}$$

Equations (12.4)–(12.8) can be assembled to give the following set of equations:

$$\begin{bmatrix} w1_1 & w2_1 & 0 & 0 & 0 \\ w2_1 & w1_1 + w1_2 & w2_2 & 0 & 0 \\ 0 & w2_2 & w1_2 + w1_3 & w2_3 & 0 \\ 0 & 0 & w2_3 & w1_3 + w1_4 & w2_4 \\ 0 & 0 & 0 & w2_4 & w1_4 + w1_5 \end{bmatrix} \begin{pmatrix} q_1 \\ q_2 \\ q_3 \\ q_4 \\ q_5 \end{pmatrix} = q_w \begin{pmatrix} w3_1 \\ -w3_1 + w3_2 \\ -w3_2 + w3_3 \\ -w3_3 + w3_4 \\ -w3_4 + w3_5 \end{pmatrix} + \begin{pmatrix} w_{T1} \\ -w_{T2} + w_{T2} \\ -w_{T2} + w_{T3} \\ -w_{T3} + w_{T4} \\ -w_{T4} + w_{T5} \end{pmatrix} \tag{12.9}$$

where

$$w1_i = w_{Bi} + w_{Si} + w_{Vi} \tag{12.10}$$

$$w2_i = -w_{Bi} - w_{Si} + w_{Vi} \tag{12.11}$$

$$w3_i = w_{Bi} + w_{Si} \tag{12.12}$$

$$w_{Bi} = \frac{5}{384}\frac{L^4}{B_i}; \quad w_{Si} = \frac{1}{8}\frac{L^2}{S_i}; \quad w_{Vi} = \frac{1}{32}\frac{B^2 L^2}{V_i} \tag{12.13a, b, c}$$

$$w_{Ti} = \frac{1}{8}\frac{(\alpha_{F2i} T_2 - \alpha_{F1i} T_1)L^2}{e_i} \tag{12.14}$$

where T_1 and T_2 are the temperatures of the outer and inner faces respectively and $q_w = Bq_0$ is a line load at the mid-line of the sandwich panel caused by a uniformly distributed pressure load q_0.

Practical methods to evaluate the transfer of the load from a completely cut sandwich panel ($B_i = S_i = V_i = 0$) to the adjacent panels have been developed by Paroc Panel Systems[12.8] as illustrated by Fig. 12.11. The sandwich panels located

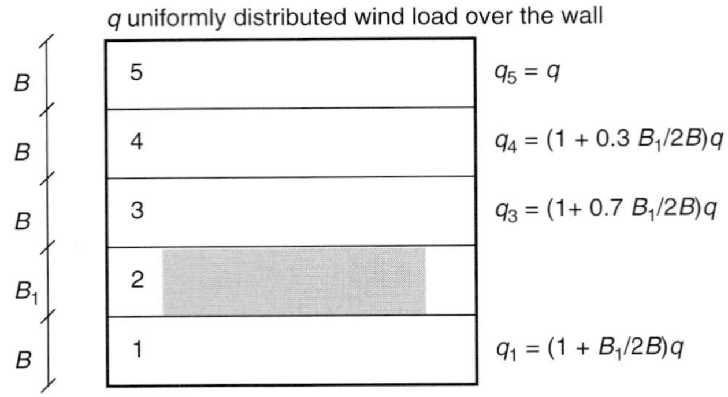

Fig. 12.11 The transfer of the wind load from the opening to the adjacent sandwich panels in a Paroc sandwich panel wall.[12.8]

immediately adjacent to the cut sandwich panel are assumed to carry 70% of the total load of the cut panel. The next pair of panels carry the remaining 30% of the load of the cut panel. This method has been shown to be valid for the specific stone wool cored sandwich panel products of this manufacturer.

If the shear loads q_i in the longitudinal joints are known, the loads and, moreover, the stresses such as the compressive stress in the face and the shear stress in the core of each sandwich panel can be calculated. The additional stress resultant, namely the torsional moment M_V, causes mainly shear stresses in the core layer. The total shear stress in the core of a complete sandwich panel with thin flat faces can be evaluated as the sum of the shear stresses caused by the shear force V and the torsional moment M_V:[12.7]

$$\tau_C = \tau_{C,S} + \tau_{C,V} \approx \frac{V}{eB} + \frac{3}{2} \frac{M_V}{2\left(\frac{2}{3}B\right)\left(\frac{1}{3}eB\right)} = \frac{V}{eB} + \frac{27}{8}\frac{M_V}{eB^2} \tag{12.15}$$

where M_V is the dominant torsional moment in the sandwich panel.

For instance, the shear stress at the support of the first sandwich panel in the example can be calculated as (Fig. 12.10):

$$\tau_C = \frac{(q_w + q_2 - q_1)L}{2eB} + \frac{27(q_1 + q_2)L}{32eB} \tag{12.16}$$

In the design, the calculated stresses have to be compared with the resistances taking into account the relevant partial safety factors for load and resistance. Methods to determine the resistances and safety factors were given in Chapters 10 and 11.

In complicated design cases such as continuous multi-span sandwich panels with large asymmetric openings, the determination of the stresses requires the use of more advanced methods of analysis such as the finite-element method. If the openings are strengthened and stiffened with additional frames, there might be complicated static interactions between the frames and the sandwich panels. In such cases, it is recommended that the frames be designed to resist the total load.

Part of the loads, q_i, are transferred through the longitudinal joints to adjacent sandwich panels and the joints have to be strong enough to carry these shear loads.

A typical failure mode of a longitudinal joint caused by shear load is shown in Fig. 12.12. The shear strength of the joint has to be determined experimentally. If the shear strength of the joint is not adequate, an additional frame has to be designed to carry the shear load.

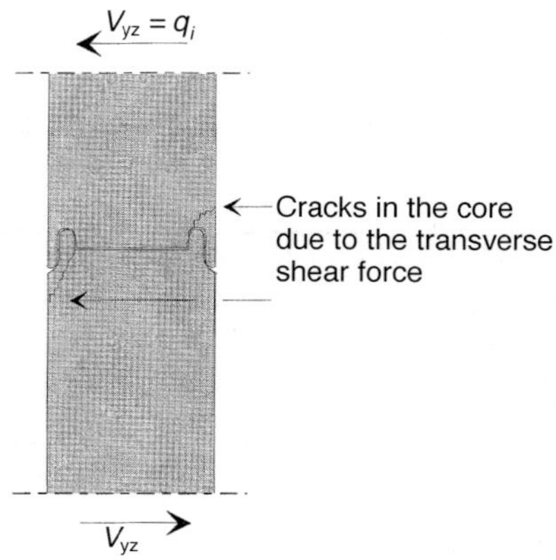

Fig. 12.12 Transverse shear forces caused by openings are transferred through the longitudinal joints of the panels which may result in shear failure in a joint.

Small openings and cut-outs may cover only a part of the width of the sandwich panel. However, small openings reduce the bending and shear strength of a sandwich panel because the area of the face and core is diminished and because the corners of the openings cause stress concentrations that may give rise to an earlier failure than for sandwich panels without openings. Similar stress concentrations are also caused by the corners of large openings. Thus, in addition to the study of the load transfer from a sandwich panel with large openings to adjacent panels, the reduction of the capacity of the sandwich panel with large openings due to the stress concentrations in the corners of the openings has also to be examined.

In the design of an opening in a sandwich panel, it is necessary to determine whether the panel is still able to carry the load, in spite of the reduction of the strength due to the opening, or whether the sandwich panel has to be strengthened. No strengthening is needed if:

- there is a structural overcapacity in the panel, perhaps because the panel has been designed to satisfy thermal insulation requirements or to avoid a fire risk, which results in higher material thicknesses than were required by the load-bearing capacity alone, or
- the openings are located outside the structurally critical cross-sections of the panel, or
- due to the reduced rigidity of the panel containing the opening, sufficient of the load is transferred to the adjacent sandwich panels so that the reduced capacity is adequate to carry the remaining load.

The reduction in the compressive strength of the face and the shear strength of the core due to openings and cut-outs is very much dependent on the individual sandwich panel product. It is not possible to give generally acceptable rules with which to evaluate this reduction at the present time.

If the sandwich panel with an opening does not have sufficient capacity to carry the stresses caused by the mechanical and temperature loads, the panel has to be strengthened by increasing the thickness of the face or the depth of the panel or by increasing the strength of the core. If, for any reason, strengthening of the sandwich panel is not possible, additional stiffeners have to be provided around the opening or additional framing to carry the whole load in the sandwich panel has to be designed.

12.5 FLEXIBILITY OF FASTENERS AND SUPPORTS

In current design calculations, sandwich panels are assumed to be supported by immovable supports which allow the panels to rotate and to move axially but which prevent any deflections normal to the panel. However, there are displacements at the supports caused by deformations in the panel itself, deformation of the fastenings and, in particular, displacements in the supporting structure. In some design cases, as for example in large cold stores, the flexibility at the supports may be utilised in order to reduce the thermal stresses in continuous multi-span sandwich panels. Figure 12.13 illustrates this by a case where the panel is free to move at a support by distances of Δw_- and Δw_+. In (a), this movement is free whereas in (b) there is a finite resistance (see Chapter 7).

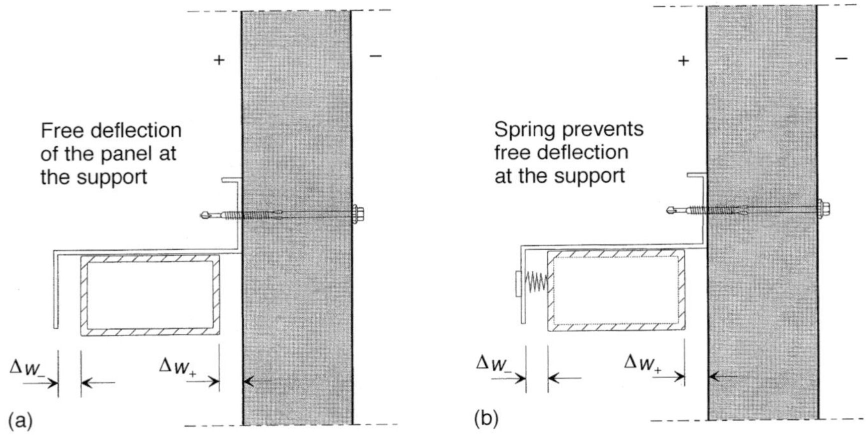

Fig. 12.13 Flexible support details.

If a sandwich panel is loaded by a positive support reaction, so that the support reaction causes compressive contact stresses between the panel and the supporting beam, the support reaction results in compressive deformations in the relatively soft core layer. The consequential deformations in the panel provide the first part of the flexibility at the support. The second part of the flexibility at the support is caused by the deflections of the supporting beams. This second part may be significant if the supporting beams are slender thin-walled cold-formed steel profiles. Figure 12.14 shows some results obtained by measuring the flexibility (a) at the end and (b) at the central support of an 85.4 mm deep and 600 mm wide sandwich panel with a polyurethane foam core in a cyclic temperature loading test. The support beams had a width of 100 mm and were supported in such a way that they could not move so that the flexibility is caused solely by the compressive deformations of the core.

Figure 12.15 shows some experimental and theoretical results obtained from a full-scale two-span temperature test on a 100 mm deep and 1200 mm wide stone wool cored sandwich panel. In this test, the support beams were again rigid so that the results show the influence of compressive deformations of the core layer on the central support reaction and deflections. The graphs show the test results compared with calculated results in which the supports have been assumed to be totally immovable in the direction transverse to the sandwich panel.

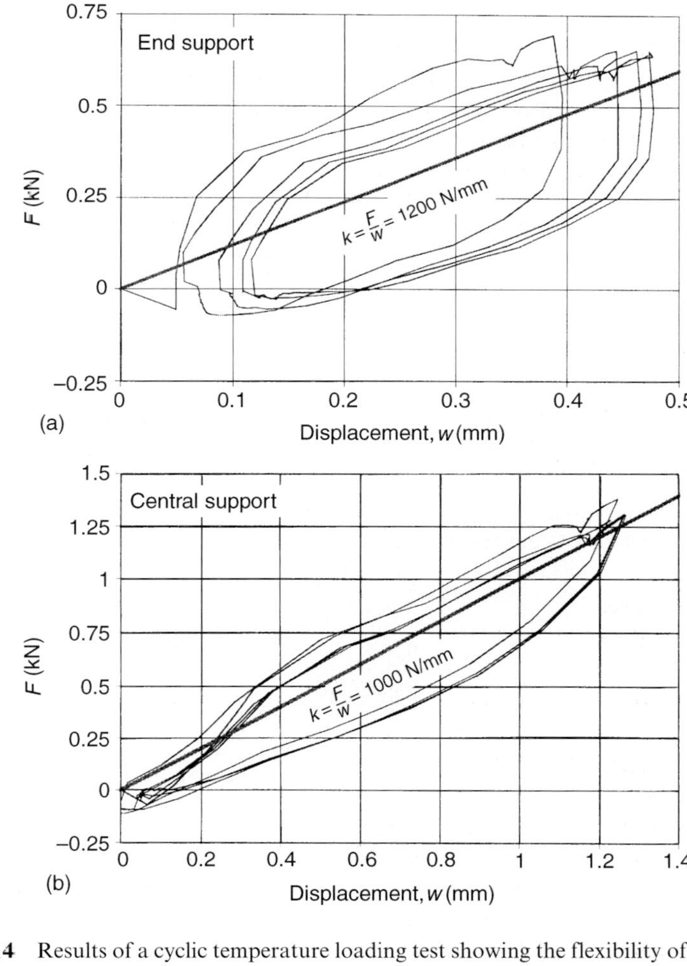

Fig. 12.14 Results of a cyclic temperature loading test showing the flexibility of a two-span polyurethane-cored sandwich panel loaded by a positive support reaction (a) at the end and (b) at the central support.[12.6]

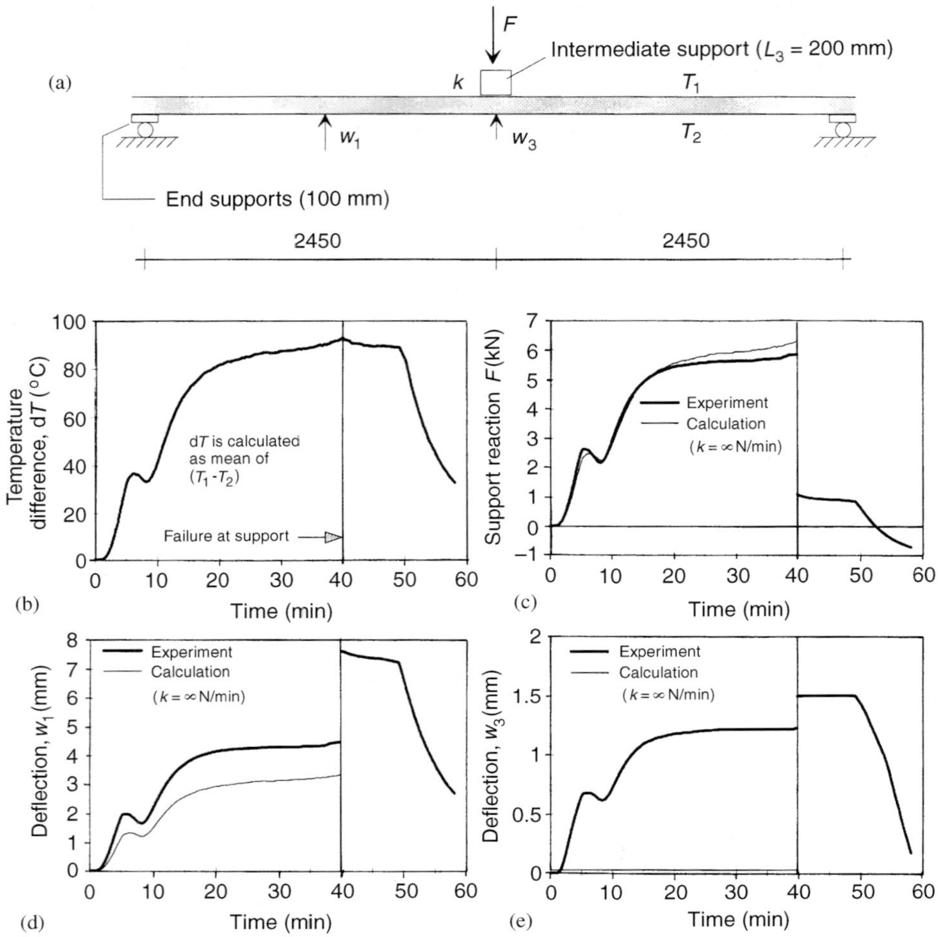

Fig. 12.15 Experimental and calculated results for full-scale two-span temperature tests on a sandwich panel with a stone wool core and flat steel sheet faces loaded by a positive support reaction: (a) static system; (b) temperature difference history; (c) support reaction at the central support; (d) deflections in the span and (e) deflections at the central support. The vertical line shows the time of local failure in the compressed external face at the central support.[12.11]

If a sandwich panel is loaded by wind suction loads, or by thermal loads which cause negative support reactions, thus causing tensile forces in the fasteners, displacements consisting of three components are produced at the supports. The first component is caused by the deformations in the core and face layers and this is very much dependent on the way in which the fastening system transmits forces into the core and face layers. The second part is caused by the deformation of the fastening system itself and the third part arises from the bending deformations of the supporting structure. The flexibility of common screw fastening systems is caused mainly by the first part, namely by the local bending deformation of the face sheet and by the local compression of the core under the heads of the screws. The flexibility then depends primarily on the number and location of the screws in the cross-section. The flexibility of a sandwich panel fixed with screw fasteners depends, of course, also on the third part, namely on the bending deformations of the supporting structure. Special roof fastening systems placed in the longitudinal joints between two sandwich panels are generally more flexible than through-fixed screw fasteners because of the tensile and bending deformations of the fastening strips, i.e., the second component of deformation. Additional flexibility arises because of the tendency of the panels to bend about a longitudinal axis.

Figure 12.16 and Table 12.1 show the tensile stiffness and resistance of a screw connection in a polyurethane cored sandwich panel. The depths of the test panels were 75 mm and 150 mm and the diameter of the screw and washer 6.3 mm and 19 mm respectively. The local screw force also causes deformations in the transverse direction of the panel.

Table 12.1 Tensile stiffness and resistance results for panel in Fig. 12.16

Depth of the panel, $e = 75\,\text{mm}$	F_{exp} (kN)	k^a (N/mm)	Depth of the panel $e = 150\,\text{mm}$	F_{exp} (kN)	k (N/mm)
End support mean value	3.9	210	End support mean value	4.6	270
Central support mean value	4.2	290	Central support mean value	4.6	350

$^a k = F/(u_1 - u_2)$

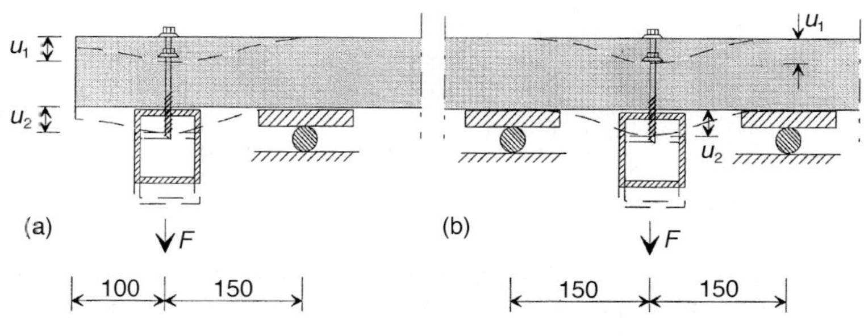

Fig. 12.16 Test arrangements for the tensile stiffness and resistance of screw fasteners drilled through polyurethane cored sandwich panels.[12.6]

Figure 12.17 shows the flexibility of a special roof panel fixture, of the type shown in Fig. 12.8(d), placed in the longitudinal joint between two 100-mm deep stone wool cored panels with flat steel sheet faces measured in a cyclic temperature loading test which caused negative support reactions at the central supports. Figure 12.18 gives some experimental results for the support reactions and deflections for a three-span full-scale sandwich panel of the type considered in Fig. 12.17. The comparison of the experimental results with the calculated results shows the influence of the flexibility of the roof panel fastening system on the support reactions and deflections. In the calculations, the supports have been assumed to be immovable ($k = \infty$) or else they have a finite tensile stiffness ($k = 300\,\text{N/mm}$) against negative support reactions.

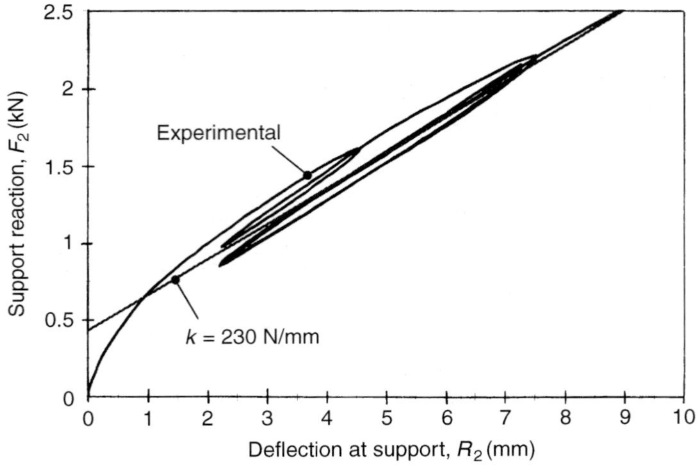

Fig. 12.17 Deflection versus support reaction at a central support of a stone wool cored sandwich panel.[12.11]

Flexibility at the supports has a large effect on the distribution of support reactions, bending moments, shear forces and deflections. Flexibility at the supports reduces the support reactions and thus the bending moments and shear forces caused by temperature differences between the external and internal faces of continuous multi-span sandwich panels. This flexibility also causes changes in the bending moment and shear force diagrams arising from wind and snow loads. Although the above effects may often be benign, increased flexibility increases the deflections caused by any load and this may be important in cases for which strict deflection limits are required.

The static behaviour of a sandwich beam fixed with flexible fastenings or supported by flexible supports can be described by a finite-element model consisting of beam elements in which the nodal points at the supports are supported by springs. Suitable finite elements for this purpose are given in Chapter 9. For the purposes of practical design calculations, a constant spring coefficient, k, can be used to describe the flexibility at the supports. Care should be taken in defining this spring constant because it is likely to require different values for positive (compressive) and negative (tensile) reaction forces.

As discussed above, the spring coefficient k may be the resultant of two or three different deformation components, namely the deformations in the core of the sandwich panel, the deformations of the fastening system and deflection of the supporting structure. If the flexibility caused by each of these deformations can be described by the spring stiffness coefficients k_1, k_2 and k_3, the resultant stiffness k can be calculated from

$$\frac{1}{k} = \frac{1}{k_1} + \frac{1}{k_2} + \frac{1}{k_3} \tag{12.17}$$

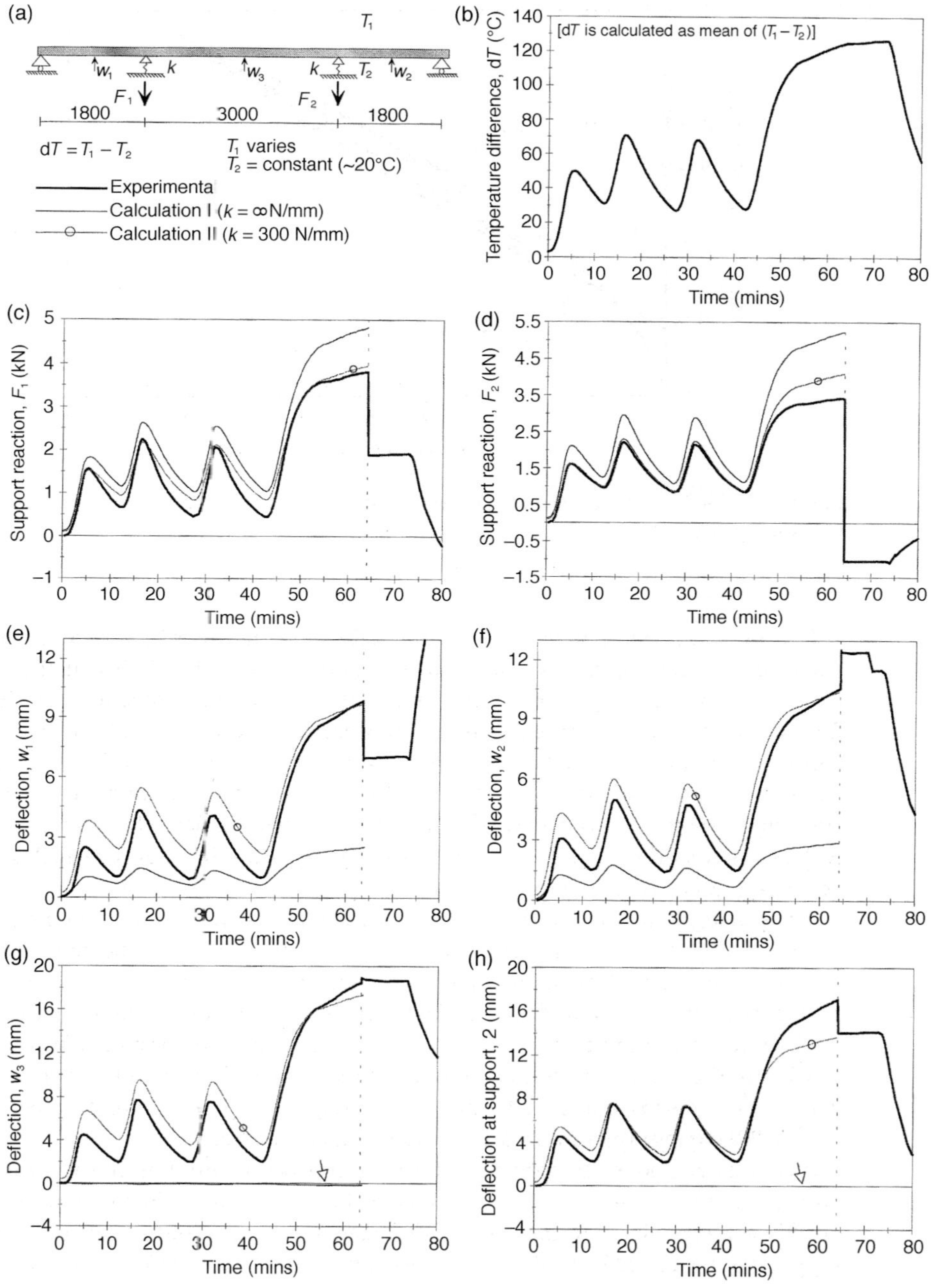

Fig. 12.18 Experimental and calculated results for a three-span full-scale sandwich panel loaded by the self-weight and a temperature difference between the face layers. (a) Static system, (b) temperature difference history, (c) and (d) support reactions at the central supports, (e) (f) and (g) deflections in the spans and (h) deflection at a central support. The broken line shows the point of local failure in the compressed external face.[12.11]

12.6 LATERAL AND ROTATIONAL STIFFNESS OF FASTENINGS

The stiffness and strength of the fastenings to sandwich panels may be utilised in order to stabilise slender support beams against lateral buckling. This form of lateral support is an especially important parameter in the design of cold-formed steel Z- and C-purlin profiles. For the purposes of analysis, the lateral support offered to a purlin is assumed to consist of two components, a transverse spring, k_S, and a rotational spring, C_D, as shown in Fig. 12.19. In practical applications, the spring coefficients are assumed to be constant, though having separate values for pressure and uplift forces. The lateral stiffness and strength of the fastenings may also have an influence in the design of long sandwich panels loaded as a consequence of a high temperature difference between the faces layers.

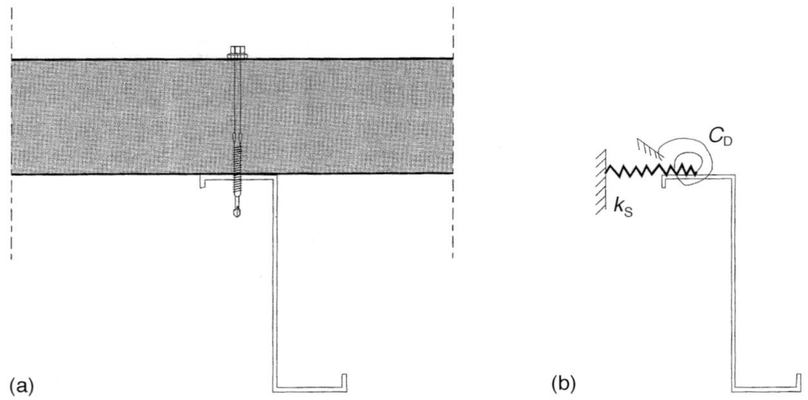

Fig. 12.19 Calculation model for the analysis of the lateral buckling strength of a Z-profile: (a) screw fastening between a sandwich panel and a cold-formed Z-profile and (b) calculation model comprising a shear spring, k_S, and a rotational spring, C_D.

The shear stiffness, k_S, of a screw fastening through a sandwich panel is mainly caused by shear deformations of the screw hole in the lower face and by bending deformation of the fastener. The shear stiffness of the special fastening systems is caused by deformations of the screw holes and the steel strips. Because the deformations of the screw holes are partially plastic, tests with repeated loads give rise to highly non-linear load–deformation curves as illustrated by Fig. 12.20. In this figure, the wall thickness of the supporting structure, which was a rectangular hollow section, was 6 mm and the steel quality was S355. The steel thickness and yield stress of the face of the panel placed against the support were 0.47 mm and 373 N/mm^2, and the corresponding values of the outer face were 0.55 mm and 398 N/mm^2, respectively. The core of the specimen was made of structural stone wool and the effective thickness of the specimen was 100 mm.

Characteristics such as the initial stiffness and the ultimate strength of the load–deformation curves depend very much on the clearance of the screw hole and on the tightening of the screw because a part of the shear force is transmitted through the friction between the face and the support structure. Therefore, the scattering

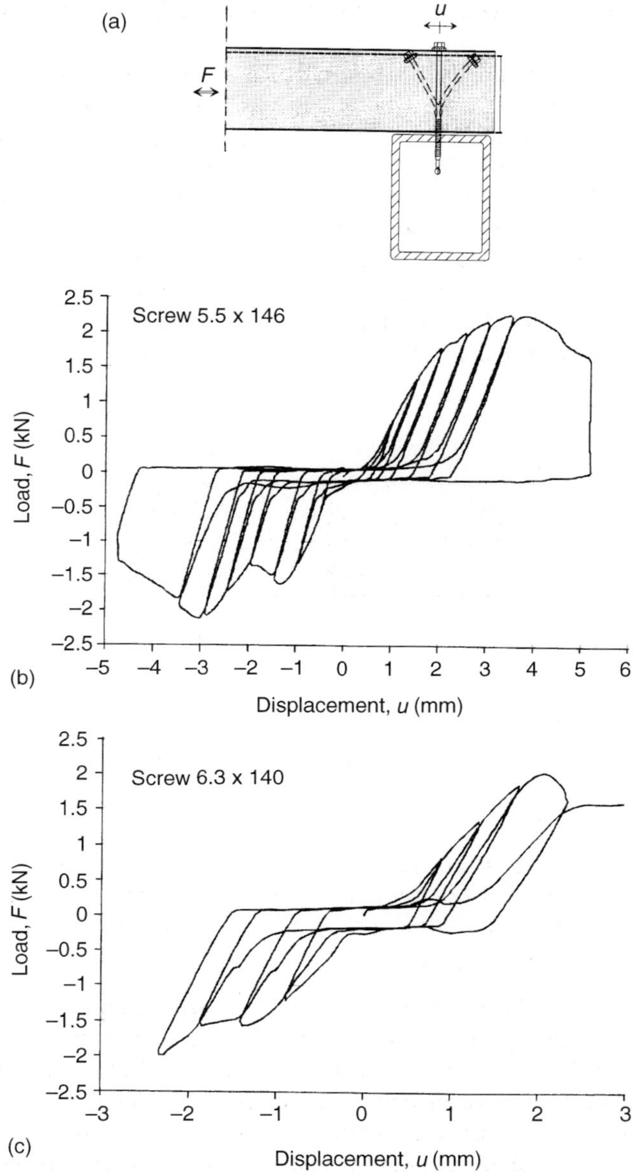

Fig. 12.20 Results of repeated loading tests on the shear resistance of self-drilling screw drilled through a sandwich panel: (a) schematic test arrangement, (b) test results for the screw connection $d = 5.5\,\text{mm}$ and (c) $d = 6.3\,\text{mm}$.

between the results of individual tests made in different laboratories might be high. In contrast to the rotational stiffness, the shear stiffness does not play an important role in the analysis of the lateral buckling of purlin profiles. However, the shear stiffness and especially the shear strength might be a useful parameter in the evaluation of the forces in the connections for loading cases in which the deflections are very large or in which the temperature expansion stresses the face against the supporting structure.

The rotational stiffness C_D provides torsional restraint to the purlin profile. This rotational stiffness depends on:

(1) the rotational stiffness, $C_{D,A}$, of the connection between the sandwich panel and the purlin profile
(2) cross-sectional distortion of the purlin profile, $C_{D,B}$, and
(3) the bending stiffness of the sandwich panel, $C_{D,C}$

The total rotational stiffness C_D can be evaluated using the equation:

$$\frac{1}{C_D} = \frac{1}{C_{D,A}} + \frac{1}{C_{D,B}} + \frac{1}{C_{D,C}} \tag{12.18}$$

The rotational flexibility of the connection depends on the width of the flange of the purlin profile, the number and location of the screws into the flange and on the bending stiffness of the lower face and the compressive stiffness of the core layer of the sandwich panel. Special fastening systems placed in the longitudinal joints of sandwich panels have additional parameters which may also have a significant influence on the rotational stiffness. Because of the large number of variables, some of which are not well understood, no design expressions have been derived for the rotational stiffness $C_{D,A}$ so that this quantity has to be determined experimentally.

The flexibility $C_{D,B}$ caused by the distortion of the purlin profile can be analysed on the basis of the geometry and the material properties of the purlin profile. It depends also on the direction of the load and on the location of the fastener in the connected flange. The third term in Expression 12.18, $C_{D,C}$, depends on the flexural rigidity of the sandwich panel and, furthermore, on the static system and the span of the sandwich panel and this too can be defined analytically.

The rotational stiffness provided by both trapezoidal sheeting and sandwich panels fastened to hot-rolled steel I-sections with self-tapping screws has been studied experimentally by Lindner.[12.9, 12.10] He has proposed expressions with which to evaluate the partial stiffnesses $C_{D,B}$ and $C_{D,C}$ and has given experimental results for $C_{D,A}$ based on tests on polyurethane foam sandwich panels.

In ENV 1993-1-3 1996,[12.4] an alternative method for the evaluation of the stabilising function provided by a trapezoidal sheeting and its connections is given, in which the rotational spring is transformed to a lateral spring connected to the unbraced flange of the purlin. In this way, the study of the lateral-torsional stability of purlin profiles braced by their fastening to trapezoidal sheeting is simplified to a study of the buckling of the elastically-supported, unbraced flange of the purlin profile. ENV 1993-1-3 1996 gives test arrangements with which to determine the rotational stiffness $C_{D,A}$ or the lateral spring stiffness experimentally.

The rotational stiffness $C_{D,C}$ provided by sandwich panels with thin flat or lightly profiled faces can be evaluated on the basis of the expressions given in ENV 1993-1-3 1996 for trapezoidal sheeting. In the case of sandwich panels with strongly profiled faces, the shear stiffness also has an influence on $C_{D,C}$ and therefore, the expressions become more complicated:

$$C_{D,C} = \frac{M}{\phi} \tag{12.19}$$

where M is the stabilizing bending moment and ϕ is the rotation at the support.

For single-span thin-faced sandwich panels, a conservative value of $C_{D,C}$ is given by

$$C_{D,C} = \frac{2B_s}{L} \tag{12.20}$$

Correspondingly, for continuous multi-span thin-faced sandwich panels, $C_{D,C}$ can be evaluated at intermediate supports using

$$C_{D,C} = \frac{4B_s}{L} \tag{12.21}$$

In Equations 12.20 and 12.21, B_s is the flexural rigidity and L the span of the sandwich panel.

In practice, the partial rotational stiffness $C_{D,A}$ has the most importance in the design calculations. In the case of thin slender cold-formed purlin profiles, the stiffness $C_{D,B}$ also has a large influence on the capacity of the purlin. In many cases, $C_{D,C}$ has only a minor effect and it can usually be neglected in the design calculations.

In the design of sandwich panels and connections it is important to bear in mind that they may also have this stabilising function. The stabilising function may be of equal importance during the erection phase of construction as during the service life.

12.7 CAPACITY UNDER AXIAL LOAD

Sandwich panels have high flexural rigidity and, therefore, have high resistance with respect to axial compressive loads. Their capacity may be limited by Euler-type global buckling of the sandwich column or by local buckling or yielding of the face. Analytical and numerical methods with which to evaluate the global Euler-type buckling load N_{cr} of sandwich columns with flat or profiled faces are given in Chapter 9.[12.3]

At the ends of sandwich panels, the axial loads are generally transferred from other parts of the structure primarily into the faces of the sandwich columns. Because of the eccentricity of the loading, the axial load is, in the worst case, transferred into one face layer only. The ends of the thin faces are very sensitive to local geometric imperfections and transverse loads and eccentricities and, therefore, they have to be stiffened to resist the local failure modes.

In order to study the local axial strength of the ends of typical sandwich panels, four different end details were used to transfer the axial load from a timber structure to the thin flat faces of the panels in a test series at Helsinki University of Technology. In the first and second cases (1 and 2 in Fig. 12.21(b)) no additional stiffeners were used but the load was transferred either to the whole cross-section or to the core layer of the test specimens. In the third and fourth cases (3 and 4 in Fig. 12.21(b)) plywood stiffeners were glued or hot-rolled steel angle stiffeners were screwed to the ends of faces. In the cross-sections of the steel angles, the holes for

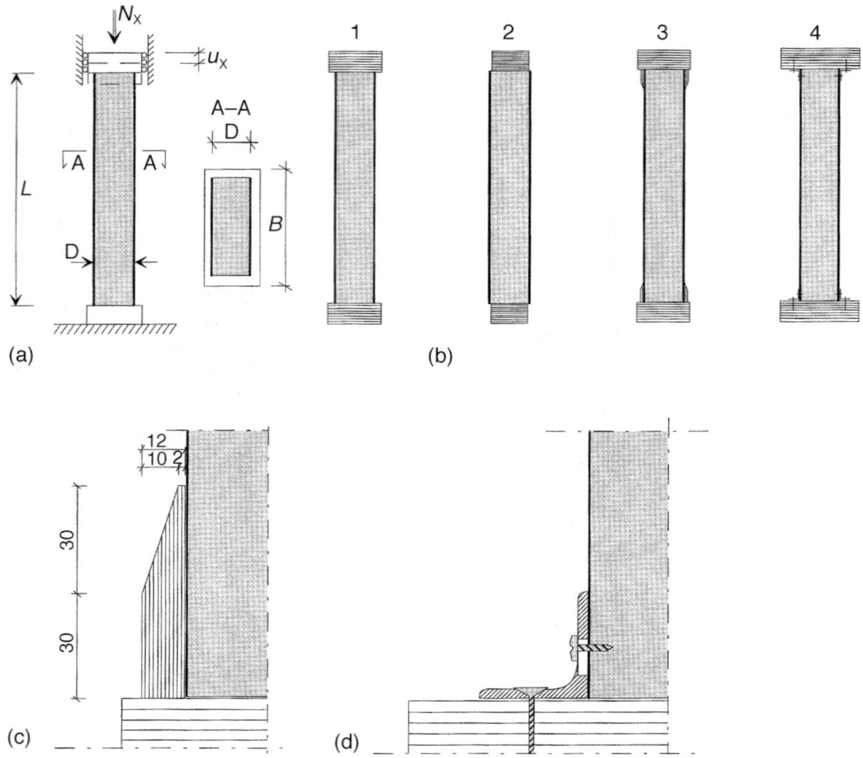

Fig. 12.21 Testing of the local compressive strength at the end of short sandwich panel columns. (a) Load arrangements and (b) four different end details for the short sandwich columns used in the tests. (c) Plywood profiles, which were glued on the steel faces, were used as stiffeners in the third end detail. (d) Stiffeners in the fourth end detail were steel angle profiles fixed with screws through enlarged screw holes into the steel faces.

the screws were enlarged in order to avoid the transfer of the axial load through the screws and thus to avoid any additional imperfections caused by the stiffeners. The specimens were loaded using displacement control with clamped boundary conditions at the ends of the specimens as shown in Fig. 12.21(a).

The width of the specimens, B, was 250 mm and the depth, D, was 100 mm. The steel thicknesses of the external and internal face sheets were 0.48 and 0.46 mm for the stone wool (rw) specimens and 0.57 and 0.57 mm expanded polystyrene (eps) specimens, respectively. The initial compressive strength of the faces of the stone wool cored specimens varied between 150 and 200 N/mm^2 and the strength of the faces of the expanded polystyrene foam cored specimens used in the tests between 100 and 150 N/mm^2. As shown in Fig. 12.22, the test results show that the stiffening system based on the steel angle sections produces local compressive strengths that are close to the initial compressive strengths between the ends of the specimens. Plywood stiffeners caused an eccentricity of local loads at the ends and thus, the strengths in case 3 were comparable to the strengths in the first case in which the load was transferred directly into the unstiffened faces. In order to

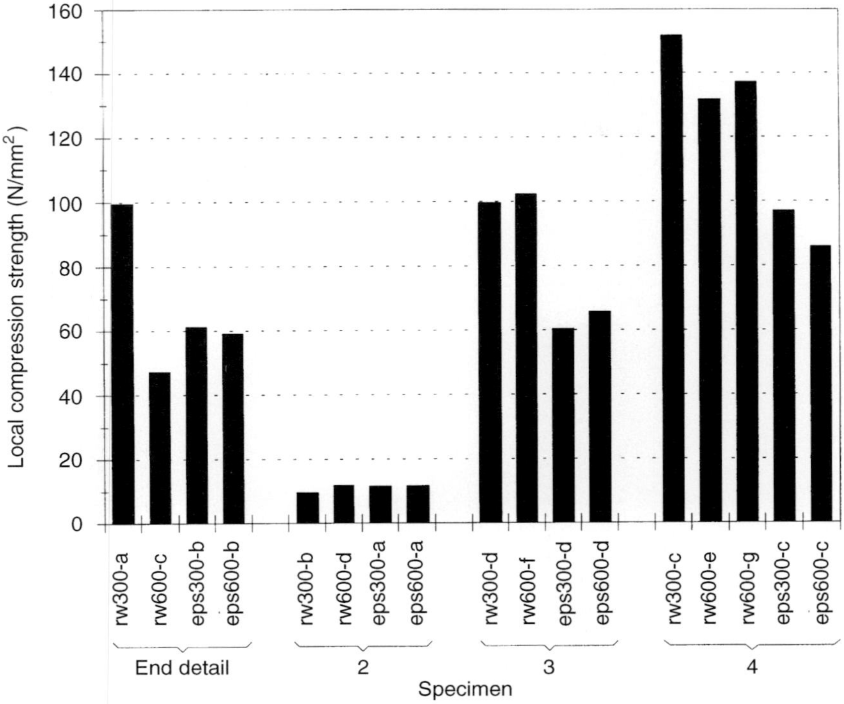

Fig. 12.22 Compression strength of the ends of axially loaded sandwich columns loaded and stiffened with different end details. The numbers in the designation of the specimens show the length of the specimen (mm).

increase the axial strength, the plywood stiffeners should be longer than those used in the tests as shown in Fig. 12.21(c). The lowest local strengths were obtained in the second case, in which the cores of the specimens failed in compressive and shear failure modes. The failure modes in cases 1, 3 and 4 were brittle with relatively little post-ultimate axial strength after the ultimate load had been reached.

The results of the test series show that there are large differences between the local axial strengths of sandwich panels provided by the different end details. It follows that the performance of any proposed end details has to be studied and their properties carefully verified. The test results show also that the stiffening of the ends of thin flat faces against the local compressive failure mode can be based on the conventional methods used in building practice.

In practical applications, axially loaded sandwich panels are usually also stressed by transverse loads. For the purposes of design, the interaction between the axial or eccentric axial compressive forces and bending moments has to be studied. Because the axial or eccentric compressive load, the bending moment and the shear force result in compressive and tensile stresses in the faces and shear stresses in the core of sandwich panels, provided that the stress resultants (N, $M + \Delta M(N)$) are known, the design of sandwich beam-columns can be based on the known failure modes introduced in Chapter 10 and the design equations given in Chapter 11 (Fig. 12.23). Hartsock[12.5] has proposed a linear interaction equation, $\sigma_F = \sigma_{FN} + \sigma_{FM}$,

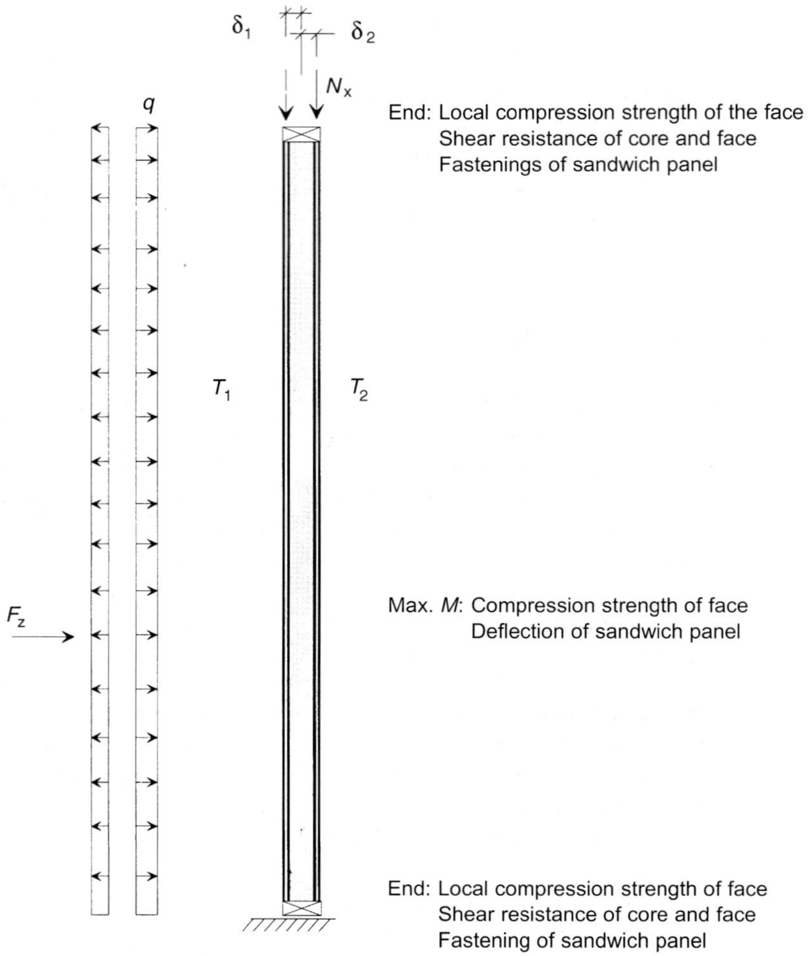

Fig. 12.23 Critical cross-sections of a single-span sandwich panel loaded with axial and transverse loads.

between the face stresses caused by axial loads σ_{FN} and bending moments σ_{FM} for sandwich beam-columns with flat or lightly profiled faces. This might be valid for axial loads which are small compared with the global Euler-type buckling loads. In such cases, $\Delta M(N) \approx 0$.

In general, the design equations for a compressive failure of the face and for a shear failure of the core of a sandwich beam-column with flat or lightly profiled faces can be written, respectively, as:

$$\sigma_{\mathrm{Fdi}} = \frac{\gamma_{\mathrm{F}} E_{\mathrm{Fi}} N_{\mathrm{k}}}{(E_{\mathrm{F1}} t_1 + E_{\mathrm{F2}} t_2) B} + \frac{\gamma_{\mathrm{F}} (M + \Delta M(N))_{\mathrm{k}}}{e B t_i} \leq \frac{f_{\mathrm{Fck}}}{\gamma_M} \tag{12.22}$$

and

$$\tau_{Cd} = \frac{\gamma_F(V + \Delta V(N))_k}{eB} \leq \frac{f_{Cvk}}{\gamma_M} \tag{12.23}$$

The geometrically non-linear interaction study can also be performed numerically using, for instance, a non-linear finite-element method together with the relevant load and material safety factors, depending on the failure mode. The design expressions given in ENV 1993-1-3 1996[12.4] for axially loaded trapezoidal steel sheets may be applicable to sandwich beam-columns with strongly profiled faces. However, no research results are available on this subject.

12.8 DIAPHRAGM ACTION

A series of sandwich panels fixed continuously together and to the supporting structure constitutes a very rigid sheet that has the potential to carry large shear loads in the plane of the sheet with small deformations. This physical function of a series of sandwich panels is called 'diaphragm' or 'stressed skin' action. The shear stiffness and the shear resistance in the plane of the sandwich panels can be utilised in stabilising the frames of buildings against either horizontal or vertical loads. They may also be utilised in stabilising structural elements such as cold-formed purlins against the different buckling failure modes. This in-plane shear stiffness may also be activated in certain types of shell structures and in folded plate roof construction. Depending primarily on the connections between adjacent sandwich panels and between the panels and the supporting structure, the in-plane shear stiffness of walls and roofs may take part in stabilising the frame of a building regardless of whether or not the sandwich panels have been designed for that function. Therefore, the in-plane shear stiffness and the resistance are important characteristics of sandwich panels mounted in large continuous areas in the walls and roofs of buildings.

Wall panels are usually only fastened to the supporting structure and are not connected to each other along the longitudinal joints. It follows that wall panels have a relatively low strength and a rather large flexibility with respect to in-plane shear forces (see Fig. 12.24). An in-plane shear wall becomes much more rigid if the

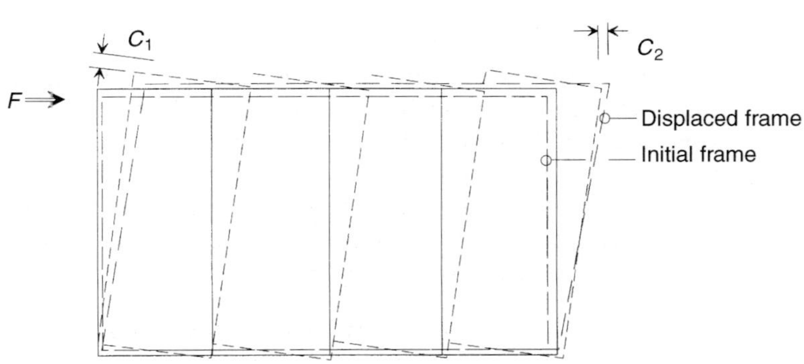

Fig. 12.24 In-plane shear deformations in the longitudinal joints between sandwich panels, C_1 and in the edge joints between the sandwich panels and the supporting structure, C_2.

adjacent sandwich panels can be connected together along the longitudinal joints. This is especially true if the span of the sandwich panels is large.

In roof panels, the typical seam fasteners in the external faces connect the adjacent panels lightly together. The in-plane shear rigidity is dependent on the location, number and type of the fasteners in the longitudinal joints between the adjacent sandwich panels and in the joints between the panels and the supporting structure. If the sandwich panels have been fixed to the supporting structure with screws passing through the full thickness of the panel, it is usually the shear strength of the shafts of the screws in the lower face which determines the in-plane stiffness and the resistance. If, as is usually the case, the lower face is relatively thin, the panel assembly is likely to have a low shear resistance. The in-plane shear stiffness and the resistance can be increased by fixing both the faces together and, furthermore, fixing them directly to the supporting structure. This generally requires additional load-bearing components in the longitudinal seams as shown in Fig. 12.25.

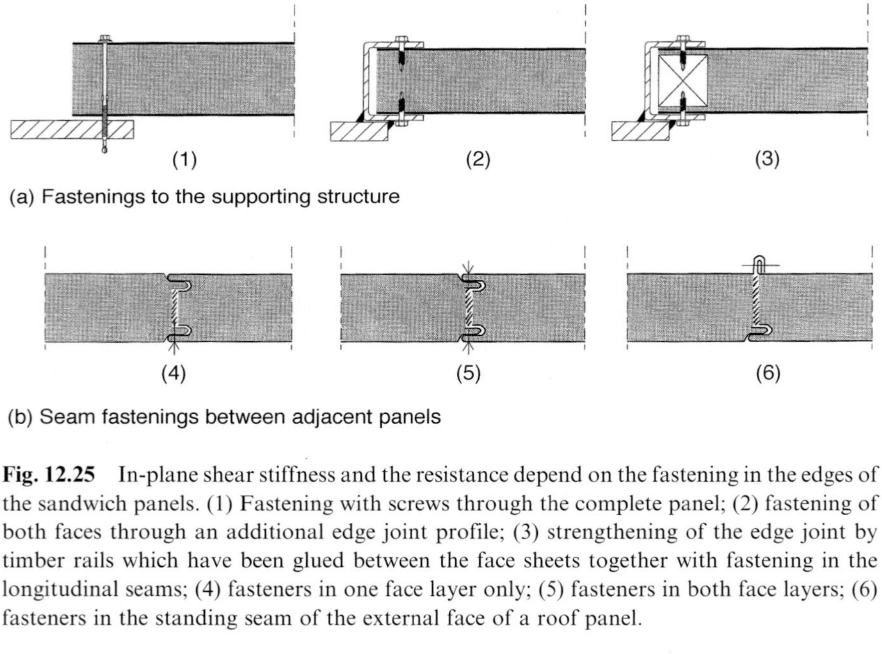

Fig. 12.25 In-plane shear stiffness and the resistance depend on the fastening in the edges of the sandwich panels. (1) Fastening with screws through the complete panel; (2) fastening of both faces through an additional edge joint profile; (3) strengthening of the edge joint by timber rails which have been glued between the face sheets together with fastening in the longitudinal seams; (4) fasteners in one face layer only; (5) fasteners in both face layers; (6) fasteners in the standing seam of the external face of a roof panel.

In order to analyse the stresses in the faces and in the fastenings of sandwich panels in shear, the displacements of each single sandwich panel relative to the supporting structure must be determined. The in plane displacements and rotations depend mainly on the shear flexibility of the fastenings in the longitudinal joints between the adjacent sandwich panels and in the joints between the sandwich panels and the supporting structure as shown in Fig. 12.24. The flexibilities of the fastenings have to be determined experimentally. From the experimental, and usually non-linear, shear load–deformation curves, tangent modulus values are determined to represent the shear flexibilities of the fastenings. In the analyses, the shear deformations have to be limited in order to avoid plastic deformations at the serviceability limit state. In some rare cases, the shear flexibility of the sandwich panel itself and the deformations in the supporting structure may also have an influence on the stresses in the face layers and in the fastenings. Calculation procedures with which to analyse the shear forces in the joints of the sandwich panels have been developed by Baehre and Ladwein.[12.15–12.18] Useful guidelines for the analysis of the shear rigidity and the resistance of sandwich panels are also given in the European Recommendations for the application of metal sheeting acting as a diaphragm.[12.14]

The in-plane shear resistance of a sandwich panel wall or roof is typically determined by a shear failure in a fastening as a result of local tearing of one of the face sheets or shear failure of the fastener itself. If the longitudinal joints have been strengthened and connected together with a large number of screws, the shear resistance may, in rare cases, also be limited either by global shear buckling of the complete sandwich panel assembly or by a local shear buckling failure mode in one of the face sheets. The possibilities to increase the in-plane shear stiffness and the resistance of a panel assembly by placing fasteners in the longitudinal seams, by adding fasteners in the edge joints and by additional strengthening of the joints depends in practice not only on using appropriate technical solutions but often also on the architectural requirements.

12.9 CONCLUSIONS

Chapter 12 introduced methods to solve practical questions in the design of sandwich panels and panel assemblies such as the influence of openings. It also introduced new ideas with which to increase the technical utilisation of sandwich panel systems by taking into account the flexibility of the fastenings and by making use of the axial loading and the in-plane shear loading capacity of the sandwich panels. These methods and ideas have not yet been developed to the level of standardised design guidelines and recommendations. More systematic research and more practical experience are needed before general design recommendations can be given. Some of the problems such as the strengthening of panels due to the presence of significant openings, due to the local failures caused by axial loads and due to the shear failures in the joints as a result of diaphragm action may be very dependent on the detailing of an individual sandwich panel product so that there are limited possibilities to draft any general design guidelines. However, these relatively new possibilities to utilise structural sandwich panels have been shown primarily for future use and design practice. Nevertheless, these ideas can already be put into use today in special design cases, in which the resistance of a particular sandwich panel product against the relevant load system can be studied on a case-by-case basis.

12.10 REFERENCES

12.1 Chong, K.P. and Hartsock, J.A. (1993) Structural analysis and design of sandwich panels with cold-formed steel facings. *Thin-Walled Structures*, **16**, 199–218.

12.2 Chong, K.P., Engen, K.Ø. and Hartsock, J.A. (1977) Thermal stresses and deflection of sandwich panels. *Journal of the Structural Division*, **103**, 35–49.

12.3. Davies, J.M. (1987) Axially loaded sandwich panels. *Journal of Structural Engineering*, **113** (11), 2212–2230.

12.4 European Committee for Standardization (1996) *Document ENV 1993-1-3. Design of Steel Structures: Part 1.3: Supplementary Rules for Cold Formed Thin Gauge Members and Sheeting*, 128 pp.

12.5 Hartsock, J. (1991) *Design of Foam-Filled Structures*, 2nd edn. Technomic Publishing Company, 89 pp.

12.6 Hassinen, P. and Helenius, A. (1990) Design of sandwich panels against thermal loads. In *Proceedings of IABSE Symposium, Mixed Structures, Including New Materials*, Brussels September 5–7, 1990. IABSE Reports, Vol. 60 (1990), pp. 605–610.

12.7 Höglund, T. (1986) Load bearing strength of sandwich panel walls with window openings. In *Proceedings of IABSE Colloquium*, Stockholm 1986, IABSE Reports Vol 49, pp. 349–356.

12.8 Paroc Panel Systems (1995) Design of Paroc walls. In *Design Guidelines of Paroc Sandwich Panels*.

12.9 Lindner, J. (1988) *Drehbettungswerte für Dachdeckungen mit untergelegter Wärmedämmung*, Studiengesellschaft für Anwendungstechnik von Eisen und Stahl e.V., Düsseldorf [in German] 143 pp.

12.10 Lindner, J. and Gregull, T. (1986) Torsional restraint coefficients of profiled sheeting. In *Proceedings of IABSE Colloquium, Thin-Walled Metal Structures*, Stockholm 1986. IABSE Reports, Vol. 49, pp. 161–168.

12.11 Martikainen, L. and Hassinen, P. (1996) Load-bearing capacity of continuous sandwich panels. Helsinki University of Technology, Department of Structural Engineering, Report 135. 221 pp.

12.12 Stamm, K. and Witte, H. (1974) *Sandwichkonstruktionen (Sandwich Structures)*, Springer-Verlag, Vienna, 337 pp. [in German].

12.13 Heselius, L. (1986) Elastically braced light gauge beams with open sections. In *Proceedings of IABSE Colloquium*, Stockholm 1986, IABSE Reports Vol. 49, pp. 27–34.

12.14 ECCS (1995) *European Recommendations for the Application of Metal Sheeting Acting as a Diaphragm*, Publication No. 88.

12.15 Baehre, R. and Ladwein, T. (1994) *Tragfähigkeit und Verformungsverhalten von Scheiben aus Sandwich-elementen mit Stahldeckschichten und PUR-Schaumkern*, Studiengesellschaft Stahlanwendung e.V. Projekt 199, 155 pp. [in German].

12.16 Baehre, R. and Ladwein, T. (1994) Diaphragm action of sandwich panels. *Journal of Constructional Steel Research*, **31**(2/3), 305–316.

12.17 Ladwein, T. (1993) Zur Schubfeldwirkung von Sandwichelementen (Teil I) (Diaphragm action of sandwich panels, Part I). *Stahlbau*, **62**(11), 342–346 [in German].

12.18 Ladwein, T. (1993) Zur Schubfeldwirkung von Sandwichelementen (Teil II) (Diaphragm action of sandwich panels, Part II). *Stahlbau*, **62**(12), 361–363 [in German].

12.19 *Good Roofing and Cladding: Guide 3: Design Detailing*, Engineered Panels in Construction (EPIC), London (undated).

Chapter 13
Mechanical Testing

13.1 INTRODUCTION

Mechanical testing plays a significant role in both the design and quality control of sandwich panels. Primarily because of the variability of core material properties, a number of the parameters which appear in the design equations can only be reliably determined on the basis of a statistically significant number of tests. Testing is then used during production in order to ensure that the values of these parameters remain within the assumptions.

In addition to the mechanical properties, a number of other properties of sandwich panels are generally determined by testing. Thus, for example, tests for fire properties have been discussed in Chapter 5 and testing to evaluate the durability of panels has been discussed in Chapter 6. This chapter deals mainly with a series of essential tests for mechanical properties which are specific to sandwich panels. These tests have been extensively discussed internationally, first of all in the drafting of the European Recommendations[13.1] and more recently in connection with the CEN product standard for sandwich panels.[13.2] This chapter draws extensively from these documents.

Tests for mechanical properties can be classified in several different ways. Design tests, sometimes termed 'type tests' are used to obtain and assess the basic design data during product development and to confirm the theoretical or practical performance of the panel. These include tests to determine such parameters as the wrinkling stress of a face, the interaction of bending moment and support reaction and the behaviour of connections as well as basic material properties such as the compressive and shear strength of the core. Quality assurance tests are based on standard methods and are used to confirm that the required mechanical and other properties are maintained during production. Tests can be made separately on face materials, core materials or a complete panel. For material tests, small specimens are generally used, whereas loading tests are usually performed using full-sized panels.

The proper functioning of a sandwich panel requires that the materials, namely those of both the skins and the core, fulfil the requirements. It is equally important that these components form an integral structure such that the bond between them is reliable and can transfer the required forces. The tests on the facing materials are standardised and there is no need to consider them in detail here. The material tests that are specific to sandwich panels are tests on core materials and tests whose outcome depends on the bond between the core and the faces.

When comparing test data for rigid plastic foam core materials from various sources, both the influence of loading arrangement and loading rate on the results and the definition of the 'strength' should be borne in mind. It should also be appreciated that the temperature and relative humidity may also influence the stress–strain behaviour of most core materials. Plastic foams are susceptible to creep and, at higher strain rates, the behaviour appears to be more rigid because insufficient time is given for the visco-elastic strains to develop. Due regard, therefore, must be given to the strain rates specified in the relevant standards which describe the test methods. Figure 13.1 shows a typical compressive stress versus strain curve derived from a material test on a polyurethane foam. The curve is highly non-linear. Typically, in both compression and shear tests, rigid plastic foams may undergo considerable non-linear deformations without rupture. The stress corresponding to a permanent deformation of 0.4% is usually denoted as the 'yield stress'. The ultimate value, i.e. the 'strength', is usually taken to be equal to the stress at a strain of 10% or to the maximum value if this is found to correspond to a strain of less than 10%. The derivation of the elastic moduli is generally based on the early approximately linear part of the stress–strain curve, but this must follow the definition given in the relevant standard.

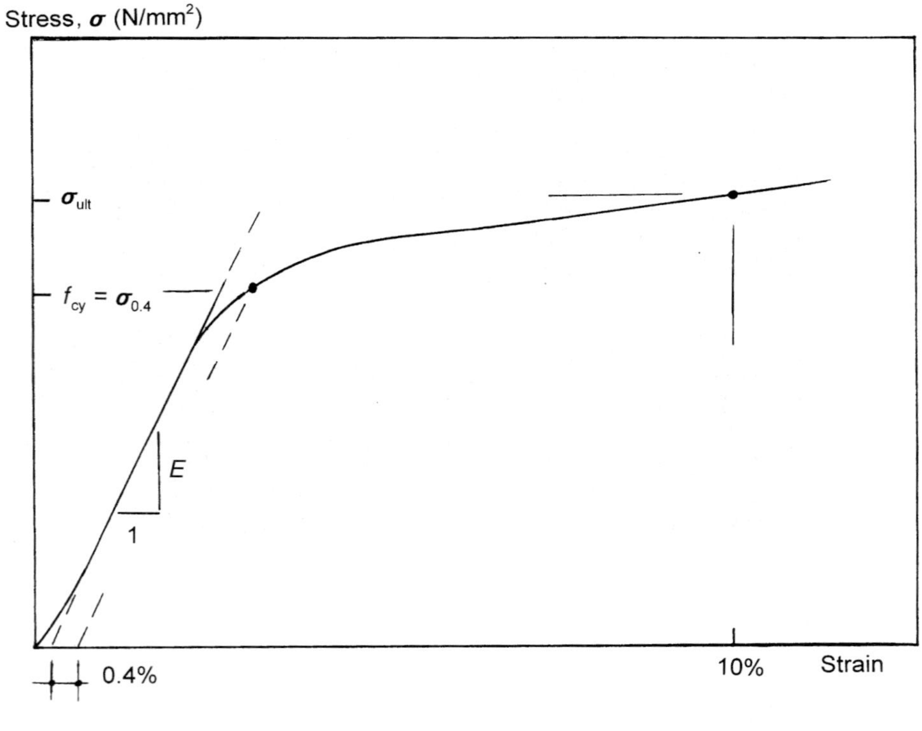

Fig. 13.1 Typical stress versus strain curve for a rigid plastic material.

13.2 TESTS ON CORE MATERIALS

The predominant requirement of the core material of a sandwich panel is the capability to resist shear stresses. Several test methods are available which may be used to measure the shear strength and shear modulus. The recommended method for rigid plastic foam core panels is to load a short strip of panel, with thin metal faces, in four-point bending. For mineral wool core panels a full-scale panel test is recommended. Sections 13.9.3.4 and 13.9.4.1 give detailed descriptions of these tests. The dimensions used in the four-point bending test should be chosen so that a shear failure is more likely to take place than wrinkling of the compressed face. Figure 13.2 shows the test arrangement and Fig. 13.3 some typical failed specimens. In these tests, however, the length of specimens was purposely chosen to be longer than in the standard shear test. As a consequence, there was the probability of both shear and wrinkling failures. In Fig. 13.3, specimens 1 and 2 have failed in a typical shear mode, whereas specimen 3 has failed by wrinkling of the face in compression.

The four-point bending test was originally developed for relatively thin panels with PUR cores. Nowadays, much thicker panels are frequently used, often with a mineral wool core. For such panels, a crushing failure may take place where the forces are transmitted to the specimen at the loading points and the supports and the test requires modification by using wider load widths at the contact points.

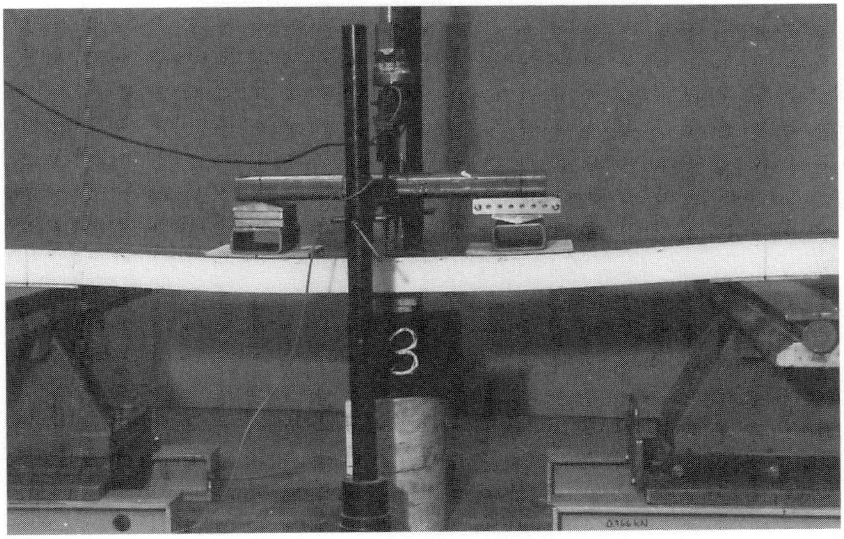

Fig. 13.2 Shear test on a specimen from PUR panel using four-point bending.

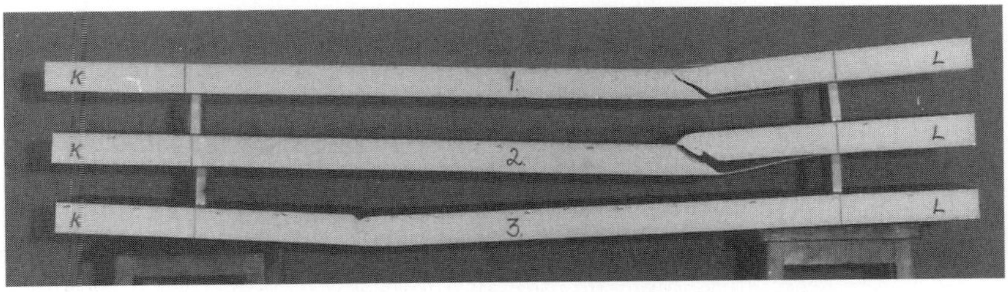

Fig. 13.3 Failed shear test specimens: shear failure in specimens 1 and 2, wrinkling failure in specimen 3.

However, wide load strips reduce the deflection and render the usual formulae for the calculation of the shear modulus invalid. For this reason, an alternative test method for thick mineral wool panels has been developed and this is described in Section 13.9.3.6.

In the so-called 'square' test, a rectangular prism is cut from the core material of the panel. This has a square cross-section which is usually 100 mm × 100 mm. The length of the prism is generally somewhat larger and of the order of 150 mm. The orientation of the prism inside the panel is such that it is generated by the axis perpendicular to the panel axis and parallel with the faces of the panel. Two of the four equal sides are parallel with the faces and the other two are perpendicular to the faces. Stiff plates, for example thick plywood pieces, are glued on to the four equal faces. When the specimen is subjected to either compression, as shown in Fig. 13.4, or to tension diagonally across the prism, the core material is subjected to the same state of shear stress as in the panel.

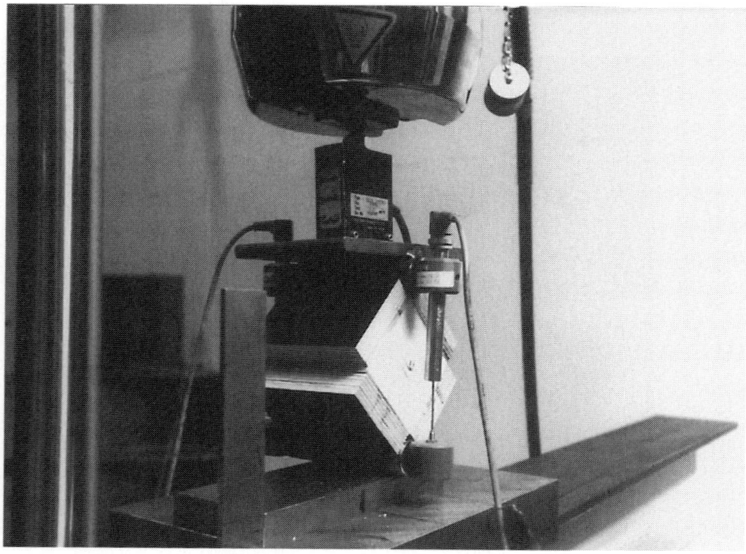

Fig. 13.4 'Square' shear test on PUR core material.

The standards ASTM C 273 and ISO R 1922 describe an alternative simple method that has been extensively used previously. A strip of core material, which is usually 25 mm × 50 mm × 250 mm, is cut from the panel and glued onto two stiff steel plates. When the specimen is placed in a testing machine and subjected to tension, the core material is approximately in pure shear, as shown in Fig. 13.5. The specimen should have an orientation corresponding to the shear plane in the panel. This test can also be carried out using three steel plates and two strips of core material as shown in Fig. 13.6. The shear strain and stress can be obtained from the force and displacement measured by transducers and a load cell. This method may be used for rigid plastic foam materials but it cannot generally be used for mineral wool. A disadvantage of this test is that steel plates tend to bend because of the eccentric force and this causes tension in the core material as well as shear.

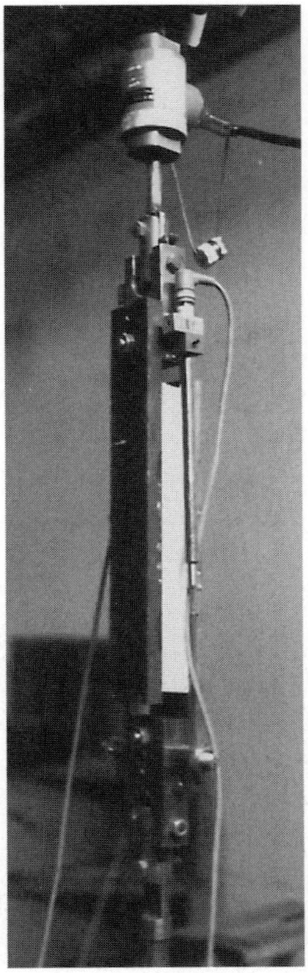

Fig. 13.5 Shear test on EPS core material in accordance with the standards ASTM C 273 and ISO R 1922.

Furthermore, the shear stress may not be uniform, particularly near the ends of the specimen.

Jungbluth and Berner[13.3] have compared the results of these alternative shear tests and have shown that the results given by the 'square' shear test agree well with those from the four-point bending test but the test with lapped steel plates underestimates real shear strength and stiffness. Some results from Jungbluth and Berner's study are shown in Fig. 13.6. The shear behaviour of a full-scale panel is usually at least as good as that of a strip of panel subject to the four-point bending test. This remains true when the strip is cut from a panel with profiled faces so that it represents the 'average' core. In some cases, for example when a panel has unsymmetrically corrugated faces or a discontinuous core (e.g. mineral wool lamellas), the shear test should be carried out with a full width of a panel.

The tensile strength of the core in a direction perpendicular to the faces and its tensile bond to the faces is measured by a test on a rectangular prismatic specimen cut from the panel. The faces are kept intact with the core because it is important in practice that the tensile bond to the faces exceeds the tensile strength of the core material. Ideally, therefore, the test measures the tensile strength of the core and, at the same time, checks that the adhesive bond between the faces and the core is

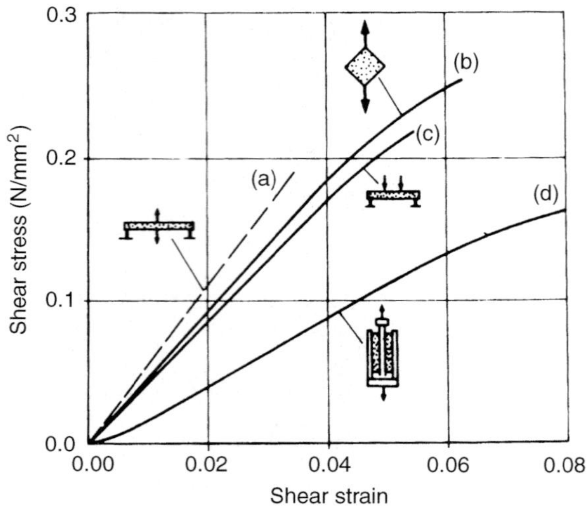

Fig. 13.6 Comparison of the results using different shear tests[13.3] for free-rise PUR foam of average density 50 kg/m^3: (a) dynamic three-point bending test specimen 35 × 110 × 2000, $G_C = 5.5\,N/mm^2$; (b) 'square' test specimen 35 × 35 × 20, $G_C = 4.6\,N/mm^2$; (c) static four-point bending test specimen 35 × 140 × 1000, $G_C = 4.3\,N/mm^2$; (d) lap shear test specimen 35 × 105 × 420, $G_C = 2.4\,N/mm^2$.

adequate. Thus, the failure should not occur in the bond but in the core. During the preparation of the specimens, special care should be taken to ensure that the bond is not damaged while cutting the specimen from the panel. Stiff plates are attached to the faces in order to obtain uniform tensile stresses in the specimen and care is taken to ensure that the tensile force is applied axially.

Specimens of discontinuous mineral wool core material should consist of a single lamella. Specimens incorporating two or more lamellas may be subjected to non-uniform loading if the stiffnesses of the lamellas differ.

Figure 13.7 shows a failed PUR specimen and Fig. 13.8, two failed rock wool specimens. Tensile failure in rigid plastic foams tends to be rather brittle whereas that of mineral wools is more ductile with a drooping force–displacement curve after the attainment of maximum load. The detailed procedure for tensile tests on core material is given in Section 13.9.3.2.

The compressive strength perpendicular to the faces is usually determined using a similar specimen to that used in the tensile test. A rectangular prismatic specimen with a height equal to that of a panel is cut from the panel and compressed centrically. The details of the test are given in Section 13.9.3.3. Figure 13.9 shows a PUR foam specimen during the test and Fig. 13.10 a failed mineral wool specimen. In order to avoid unnecessary damage to the specimen, the test is usually performed with the faces intact to the core. In order to achieve uniform compression in the core, thick plates (plywood) are usually glued on to the faces.

It should be noted that, in both tension and compression tests on small specimens, the stress state is not the same as that inside a panel because the edges of a small specimen are free and not supported by the surrounding material. In general, this difference can be neglected but, in compression tests on specimens of

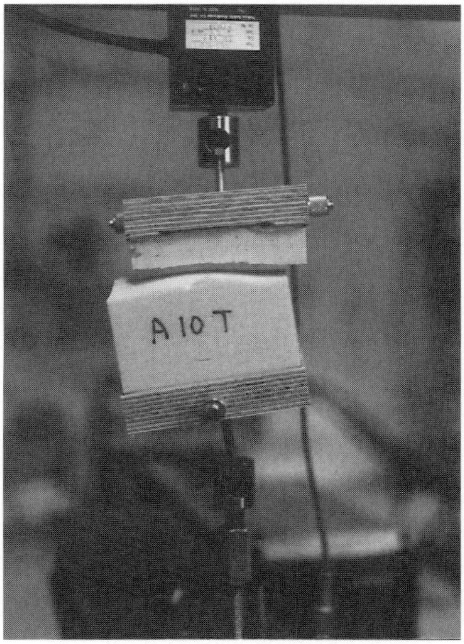

Fig. 13.7 Failed PUR foam specimen after a tension test.

Fig. 13.8 Failed mineral wool specimens after a tension test.

Fig. 13.9 Compression test on a PUR core material.

Fig. 13.10 Failed rock wool specimen after a compression test.

mineral wool core material, the size of the specimens has some effect on the results because the strength of slender compressed fibres is highly dependent on the support that they receive from the adjacent fibres. The compressive force is generally applied to the specimen through a pin and stiff platens. In tests on mineral wool specimens, this loading arrangement may be significant because axial compression changes to combined compression and bending once the fibres have started to fail at one edge.

13.3 TESTS TO DETERMINE THE WRINKLING STRESS

The wrinkling stress should be determined from the ultimate moment in bending tests using simply supported single-span panels. In such tests, the panels may be loaded with line loads or a uniformly distributed load. At least two line loads should be used but four are preferable. Particularly with thick panels, where a concentrated line load may tend to penetrate the face thus reducing the wrinkling stress, an alternative and more suitable method of loading is the so-called vacuum method,[13.4] shown in Fig. 13.11. In this method the panel is seated inside the top of an air-tight box and a plastic sheet is laid over it and sealed to the edge of the box (see Section 13.9.4.2). An almost prefect uniformly distributed load is achieved when the air is pumped out of the box. Another very similar method uses airbags which are overpressurised[13.5] in order to apply the load.

Fig. 13.11 Test arrangement for the vacuum method.

Figure 13.12 shows a typical wrinkling failure in the compressed face of a lightly profiled panel. The wrinkle is sharply defined and deforms into the panel. This is because, with most core materials, the compressive properties are smaller than the tensile properties. In a wrinkling test, a poorly defined wrinkle, or a wrinkle

Fig. 13.12 Typical wrinkling failure in a sandwich panel.

outwards rather than inwards, is generally an indication of poor core properties or poor bond between the core and the faces.

13.4 TESTS TO DETERMINE THE INTERACTION BETWEEN BENDING MOMENT AND SUPPORT FORCE

At an internal support in a multi-span panel, the wrinkling stress (or apparent yield stress) of the face of the panel which is in contact with the support must be reduced in order to take account of the influence of the support force which causes local deformation of this face. This deformation has a significant influence on the wrinkling stress of the face as shown in Fig. 13.13.

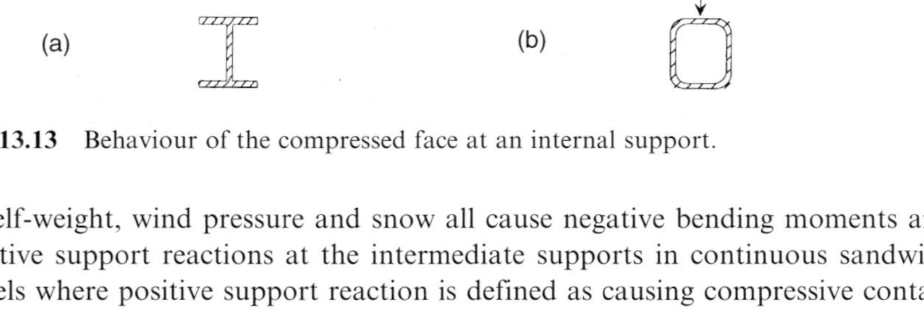

Fig. 13.13 Behaviour of the compressed face at an internal support.

Self-weight, wind pressure and snow all cause negative bending moments and positive support reactions at the intermediate supports in continuous sandwich panels where positive support reaction is defined as causing compressive contact stresses between the panel and the supporting structure. In this loading case, and particularly when considering the serviceability limit state, the local compressive and bending strengths of the lower face and the compressive strength of the core are critical.

The load case of wind suction, and some thermal loading cases, cause negative support reactions. Here, the support reaction force causes local tensile forces in the connections between the panel and its supports and these interact with the compressive stresses in the outer face of the panel.

In order, therefore, to determine the wrinkling stress at an intermediate support, it is necessary to carry out two types of line load test on a simply supported panel (see Section 13.9.4.4). Both of these tests should be carried out on a full panel width.

13.5 CREEP AND LONG-TERM SHEAR STRENGTH TESTS

Panels with a core material of rigid plastic foam and that are subjected to long-term loading should be tested to determine their visco-elastic behaviour. Creep tests can be performed using the same test arrangement that is used to determine the wrinkling strength. The determination of creep parameter ϕ_t (see Section 10.2.7 *Load-bearing capacity*) for panels with flat faces is given in Section 13.9.4.3. For panels with bending stiff faces see Reference 13.1. Long-term shear strength tests (see Chapter 6 *Durability*) can be carried out using the four-point bend test.

13.6 TESTS ON FASTENERS

The fastening systems used to connect sandwich panels to the supporting structure are often specific to the particular manufacturer and vary greatly between different panel systems. The test arrangement should reflect the true behaviour of the fastening system. One specific test related to sandwich panels is the repeated bending test on a fastener. This test is needed to ensure that a fastener is capable of resisting the cyclic bending caused by thermal movements of the faces (see Section 13.9.5).

13.7 TESTS USED IN QUALITY CONTROL

The quality control procedure in a production plant must first ensure that the prefabricated components, namely the facing materials and also the core materials when slab stock material is used meet the design requirements. Secondly, the required bond between the faces and core must be checked. In the case of panels with polyurethane core, which are foamed in the manufacturing process, the properties of the PUR-core can only be controlled after the manufacture. Table 13.1[13.1] summarises typical quality assurance procedures for continuously foamed PUR-panels.

The tests themselves are, of course, the same as those already described in the context of design tests.

A typical minimum quality assurance procedure for panels formed from slabstock may include testing according to Table 13.2.

Table 13.1 Typical quality assurance procedure for continuously-foamed panels

Type of test	Number of specimens	Frequency
Core material		
Density	3	1 per shift
Tensile strength and modulus (including bond to faces)	3	1 per shift
Compressive strength and modulus	3	1 per shift
Shear strength and modulus	3	1 per week
Dimensional stability under temperature attack	1	1 per week
Face material		
Tensile strength		All deliveries unless attested by material certificates
Thickness		
Complete panel		
Strength and stiffness (Sections 13.9.4.1 and 13.9.4.2)	1	1 per week
Dimensional control	1	1 per shift
Ageing	1	1 per year

Table 13.2 Typical quality assurance procedure for panels formed from slabstock

Type of test	Number	Frequency
Discrete core material		
Density	3	1 per shift
Tensile strength and modulus (incl. bond to faces)	3	1 per shift
Compressive strength and modulus	3	1 per shift
Shear strength and modulus	3	1 per week
Dimensional stability under temperature attack	1	1 per week
Face material		All deliveries unless attested by material certificates
Tensile strength	3	
Thickness		
Adhesive		
Shear strength	3	1 per week
Wedge test	3	1 per week
Complete panel		
Strength and stiffness (Sections 13.9.4.1 and 13.9.4.2)	1	1 per week
Dimensional control	1	1 per shift
Ageing	1	1 per year
Reaction to fire	1	1 per year

13.8 STATISTICAL INTERPRETATION OF TEST DATA

The scatter in any series of test results reflects, among other things, the variation of material properties. The test data may vary considerably, even if the same test is repeated on nominally identical specimens. This data should therefore be processed with statistical means in order to obtain reliable design or quality assurance values. Usually, the distribution of results is assumed to be normal. A normal distribution is uniquely determined by the mean $\bar{x}$ and the standard deviation σ. One may calculate these from a family of test results x_i ($i = 1, \ldots n$ where n is the number of tests) using the following equations:

mean:

$$\bar{x} = \frac{\sum x_i}{n}$$

standard deviation:

$$s = \sqrt{\frac{\sum(x - x_i)^2}{n - 1}}$$

From these estimates one can derive a fractile, i.e. the value below which only a chosen part of values will fall below a certain confidence level. The fractile can be expressed as

$$\bar{x} - k_\sigma s$$

where k_σ is a coefficient which depends on the number of tests and the chosen confidence level. Naturally the larger the number of tests, the more reliable is the estimate. The fractal of 5% is commonly used with the confidence level of 75%. This means that there is a probability of 75% that only 5% of the values will be below the fractal. The ISO-standard[13.6] gives rigorous statistical methods to deal with data. In Table 13.3 some values of k_σ are given for the 5% fractile with a confidence level of 75%.

Table 13.3 Value of k_σ for a 5% fractile with a confidence limit of 75%

Number of specimens, n	3	4	5	6	7	8	9	10	15	20	30	60	100
k_σ	3.15	2.68	2.46	2.34	2.25	2.19	2.14	2.10	1.99	1.93	1.87	1.80	1.76

13.9 DETAILED TEST PROCEDURES

13.9.1 General

The procedures which follow describe the principal mechanical tests that are used in the design and quality control of sandwich panels and are generally taken, without comment, from the draft update of Reference 13.1 (Summer 1999). Some notes, comments and points of minor detail have been omitted. It should also be noted that the important topic of the ageing of core material, and the influence of this on the mechanical properties, is dealt with in Chapter 6 where procedures for accelerated ageing are described.

In general, and unless stated otherwise, each property shall be determined on the basis of a minimum of three tests. Where a number of different core thicknesses are to be produced for a range of otherwise similar panels, it shall be permissible to determine the core-dependent properties for the thinnest and thickest panels to be produced together with a panel near the middle of the range and to use linear interpolation.

Material properties (tensile, compression and shear strengths and the influence of ageing) based on specimens cut from full panels shall generally be determined from a considerably larger number of tests with a minimum of five.

All test specimens shall be taken from the middle of the panel at a minimum distance from any outside edge of 20% of the cover width of the panel. Specimens shall be carefully cut in such a manner as to avoid damage using a band saw or similar device.

Unless otherwise noted, both the equipment applying loads and the devices measuring the load applied shall be capable of working to an accuracy of at least 1.0% and all deflections shall be measured with an accuracy of at least 0.1 mm.

13.9.2 Test environment and conditioning of test specimens

In the procedures that follow, unless stated otherwise, all testing shall be carried out in a laboratory under normal indoor conditions.

The minimum age of specimens for design tests shall be 48 hours. Specimens for quality control tests may be taken immediately after production. The date, time, temperature and relative humidity shall be recorded at the time of sampling.

In cases of dispute, or where the temperature or relative humidity are considered to be of particular significance, control tests shall be carried out after a minimum of 48 hours conditioning with:

Temperature:	$(20 \pm 5)^\circ C$
Relative humidity:	$(55 \pm 10)\%$

Specimens for design tests shall be stored under the specified test conditions for at least 48 hours prior to the commencement of the tests. Specimens for quality control tests may be tested immediately after manufacture.

For identification purposes, the core density of all tested specimens shall be recorded with the test results. For this purpose, it is sufficient to weigh three small prismatic core specimens cut from different regions of each complete panel used for the preparation of test specimens.

13.9.3 Test procedures for material properties

13.9.3.1 Tensile test on the face material

Tensile tests to determine the yield strength and other properties of the face material should be carried out in accordance with the standard EN 10002-1 (EN 10002-1 *Metallic Materials. Tensile Testing. Part 1: Method of Test*).

13.9.3.2 Tensile test on the core material

This test shall be performed in one of the following two ways:

(1) with the faces of the panel intact in order to determine the tensile bond strength between the faces and the core or to demonstrate adequate bond while determining the tensile strength of the core material
(2) before the faces are attached in order to determine the tensile strength of the core

In general, the bond with the faces is of fundamental importance and the test should be carried out with the faces intact and failure should not take place in the bond layer.

Specimens of square cross-section shall be prepared to the dimensions shown in Fig. 13.14 and bonded using a suitable adhesive to platens of sufficient stiffness to ensure a uniform tensile stress over the area of the specimen. With lightly profiled faces, special measures may be required in order to ensure full adhesion between the platens and the faces.

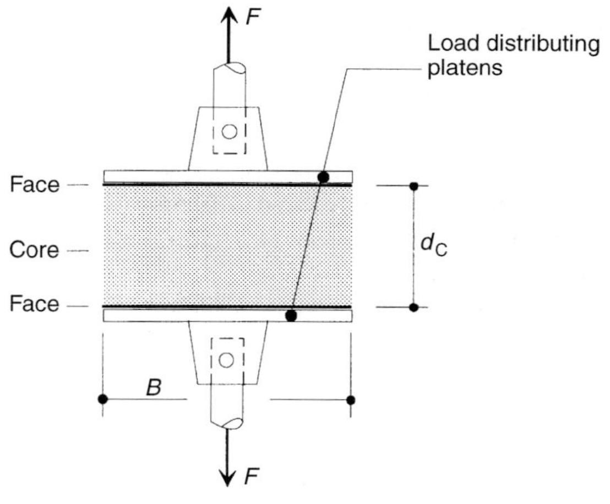

Fig. 13.14 Tensile test on core material ($0.5\,d_C \leq B \leq 1.5\,d_C$ but not less than 50 mm where d_C = thickness of the core of the test specimen and B = width of test specimen).

For panels with profiled faces the specimens should be cut from the predominant thickness.

Better results are generally obtained with larger specimens and it is recommended that, where possible, specimens with width B of at least 100 mm should be used.

The test shall be carried out by loading the specimen in increments in a suitable tensile testing machine. The strain rate shall have a minimum value of 1% per minute and should not exceed 3% per minute. At each increment of load, the extension shall be measured and a load–deflection curve drawn.

Tensile strength, f_{Ct} shall be given by:

$$f_{Ct} = F_u / B^2$$

Tensile modulus, E_{Ct}, shall be given by:

$$E_{Ct} = F_u \, d_C / w_u \, B^2$$

where F_u and w_u are defined in Fig. 13.15.

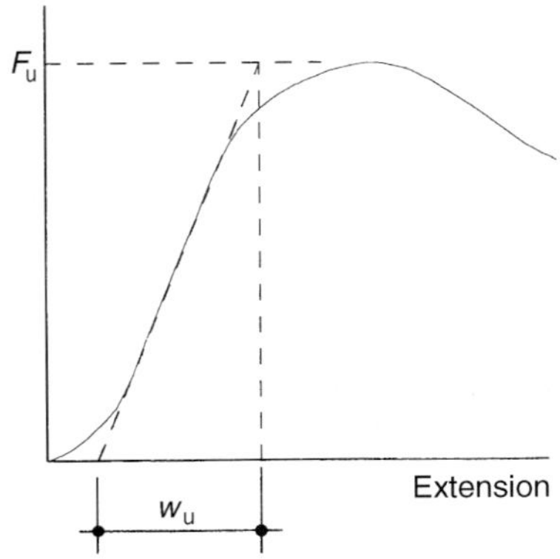

Fig. 13.15 Load–deflection curve from a tensile test on core material.

For specimens which do not exhibit a well-defined ultimate load, F_u may alternatively be defined as the load at a specified relative deformation. For polyurethane foams, 10% relative deformation is an appropriate limit. For materials with a more rigid cell structure or of non-cellular structure, a lower value may be used.

For design, but not for quality assurance, this test shall also be carried out on specimens heated to 80 ± 3°C. This test at elevated temperature is usually carried out by heating the specimen to a temperature a little above 80°C and then carrying out the test immediately in a conventional testing machine before it has cooled below 80°C.

13.9.3.3 Compression test on the core material

Specimens of square cross-section shall be prepared to the dimensions shown in Fig. 13.16. The metal faces need not be removed but, if the faces are removed, care shall be taken not to reduce the height of the core.

For panels with profiled faces, the specimens shall be cut from the predominant thickness as previously for tensile tests on core materials.

Better results are generally obtained with larger specimens and it is recommended that, where possible, specimens with width B of at least 100 mm shall be used.

The specimen shall be placed between the two parallel stiff loading plates of a suitable compression testing machine and loaded in increments. The strain rate shall have a minimum value of 1% per minute and should not exceed 3% per minute. At each increment of load, the displacement shall be measured and a load–deflection curve drawn.

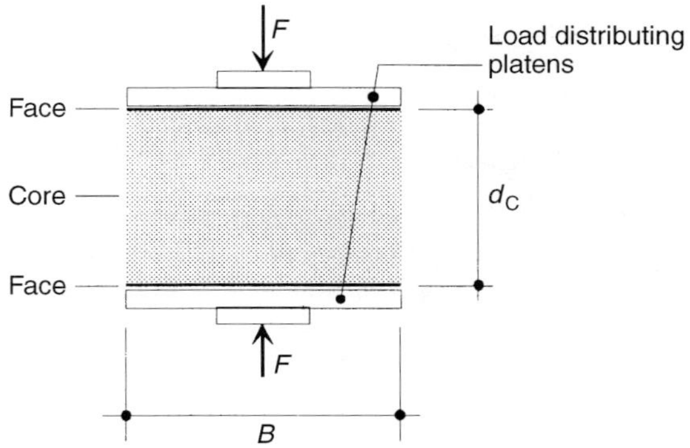

Fig. 13.16 Compression test on core material ($0.5\,d_C \leq B \leq d_C$ but not less than 50 mm where; d_C = thickness of the core of the test specimen and B = width of test specimen).

Compressive strength, f_{Cc}, of the core material shall be given by:

$$f_{Cc} = F_u/B^2$$

Compressive modulus, E_{Cc}, of the core material shall be given by:

$$E_{Cc} = F_u\,d_C/w_u\,B^2$$

For specimens which do not exhibit a well-defined ultimate load, F_u may alternatively be defined as the load at a specified relative deformation. For polyurethane foams, 10% relative deformation is an appropriate limit. For materials with a more rigid cell structure or of non-cellular structure, a lower value may be used (see Section 13.1 and Fig. 13.1).

For design, but not for quality assurance, this test shall also be carried out on specimens heated to 80 ± 3°C.

13.9.3.4 Shear test on the core material

The shear strength and shear modulus of the core material under short-term loading shall be determined using the four-point bending test as shown in Fig. 13.17.

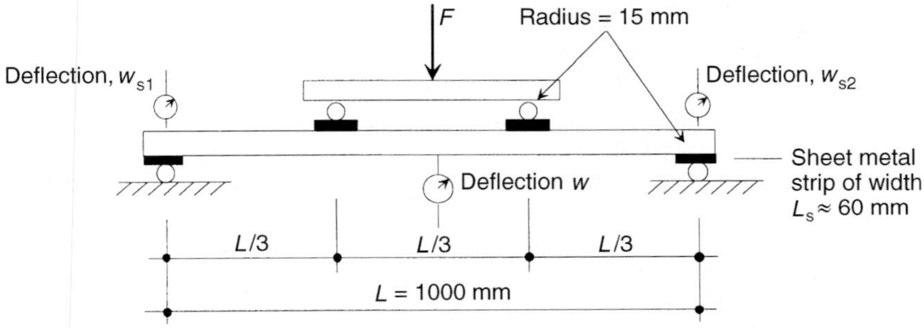

Fig. 13.17 Shear test on core material.

If the recommended span does not result in a shear failure, the span may be reduced in increments of 100 mm until a shear failure is obtained. Subsequent tests shall then be carried out at the reduced span.

The width B of the test specimen shall generally be chosen to give flat faces without any stiffening ribs. There is no upper limit to the width B of the test specimen. In order to ensure the stability of the specimen during testing, B shall be at least equal to the depth D of the panel and preferably at least $1.5D$.

For panels with discontinuous core material, tests shall generally be carried out on the full cover width of the panel with joints in the core material in the worst arrangement that may arise in practice. The joint arrangement used in the tests shall be described in the test report.

The width L_s of the sheet metal strips at the support and load points shall typically be 60 mm. This value may be increased, if necessary, in order to avoid local crushing of the core.

The loading rate shall be such as to result in failure between 5 and 10 minutes after the commencement of the test.

The net metal thickness of both faces of each test specimen shall be measured.

For each test specimen, the shear modulus G_C shall be calculated from the slope of the straight part of the load–deflection ($\Delta F/\Delta w$) curve as follows:

Flexural rigidity $$B_S = \frac{E_{F1} A_{F1} E_{F2} A_{F2}}{E_{F1} A_{F1} + E_{F2} A_{F2}} e^2$$

Bending deflection $$\Delta w_B = \frac{\Delta F L^3}{56.34 B_S}$$

Shear deflection $$\Delta w_S = \Delta w - \Delta w_B$$

Shear modulus $$G_C = \frac{\Delta F L}{6 B d_C \Delta w_S}$$

where E_{F1} = Young's modulus of the top face
A_{F1} = area of cross-section of the top face
A_{F2} = area of cross-section of the bottom face
E_{F2} = Young's modulus of the bottom face
e = depth between the centroids of the faces
Δw = deflection at mid-span for a load increment ΔF taken from the slope of the linear part of the load–deflection curve
d_C = depth of the core material

The ultimate shear stress f_{Cv} of the core material shall be calculated from the maximum load attained in a specimen failing in shear as follows:

$$f_{Cv} = \frac{F_u}{2 B e}$$

where F_u = ultimate load carried by the specimen failing in shear.

13.9.3.5 *Long-term shear loading*

In order to determine the long-term creep-rupture strength, long-term loading tests using procedures similar to those in Section 13.9.3.4 shall be carried out at a temperature of approximately +20°C in such a way that $n \geq 10$ samples fail within the time interval $0.1\,\text{h} \leq t \leq 10^3\,\text{h}$. Ideally, they should be equally spread out within this range.

Deformation measurements are not required.

Based on the test results, a regression line shall be drawn, as shown in Fig. 13.18, in order to show the relationship of the mean long-term shear strength to the initial shear strength (short-term strength) as a function of the loading time.

The long-term shear strength (e.g. for 2000 or 100 000 hours) shall be calculated using an expression based on the mean-value regression line.

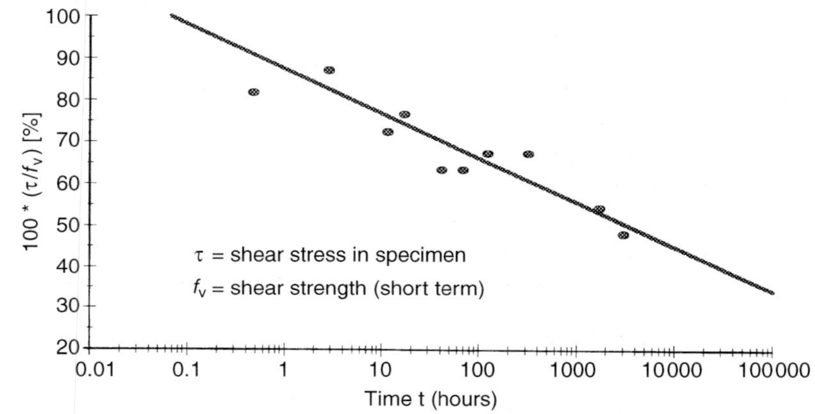

For example:
Regression line $100 * (\tau/f_v) = -4.58911 * \ln(t) + 87.5289$
$100 * (\tau/f_v) = -4.58911 * \ln(100\,000) + 87.5289 = 34.7\%$

Fig. 13.18 Determination of the long-term shear strength.

13.9.3.6 Alternative shear test for thick panels

An alternative shear test which has been developed for thick mineral wool panels is shown in Fig. 13.19. This takes account of the tendency of the core to crush at the supports and points of application of the loads by using much wider loading platens. However, a consequence of the use of wider platens is that the equations given in Section 13.9.3.4 for the calculation of the shear modulus become invalid. Consequently, a different approach to the determination of the shear modulus is required.

The length d of the platens and the total length of the specimen are chosen so that there is no significant crushing of the specimen and failure takes place in shear within the relatively narrow 'shear areas'. A suitable length for a shear area is typically 200 mm. Two displacements at points approximately 100-mm apart are measured within the shear area in order to determine the shear strain. These measurements require the use of accurate equipment. The loading time to failure should be between 5 and 10 minutes.

The width of the specimen should be at least the height of the beam. No butt joints in the core slabstock (mineral wool lamellas) should be within the shear areas or close to them.

The shear stress τ within the shear area is calculated in the conventional way:

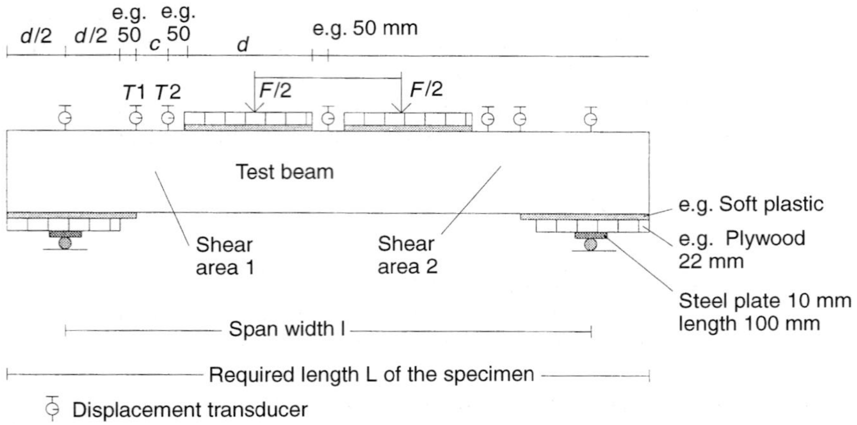

Fig. 13.19 Alternative shear test for thick panels.

$$\tau = F/2A$$

where A is the cross-sectional area of the test beam.

The shear strain in the shear area is:

$$\gamma = (T2 - T1)/c$$

where c is the distance between the transducers T1 and T2.

13.9.4 Tests on complete panels

13.9.4.1 Test to determine the shear strength of a complete panel

This test offers a more reliable method of determining the shear strength of panels with slabstock cores where joints between the core elements may affect the shear properties. If the panel has flat or lightly profiled faces, this test may be used as an alternative to that described in Section 13.9.3.4.

For panels with discontinuous core material, tests shall generally be carried out on the full cover width of the panel with joints in the core material in the worst arrangement that may arise in practice. The joint arrangement used in the tests shall be described in the test report. The test shall be carried out by subjecting a short simply supported panel to two line loads either equally spaced at the 1/3 points or applied at the 1/4 points or to air pressure caused by either a partial vacuum chamber test apparatus or air bags. The panel shall be loaded in increments up to failure and the failure load noted.

This test is not necessary if the test described in Section 13.9.4.2 is carried out on sufficiently small spans.

The span shall be sufficiently short to ensure a shear failure.

The loading rate shall be such as to result in failure between 5 and 10 minutes after the commencement of the test.

When using air pressure loading, the load shall be measured by means of load cells, not air pressure.

The net metal thickness of both faces of each test specimen shall be measured.

The load F_u at failure gives the shear strength of the complete panel including the contribution of both the core and faces.

For panels with flat or lightly profiled faces, it may be assumed that all of the shear force is carried by the core so that the ultimate shear strength f_{Cv} of the core is given by:

$$f_{Cv} = \frac{F_u}{2\,Be}$$

where B = width of the specimen
F_u = ultimate load carried by the specimen failing in shear
e = depth between centroids of the faces

13.9.4.2 *Test to determine the bending strength and stiffness of a simply supported panel*

This test is generally used to determine the wrinkling stress and, in such cases, the span L shall be sufficiently large to ensure a bending (wrinkling or face buckling) failure. The necessary span is dependent on several factors including the depth D of the specimen. The following values are offered for guidance:

$D \leq 40$ mm	$L = 3.0$ m
$40 < D \leq 60$ mm	$L = 4.0$ m
$60 < D \leq 100$ mm	$L = 5.0$ m
$D > 100$ mm	$L = 6.0$ m

If the above values of span are found to give rise to shear failure, they shall be increased in increments of 1 metre until a bending failure is obtained.

The test shall be carried out by subjecting a simply supported panel to four line loads, as shown in Fig. 13.20, extending across the full width of the panel or to air pressure caused by either a partial vacuum chamber test apparatus or air bags. The panel shall be loaded in increments up to failure and the failure load noted.

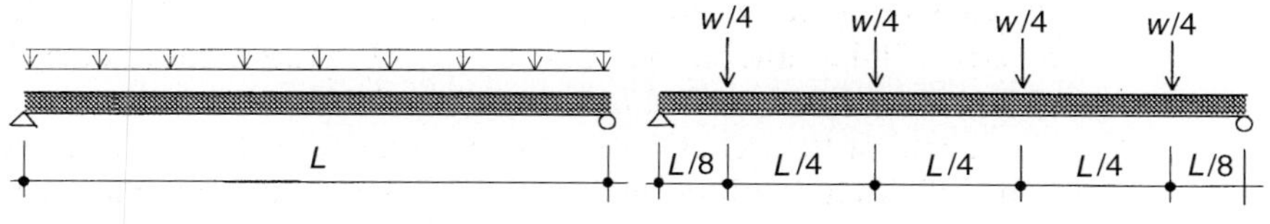

Fig. 13.20 Alternative loading systems for bending tests.

If line loads are applied to a profiled face, they shall be applied through timber or steel transverse loading beams together with timber loading platens placed in the troughs of the profile as shown in Fig. 13.21. A layer of felt, rubber or other similar material may be placed between the loading platens and the panel in order to reduce the possibility of local damage.

If the trough of the profile includes rolled-in stiffeners, the loading platens may be shaped appropriately.

The loads shall be maintained perpendicular to the panel throughout the test.

After completion of the test, the thickness and yield stress of each face shall also be determined and recorded.

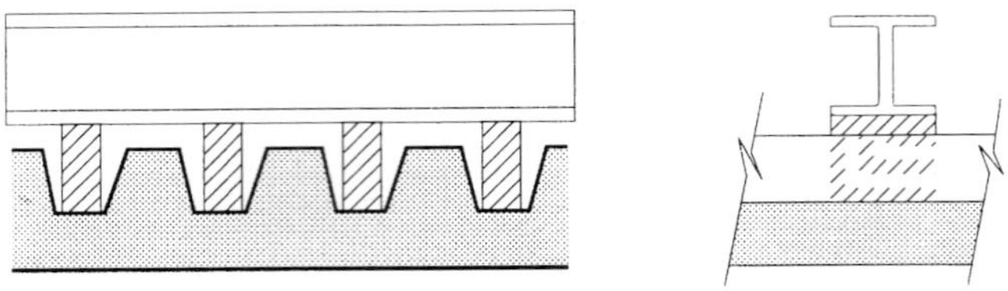

Fig. 13.21 Method of applying load to a profiled face.

Support conditions
In general, the support width shall be within the range 50–100 mm. Timber blocks may be used to avoid deformation of a side rib which does not contain foam.

The tested panel may be attached to the supports through either the profile valleys or crests as in practice.

When this test is used to determine the wrinkling stress for use in design calculations, the support conditions shall be such as to apply no restraint to the rotation of the panel about the line of support. A suitable support detail is shown in Fig. 13.22.

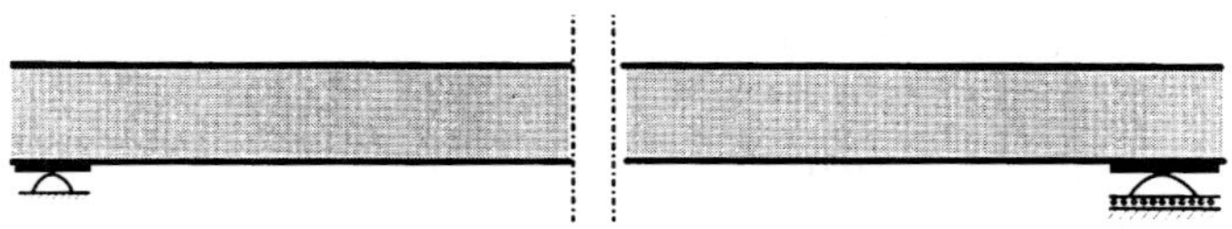

Fig. 13.22 Support detail for simply supported tests.

Loading of specimens
It is preferable to carry out this test by controlling the deflection rather than the load (i.e. using a constant deflection speed). However, either procedure may be used provided that the deflection speed does not exceed 1/50 of the span per minute at any time during the test. The load shall be increased steadily until failure occurs. The failure load and the nature and location of the failure and the relationship between load and deflection shall be recorded.

It is preferable to preface the formal test by a small preload.

Determination of the wrinkling stress
Although, in most cases, the design value of the wrinkling stress may be calculated, more favourable values of the wrinkling stress will generally be obtained by testing. In this case, the wrinkling stress shall be obtained by determining the ultimate moment of resistance using the test procedure described above and then the face stress at failure shall be obtained by calculation.

For symmetrical or nearly symmetrical panels, it is essential to carry out this test with both orientations of the panel because the wrinkling stress may be greatly influenced by whether the face was at the top or bottom of the panel during manufacture. For symmetrical panels, the design shall be based on the least favourable wrinkling stress.

The wrinkling stress σ_w is only directly relevant for panels with flat or lightly profiled faces. For such cases, σ_w is given by:

$$\sigma_W = \frac{M_u}{eBt_1}$$

where M_u = the ultimate bending moment recorded in the test, including the effect of the self-weight of the panel and the weight of the loading equipment
B = width of the test specimen
e = depth between centroids of the faces
t_1 = core metal thickness of the face in compression

If a partial vacuum chamber or air bag test apparatus is used in order to provide a uniformly distributed load over the surface of the specimen, the wrinkling stress is:

$$\sigma_W = \frac{(F_G + F_u)L}{8Bt_1 e}$$

where F_G = self-weight of the test panel
F_u = $q_u BL$ = ultimate applied load

If the load is applied as four equal line loads at positions 1/8, 3/8, 5/8, 7/8 of the span, the wrinkling stress is:

$$\sigma_W = \frac{(F_G + F_u)L}{8Bt_1 e}$$

where F_G = self-weight of the test panel (negative for air bag tests)
F_u = $q_u BL$ = ultimate applied load plus the weight of the loading equipment

If the face under tension in this test is profiled, the wrinkling stress of the flat or lightly profiled face in compression may be determined using

$$\sigma_W = \frac{M_u - M_{F2}}{Bt_1 e}$$

where M_u = the ultimate bending moment recorded in the test including the effect of the self-weight of the specimen and the weight of the loading equipment
M_{F2} = the bending moment carried by the profiled face. The value of M_{F2} shall be determined by calculation

Determination of the shear modulus of the core material

This bending test may also be used in order to determine a reliable value for the shear modulus of the core material.

If both faces of the test panel are flat or lightly profiled, the total deflection at the centre of the test panel shall be divided into two parts:

$$w = w_B + w_S$$

where w = measured deflection at mid-span of the test panel
w_B = deflection due to axial deformation in the faces
w_S = deflection due to shear deformation of the core material

The shear modulus of the core shall be determined from w_S.

If a partial vacuum chamber or air bag test apparatus is used in order to provide a uniformly distributed load over the surface of the specimen, the deflections at mid-span shall be expressed as:

$$w_B = \frac{5}{384}\frac{FL^3}{B_s} \quad \text{and} \quad w_S = \frac{1}{8}\frac{FL}{G_{eff}A_C}$$

where G_{eff} = effective shear modulus of the core = $G_C\,(e/d_C)$.

The shear modulus of the core shall then be expressed as:

$$G_C = \frac{FL}{8A_C(w - w_B)}\frac{d_C}{e}$$

In these expressions, the deflection w is taken from the linear part of the load–deflection curve, F is the corresponding applied load and

$$B_S = \frac{E_{F1}A_{F1}E_{F2}A_{F2}}{E_{F1}A_{F1} + E_{F2}A_{F2}}e^2; \quad A_C = Be$$

If the load is applied as four equal line loads at positions 1/8, 3/8, 5/8, 7/8 of the span, the expressions for the deflection components at mid-span become

$$w_B = \frac{41}{3072}\frac{FL^3}{B_S} \quad \text{and} \quad w_S = \frac{1}{8}\frac{FL}{G_{eff}A_C}$$

The calculation procedure is otherwise unchanged.

13.9.4.3 *Test to determine the creep coefficient ϕ_t*

A single test shall be sufficient to determine the creep coefficient for a particular core material.

The test shall be carried out on a complete panel of span equal to that used for the bending test in Section 13.9.4.2. The core thickness shall be the maximum in the test series.

The test shall be carried out under a constant load which shall be sustained undisturbed for a minimum of 1000 hours. During this time, the deflection shall be regularly monitored to give a continuous relationship between deflection and time. The load used for the creep test shall correspond to approximately 30% of the average load for shear failure at ambient temperature determined from the tests carried out according to Section 13.9.3.4 or 13.9.4.1.

The test shall be carried out by subjecting a simply supported panel to a uniformly distributed dead load. During the placing of the load, the panel shall be propped from below in such a way that the propping can be removed quickly and smoothly in order to initiate the test. Deflection measurements shall commence the instant that the full load is applied.

Alternatively, the initial deflection shall be calculated from the slope of the load deflection curve obtained during the corresponding bending test in Section 13.9.4.2. In this case, the dead load may be applied more gradually in the conventional manner.

The creep coefficient for the core of a thin-faced sandwich panel shall be determined using the expression:

$$\phi_t = \frac{w_t - w_0}{w_0 - w_B}$$

where w_t is the deflection measured at time t,
w_0 is the initial deflection at the time $t = 0$
w_B is the deflection caused by the elastic extension of the faces

The deflections caused by the bending and shear deformations of a sandwich panel with strongly profiled faces cannot be separated in the expression for the deflection because the distribution of the bending moment into the sandwich component M_S and the flange components M_{F1}, M_{F2} depends on the shear stiffness of the core. Therefore, the creep coefficient shall be evaluated iteratively on the basis of the measured deflections as a function of the time.

On the basis of the results of the tests within a total time $t \geq 1000$ h, creep coefficients shall be extrapolated using a semi-logarithmic diagram in order to determine the creep coefficients required in the design (i.e. $\phi_t(t = 2000\,\text{h})$ and $\phi_t(t = 100\,000\,\text{h})$).

The creep test may alternatively be carried out on a complete panel using a uniformly distributed load.

More exact and usually advantageous values of the creep coefficient ϕ_t will be obtained if the duration of the test is extended to (say) 2000 hours.

The development of new core materials may give rise to increased susceptibility to creep and may therefore require longer tests.

The load used for creep tests is not unduly critical and similar results will be obtained for any load in the range 30–40% of the failure load. However, if the load level is significantly greater than 40% of the failure load, shear creep may no longer be a linear visco-elastic phenomenon and the results obtained from the test may no longer correspond to those arising in practice.

13.9.4.4 *Interaction between bending moment and support force*

The test arrangement for the interaction between bending moment and support reaction force shall be a single-span panel subject to a line load. This is often referred to as the 'simulated central support test' because it simulates the conditions in the central support of a two-span beam.

In order to determine the wrinkling stress at an intermediate support it is necessary to carry out two types of test as shown in Fig. 13.23:

(1) tests which simulate downward load
(2) tests which simulate uplift load

Both of these tests shall be carried out on a full panel width.

It is important that the span (which may in some cases be 5 metres or more) shall be sufficient to ensure that:

- for tests (1), the compressive stress between the panel and the support (under the line load) at the time of wrinkling failure is less than the compressive strength of the panel core material
- and for test (2), the forces in the fasteners at wrinkling failure of the panel are less than their design values

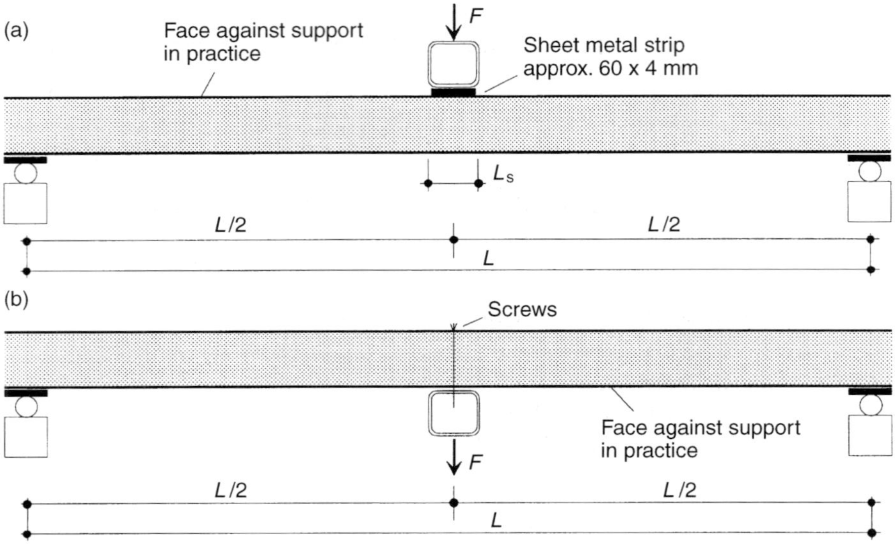

Fig. 13.23 Simulated central support test.

This ensures that, for the prototype panel, all failure modes (wrinkling of the face, compressive failure of the core and tensile failure of the connection) are designed for approximately equal levels of safety.

If the test is carried out on a shorter specimen than that described above, the failure mode is likely to be dominated by core crushing and a conservative value of the wrinkling stress will be obtained.

The wrinkling stress σ_w is only directly relevant for panels with flat or lightly profiled faces. For such cases, σ_w shall be given by:

$$\sigma_w = \frac{F_u L}{4 B t_1 e}$$

where B = width of the specimen
F_u = ultimate load carried by the specimen
t_1 = core metal thickness of the face in compression
e = depth between the centoids of the faces

13.9.4.5 *Test on a two-span panel*

This test may be used as an alternative to the test described in Section 13.9.4.4 in order to investigate the interaction between the bending moment and reaction force at an internal support.

In carrying out this test, one of the arrangements shown in Fig. 13.24 shall be used following the principles described in Section 13.9.4.4.

Care shall be taken to identify the onset of permanent deformations corresponding to buckling or yielding of the face or crushing of the core at the internal support. This will generally require that the panel is unloaded after certain increases in the applied load in order to determine the residual deflection.

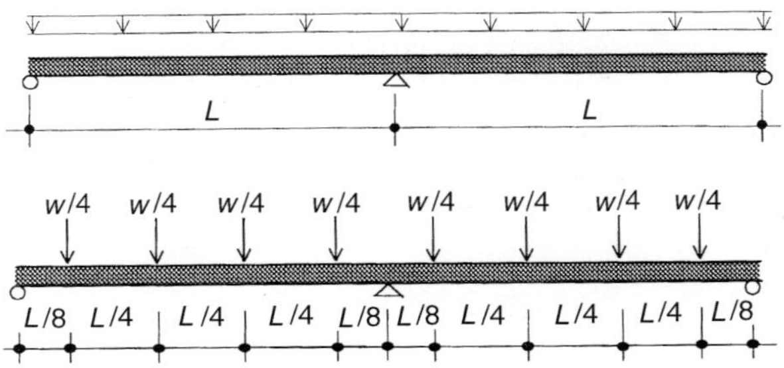

Fig. 13.24 General arrangements for two-span tests.

13.9.4.6 *Support reaction capacity*

A suitable test with which to determine the capacity with respect to an end support reaction is shown in Fig. 13.25. The end support reaction capacity shall be defined as

$$F_{R1} = \frac{L_2}{(L_1 + L_2)} F$$

where F is the maximum load measured in the test or the load corresponding to a deflection $w = 0.1e$ if this is lower than the maximum load and on the rising part of the load deflection curve as shown in Fig. 13.26.

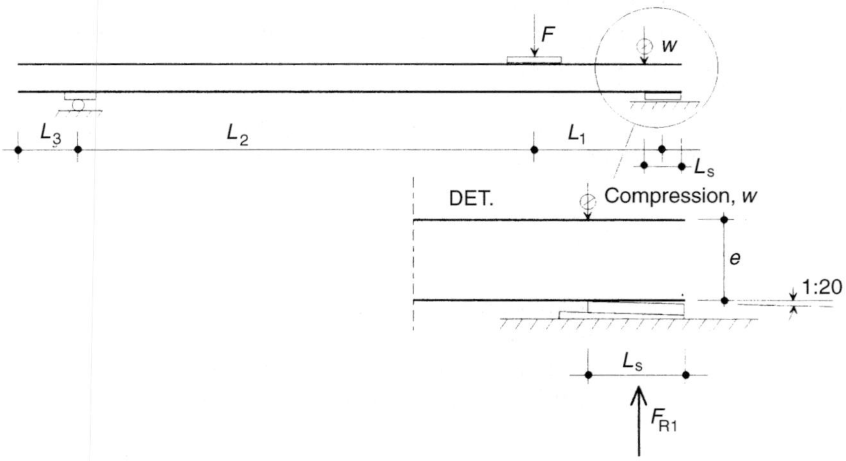

Fig. 13.25 Test arrangements for the determination of the end support reaction capacity.

The dimensions L_1, L_2 and L_3 shall be chosen in such a way that the failure mode of the specimen is a compression failure at the support. If the failure mode is a shear failure between the loading plate, F, and the support plate, L_s, the end support reaction capacity may be taken to be the value of the support reaction force at the time of shear failure.

The support reaction capacity at an intermediate support shall be defined as follows.

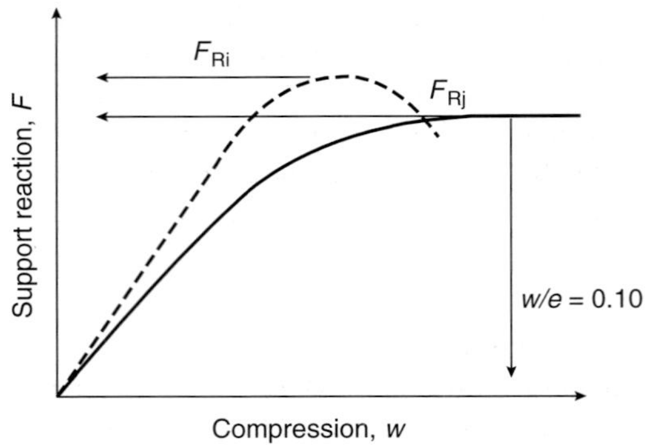

Fig. 13.26 Definition of the ultimate load in an end support reaction test.

The compression strength f_{Cc} of the core shall be determined according to Section 13.9.3.3.

$$F_{R2} = f_{Cc}B(L_s + ke) \quad \text{where } k = 2\frac{(F_{R1} - f_{Cc}BL_s)}{f_{Cc}Be}$$

The loading rate shall be such as to result in failure between 5 and 10 minutes after the commencement of the test.

13.9.5 Tests on fasteners and connections

13.9.5.1 General

Many sandwich panels have proprietary fastening systems which may require specially designed test arrangements. It is not, therefore, easy to define general test procedures which can be used in all circumstances. The two general principles which should be followed are as follows.

(1) The test procedure must reproduce all significant influences on the performance of the connections which will arise in practice.
(2) If in doubt, test connections in their actual locations within a system of full sized panels.

Having said this, the principal force to which the connections are subjected is tension due to wind suction. The test described in Section 13.9.4.4 for the interaction between bending moment and support force does indeed test the fasteners at an internal support in tension. It is implicit that, in this test, there will be sufficient fasteners to ensure that the failure mode is in the panel rather than in the fasteners. If this is not the case, this test is a valid test for the tensile capacity of the fasteners at an internal support, which will often be the critical case. However, it is also desirable to have a rather simpler test arrangement which is capable of testing an individual fastener or connection and such testing is described in Section 13.9.5.2.

A particularly simple method of connecting sandwich panels to the supporting structure, which is widely used in practice, is by means of screw fasteners passing right through the panel. When panels fastened in this way undergo cyclic changes in temperature, such screws are bent back and forth by the relative longitudinal movement of the faces of the panel. The screw therefore suffer fatigue loading. There is a specific test to evaluate their response to this rather unusual load case which is described in Section 13.9.5.4.

Calculable shear forces in the connections generally only arise when advantage is taken of the stressed skin (shear diaphragm) effect in order to stabilise the primary structure.

13.9.5.2 Tensile test on a connection

The definitive document which describes alternative procedures for testing fasteners in tension is rather old.[13.8] However, it is still valid and has not been superseded. The European recommendations[13.1] include the statement that testing of fastenings shall generally be in accordance with this document and do not go into very much detail regarding the appropriate test assemblies for fasteners in sandwich panels.

In general, a minimum of five tests should be carried out on any given fastening arrangement.

Separate tests shall be carried out for fastenings at the end of a panel and those at an intermediate support. In tests to determine the tensile resistance at an end support, the minimum distance between the fastener and the end of the panel shall be used.

Tests shall be carried out for the smallest and largest panel depths to be used in practice.

The load or displacement shall be increased monotonically up to the ultimate load. The use of a displacement-controlled testing machine is to be preferred. The local displacement at the fastening shall be measured in order to determine the tensile stiffness and the displacement corresponding to the maximum load in the fastening.

The ultimate failure load and the mode of failure (pull through, pull out, failure of fastener itself, etc.) shall be recorded.

13.9.5.3 Testing of fasteners in shear

If the distance between the end of the panel and the fastener is less than 30 mm, separate tests shall be carried out for fastenings at the end of a panel and those at an intermediate support.

Tests shall be carried out for the largest panel depth to be used in practice.

The load or displacement shall be increased monotonically up to the ultimate load. The use of a displacement-controlled testing machine is to be preferred. The shear displacement at the fastening shall be measured in order to determine the shear stiffness and the shear displacement corresponding to the maximum load in the fastening.

The ultimate load is defined to be the smallest of:

- the maximum load recorded during the test
- the load at which the first fall in load is observed on the load deflection curve
- the load corresponding to a displacement of 3 mm if this occurs on the rising portion of the load deflection curve

The ultimate failure load and the mode of failure (pull through, pull out, failure of fastener itself, etc.) shall be recorded.

13.9.5.4 Repeated bending test on a screw fastener

A suitable arrangement for this test is shown in Fig. 13.27. The fastener shall be tested as a cantilever of length l equal to the depth of the panel to the point of connection. The details of the test arrangement shall correspond as closely as possible to those in the actual structure, in particular, it shall be fixed into material of the same thickness t as the supporting member in the actual structure. The head of the fastener shall not be restrained during the test.

The fastener shall be tested in a displacement-controlled testing machine and shall be subjected to the following unilateral displacement spectrum where u is the maximum lateral displacement calculated at the point of attachment:

(1) 20 000 cycles at $(4/7)u$
(2) 2000 cyles at $(6/7)u$
(3) 100 cycles at u

The frequency shall not exceed 5 Hz.

After application of the above displacement spectrum, the fastener shall be tested to failure in tension.

By varying l and u, this test can be used to determine the maximum allowable value of u for different panel thicknesses. In systematic testing of a particular fastener, the thickness t of the sub-structure material shall also be varied because a greater thickness results in increased bending restraint and therefore reduced deformation capacity.

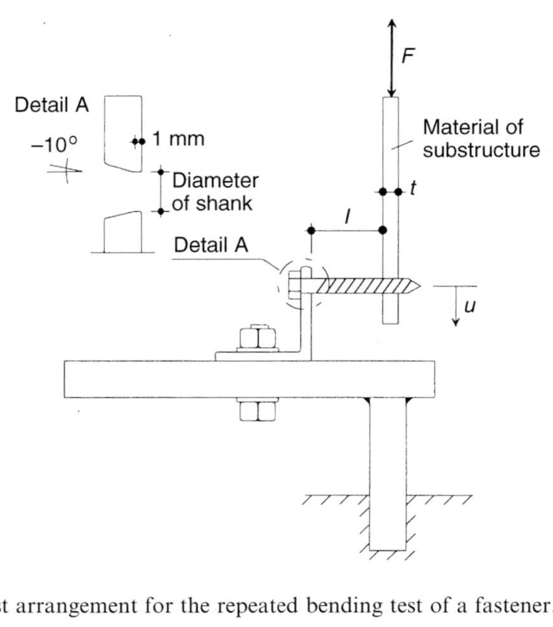

Fig. 13.27 Test arrangement for the repeated bending test of a fastener.

13.10 REFERENCES

13.1 European Convention for Constructional Steelwork (1992) European recommendations for sandwich panels with additional recommendations with mineral wool core material. Part 1: Design. CIB Report. Publication 148. TWG 7.4, International Council for Building Research Studies and Documentation W56. Also, update of this document published as CIB Publication 257 (2000).

13.2 CEN standard (1999) *Double Skin Metal Faced Insulating Sandwich Panels for Roofing and Cladding*, 5th Draft, Doc. CEN/TC128/SC11/N95, October 1999.

13.3 Jungbluth, O. and Berner, K. (1986) *Verbund- und Sandwichtragwerke*. Springer-Verlag, Berlin, Heidelberg.

13.4 (1981) *Floor and Roof Elements: Uniformly Distributed Loading with Air Pressure or Vacuum*, Nordtest remiss No. 267–81.

13.5 König, J. (1984) Der Einfluss der Verformungen bei der Prüfung von dünnwandigen Bauteilen nach der Luftsackmethode. *Stahlbau*, **4**, 105–110.

13.6 ISO Standard 3207 (1975) *Statistical Interpretation of Data – Determination of a Statistical Tolerance Level.*

13.7 European Convention for Constructional Steelwork (1978) *The Testing of Profiled Metal Sheets*. Publication No. 20.

13.8 European Convention for Constructional Steelwork (1983) *The Design and Testing of Connections in Steel Sheeting and Sections*. Publication No. 21, 176 pp.

Chapter 14
Examples of Calculation Procedures

14.1 GENERAL

The two examples which follow have been chosen to illustrate the design procedures used in the draft European Recommendations for the Design of Sandwich Panels[14.1] which have been further expounded in this book. These examples indicate the complete design procedure for a two-span wall panel with quasi flat steel faces and a two-span roof panel with an outer profiled face, both subject to external climatic loading. There are, of course, a large number of possible permutations of span condition, type of face and type of loading but the two that have been chosen illustrate all of the main considerations. In particular, it should be noted that the single-span case is a considerable simplification of the two-span case whereas the three-span case is very similar to the two-span case considered.

14.2 TWO-SPAN WALL PANEL

The structural system is shown in Fig. 14.1.

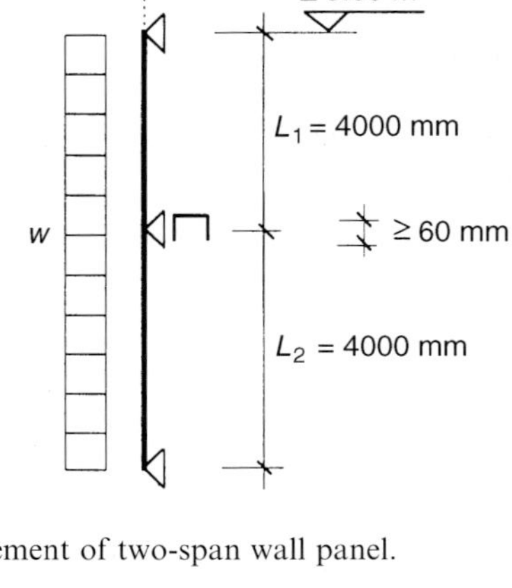

Fig. 14.1 General arrangement of two-span wall panel.

The following points may be noted:

- The panel is designed for the two load cases:

  (1) wind pressure and winter temperature
  (2) wind suction and summer temperature

- In the ultimate limit state, it is conservatively assumed that 'plastic hinges' of zero moment capacity exist over the internal supports. A consequence of this is that the temperature-induced stresses are zero at the ultimate limit state and the calculation becomes simply that for a single-span panel subject to wind pressure or suction.
- At the serviceability limit state, the combination of wind load and the action of temperature must be considered, bearing in mind the continuity over the internal support. The calculations for this limit state are, therefore, rather more complicated.
- There are no stresses due to self-weight in a wall panel and, therefore, no influence of creep.

14.2.1 Actions (loads)

The panel is designed for the following actions:

Wind pressure: $w_D = +0.50\,\text{kN/m}^2$
Wind suction: $w_S = -0.25\,\text{kN/m}^2$ (typical region of wall)

Note Higher wind suction loads near the corners and eaves, which vary from country to country, are not taken into account in this example. They are generally taken into account by locally reducing the purlin spacing and/or increasing the number of fasteners.

Temperature: Winter: $T_1 = -20°\text{C}$ $T_2 = +20°\text{C}$
Summer: $T_1 = +65°\text{C}$ $T_2 = +25°\text{C}$
(colour group 2: medium colours such as sandy yellow or light blue)

14.2.2 Cross-section of the panel

The cross-section of the panel is shown in Fig. 14.2
The relevant dimensions are as follows:

$t_{nom.1} = 0.60\,\text{mm}$ $t_{d.1} = 0.56\,\text{mm}$ $A_{F1} = 584\,\text{mm}^2/\text{m}$ $I_{F1} \approx 0$
$d_{11} = 0.63\,\text{mm}$ $d_{12} = 4.37\,\text{mm}$

$t_{nom.2} = 0.50\,\text{mm}$ $t_{d.2} = 0.46\,\text{mm}$ $A_{F2} = 469\,\text{mm}^2/\text{m}$ $I_{F2} \approx 0$
$d_{21} = 2.45\,\text{mm}$ $d_{22} = 0.20\,\text{mm}$

nominal depth $d = 60.0\,\text{mm}$

$$e = 60 - \tfrac{1}{2}(t_{d.1} + t_{d.2}) - d_{11} - d_{22} = 60 - \tfrac{1}{2}(0.56 + 0.46) - 0.63 - 0.20 = 58.66\,\text{mm}$$

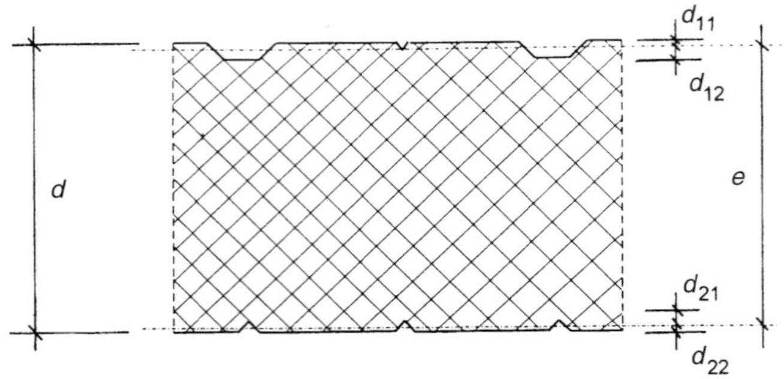

Fig. 14.2 Cross-section of wall panel.

14.2.3 Design values for the panel

The following design values have been determined on the basis of the test procedures described in this book:

Shear modulus of the core:	$G_C = 3.1\,\text{N/mm}^2$
Shear strength of the core:	$f_{Cv} = 0.11\,\text{N/mm}^2$
Compressive strength of the core:	$f_{Cc} = 0.10\,\text{N/mm}^2$
Yield strength of the faces:	$f_{ty} = 320\,\text{N/mm}^2$
Wrinkling stress of the faces:	f_{Fc} or $\sigma_w = 120\,\text{N/mm}^2$ (both faces)

The following additional values are required for the calculation:

Elastic modulus of steel:	$E_s = 210\,\text{kN/mm}^2 = E_{F1} = E_{F2}$
Coefficient of thermal expansion of steel:	$\alpha_s = 12 + 10^{-6}$ per $^\circ\text{C} = \alpha_{F1} = \alpha_{F2}$

14.2.4 Load, material and combination factors

The relevant material factors are given in Table 11.2. The relevant load and combination factors are given in Table 11.3.

14.2.5 Ultimate limit state

In the ultimate limit state, the individual spans are considered to be simply supported over a span of $L = 4.0$ metres with 'hinges' with zero rest moment at the internal supports.

14.2.5.1 Wind pressure case (per metre width of panel)

Unfactored bending moment: $M_S = \dfrac{w_D L^2}{8} = \dfrac{0.50 \times 4.0^2}{8} = 1.0\,\text{kN m/m}$

Design bending stresses: $\sigma_{F1} = (-)\dfrac{\gamma_w M_S}{e A_{F1}} \leq \dfrac{\sigma_w}{\gamma_M}$ (compression)

$= (-)\dfrac{1.5 \times 1.0 \times 10^6}{58.66 \times 584.0} \leq \dfrac{120}{1.25}$

$= (-)43.8 \leq 96.0\,\text{N/mm}^2$

$\sigma_{F2} = \dfrac{\gamma_w M_S}{e A_{F2}} \leq \dfrac{f_{ty}}{\gamma_M}$ (tension)

$= \dfrac{1.5 \times 1.0 \times 10^6}{58.66 \times 469.0} \leq \dfrac{320}{1.1}$

$= 54.52 \leq 290.9\,\text{N/mm}^2$

Unfactored shear force: $V = \dfrac{w_D L}{2} = \dfrac{0.5 \times 4.0}{2} = 1.0\,\text{kN/m}$

Design shear stress: $\tau_C = \dfrac{\gamma_w V}{e} \leq \dfrac{f_{Cv}}{\gamma_M}$

$= \dfrac{1.5 \times 1.0 \times 10^3}{58.66 \times 1000} \leq \dfrac{0.11}{1.25}$

$= 0.0256 \leq 0.088\,\text{N/mm}^2$

At the intermediate support:

Unfactored support reaction: $F_2 = w_D L = 0.5 \times 4.0 = 2.0\,\text{kN/m}$

Design compressive stress: $\sigma_{Cc} = \dfrac{\gamma_w F_2}{L_s} \leq \dfrac{f_{Cc}}{\gamma_M}$

$= \dfrac{1.5 \times 2.0 \times 10^3}{60 \times 1000} \leq \dfrac{0.10}{1.25}$

$= 0.050 \leq 0.080\,\text{N/mm}^2$

14.2.5.2 Wind suction case

Unfactored bending moment: $M_S = \dfrac{w_S L^2}{8} = \dfrac{-0.25 \times 4.0^2}{8} = -0.5\,\text{kN m/m}$

Design bending stresses: $\sigma_{F1} = -\dfrac{\gamma_w M_S}{e A_{F1}} \leq \dfrac{f_{ty}}{\gamma_M}$ (tension)

$= -\dfrac{1.5 \times (-0.5) \times 10^6}{58.66 \times 584.0} \leq \dfrac{320}{1.1}$

$= 21.9 \leq 290.9\,\text{N/mm}^2$

$\sigma_{F2} = \dfrac{\gamma_w M_S}{e A_{F2}} \leq \dfrac{\sigma_w}{\gamma_M}$ (compression)

$= (-)\dfrac{1.5 \times 0.5 \times 10^6}{58.66 \times 469.0} \leq \dfrac{120}{1.25}$

$= (-)27.3 \leq 96.0\,\text{N/mm}^2$

Design shear stresses: Not critical because $|w_S| < w_D$

14.2.5.3 *Design of fasteners for the ultimate limit state*

The panels are fastened to the supporting structure using 5.5-mm diameter steel screws with 22-mm diameter washers. These screws pass through the complete panel.

The characteristic tensile strength of a single screw fastening is $F_{Rt} = 1.95\,\text{kN}$.

There are two screws per panel at both the internal and end supports. It follows that the internal support case is always critical and there is no need to check the end supports.

Under the load case of wind suction (conservatively)

$$F_2 = (-)1.25 w_s L = (-)1.25 \times 0.25 \times 4.0 = (-)1.25\,\text{kN/m}$$

Design forces: $\gamma_w F_2 \leq \dfrac{n F_{Rt}}{\gamma_M}$

$$1.5 \times 1.25 \leq \frac{2 \times 1.95}{1.33}$$

$$1.875 \leq 2.98\,\text{kN}$$

14.2.5.4 *Concluding remark*

The conservative assumption that there are 'hinges' over the internal supports with zero rest moment removes the temperature stresses from the ultimate limit state calculations and means that a two-span panel acts as two single-span panels. As illustrated by the above calculation, the resulting stresses caused by wind pressure and wind suction acting alone are rather small and can be easily carried by the panel. This is typically the case and justifies this simple and safe approach. It follows that it is usually the serviceability limit state which is critical and governs the design of panels which are continuous over two or more spans.

14.2.6 Serviceability limit state

The serviceability limit state requires proper consideration of the two-span panel. The equations used in the following calculations can be found in Table 9.1.

14.2.6.1 *Calculation of the initial design parameters*

$$B_D = E_{F1} I_{F1} + E_{F2} I_{F2} = 0$$

$$B_S = \frac{E_{F1} A_{F1} E_{F2} A_{F2} e^2}{E_{F1} A_{F1} + E_{F2} A_{F2}} = \frac{210^2 \times 584 \times 469 \times 58.66^2}{210\,(584 + 469)} = 188.0 \times 10^6\,\text{kN}\,\text{mm}^2/\text{m}$$

$$k = \frac{3 B_S}{A_C G_{eff} L^2} = \frac{3 \times 188.0 \times 10^9}{58.66 \times 10^3 \times 3.1 \times 4000^2} = 0.194$$

Notes

(1) There is no bending moment in the faces so that $M_D = 0$ in all of the subsequent calculations.

(2) The effective shear modulus of the core:

$$G_{eff} = \frac{e\,G_C}{d_C} = \frac{58.66\,G_C}{60 - 0.56 - 0.46} \approx G_C \text{ in this example}$$

14.2.6.2 Load case: wind pressure and winter temperature

Temperature effect: $\theta = \frac{\alpha_{F2}\,T_2 - \alpha_{F1}\,T_1}{e} = \frac{12 \times 10^6 \times [20 - (-20)]}{58.66} = 8.183 \times 10^{-6}/\text{mm}$

Stresses at the internal support due to wind pressure

Unfactored bending moment:

$$M_S = -\frac{w_D\,L^2}{8(1+k)} = -\frac{0.5 \times 4.0^2}{8(1+0.194)} = -0.838\,\text{kN m/m}$$

Stresses: $\sigma_{F1} = -\frac{M_S}{e\,A_{F1}} = \frac{0.838 \times 10^6}{58.66 \times 584.0} = 24.5\,\text{N/mm}^2$ (tension)

$$\sigma_{F2} = \frac{M_S}{e\,A_{F2}} = -\frac{0.838 \times 10^6}{58.66 \times 469.0} = -30.5\,\text{N/mm}^2 \text{ (compression)}$$

Unfactored shear force:

$$V = \frac{w_D\,L}{2}\left[1 + \frac{1}{4(1+k)}\right] = \frac{0.5 \times 4.0}{2}\left[1 + \frac{1}{4(1+0.194)}\right] = 1.209\,\text{kN/m}$$

Shear stress: $\tau_C = \frac{V}{e} = \frac{1.209 \times 10^3}{58.66 \times 1000} = 0.0206\,\text{N/mm}^2$

Unfactored support reaction:

$$F_2 = w_D L\left[1 + \frac{1}{4(1+\text{k})}\right] = 0.5 \times 4.0\left[1 + \frac{1}{4(1+0.194)}\right] = 2.418\,\text{kN/m}$$

Compressive stress: $\sigma_{Cc} = \frac{F_2}{L_s} = \frac{2.418 \times 10^3}{60 \times 1000} = 0.0403\,\text{N/mm}^2$

Stresses at the internal support due to winter temperature

Unfactored bending moment:

$$M_S = -\frac{3\,B_S\,\theta}{2(1+k)} = -\frac{3 \times 188.0 \times 10^6 \times 8.183 \times 10^{-6}}{2(1+0.194)} = -1932\,\text{kN mm/m}$$

Stresses: $\sigma_{F1} = -\frac{M_S}{e\,A_{F1}} = \frac{1932 \times 10^3}{58.66 \times 584.0} = 56.4\,\text{N/mm}^2$ (tension)

$$\sigma_{F2} = \frac{M_S}{e\,A_{F2}} = -\frac{1932 \times 10^3}{58.66 \times 469.0} = -70.2\,\text{N/mm}^2 \text{ (compression)}$$

Fig. 14.3

Unfactored shear force:

$$V_T = \frac{M_S}{L} = \frac{1932}{4000} = 0.483\,\text{kN/m}$$

Shear stress: $\tau_C = \frac{V_T}{e} = \frac{0.483 \times 10^3}{58.66 \times 1000} = 0.0082\,\text{N/mm}^2$

Unfactored support reaction:

$$F_2 = \frac{2M_S}{L} = \frac{3864}{4000} = 0.966\,\text{kN/m}$$

Compressive stress: $\sigma_{Cc} = \frac{F_2}{L_s} = \frac{0.966 \times 10^3}{60 \times 1000} = 0.0161\,\text{N/mm}^2$

Load combinations

$$\sum \sigma_{F1} = \gamma_w \sigma_{F1.wD} + \gamma_T \psi \sigma_{F1.TW} \leq \frac{f_{ty}}{\gamma_M}$$

$$= 1.0 \times 24.5 + 1.0 \times 0.6 \times 56.4 \leq \frac{320}{1.0}$$

$$= 58.3 \leq 320\,\text{N/mm}^2$$

$$\sum \sigma_{F1} = \gamma_w \psi \sigma_{F1.wD} + \sigma_T \sigma_{F1.TW} \leq \frac{f_{ty}}{\gamma_M}$$

$$= 1.0 \times 0.6 \times 24.5 + 1.0 \times 56.4 \leq \frac{320}{1.0}$$

$$= 71.1 \leq 320\,\text{N/mm}^2$$

$$\sum \sigma_{F2} = \gamma_w \sigma_{F2.wD} + \gamma_T \psi \sigma_{F2.TW} \leq \frac{\sigma_{w.support}}{\gamma_M}$$

$$= 1.0 \times 30.5 + 1.0 \times 0.6 \times 70.2 \leq \frac{0.9 \times 120}{1.1}$$

$$= 72.6 \leq 98.2\,\text{N/mm}^2$$

$$\sum \sigma_{F2} = \gamma_w \psi \sigma_{F2.wD} + \gamma_T \sigma_{F2.TW} \leq \frac{\sigma_{w.support}}{\gamma_M}$$

$$= 1.0 \times 0.6 \times 30.5 + 1.0 \times 70.2 \leq \frac{0.9 \times 120}{1.1}$$

$$= 88.5 \leq 98.2\,\text{N/mm}^2$$

Note The factor of 0.9 on the right-hand side of the above equations is applied to the wrinkling stress in order to take account of the interaction with the support reaction force, see Section 10.2.2.3.

$$\sum \tau_C = \gamma_w \tau_{C.wD} + \gamma_T \psi \tau_{C.TW} \le \frac{f_{Cv}}{\gamma_M}$$

$$= 1.0 \times 0.0206 + 1.0 \times 0.6 \times 0.0082 \le \frac{0.11}{1.0}$$

$$= 0.0255 \le 0.11\,\text{N/mm}^2$$

$$\sum \tau_C = \gamma_w \psi \tau_{C.wD} + \gamma_T \tau_{C.TW} \le \frac{f_{Cv}}{\gamma_M}$$

$$= 1.0 \times 0.6 \times 0.0206 + 1.0 \times 0.0082 \le \frac{0.11}{1.0}$$

$$= 0.0206 \le 0.11\,\text{N/mm}^2$$

$$\sum \sigma_{Cc} = \gamma_w \sigma_{Cc.wD} + \gamma_T \psi \sigma_{Cc.TW} \le \frac{0.71 f_{Cc}}{\gamma_M}$$

$$= 1.0 \times 0.0413 + 1.0 \times 0.6 \times 0.0161 \le \frac{0.71 \times 0.10}{1.0}$$

$$= 0.0510 \le 0.071\,\text{N/mm}^2$$

$$\sum \sigma_{Cc} = \gamma \psi \sigma_{Cc.wD} + \gamma_T \sigma_{Cc.TW} \le \frac{0.71 f_{Cc}}{\gamma_M}$$

$$= 1.0 \times 0.6 \times 0.0413 + 1.0 \times 0.0161 \le \frac{0.71 \times 0.10}{1.0}$$

$$= 0.0409 \le 0.071\,\text{N/mm}^2$$

Note The factor of 0.71 on the right-hand side of the above equations is applied to the compressive stress in order to take account of the interaction of the bending moment and the support reaction force, see Section 10.2.2.3.

Deflection
Under wind pressure w_D:

$$w_{max} = \frac{w_D L^4}{48 B_S (1 + k)} (0.26 + 2.6k + 2k^2)$$

$$= \frac{0.5 \times 4000^4 \times 10^{-3}}{48 \times 188.0 \times 10^6 (1 + 0.194)} (0.26 + 2.6 \times 0.194 + 2 \times 0.194^2)$$

$$= 9.98\,\text{mm}$$

Under winter temperature:

$$w_{max} = \frac{\theta L^2}{32} \frac{1.089 + 3.96k}{1 + k}$$

$$= \frac{8.183 \times 10^{-6} \times 4000^2}{32} \frac{1.089 + 3.96 \times 0.194}{1 + 0.194}$$

$$= 6.36\,\text{mm}$$

Load combinations:

$$\sum w_{max} = 0.75 w_{max.wD} + \psi w_{max.TW} = 0.75 \times 9.98 + 0.6 \times 6.36 = 11.30\,\text{mm} = L/354$$

$$\sum w_{max} = 0.75 \psi w_{max.wD} + w_{max.TW} = 0.75 \times 0.6 \times 9.98 + 6.36 = 10.85\,\text{mm} = L/369$$

14.2.6.3 Load case: wind suction and summer temperature

$$\text{Temperature effect: } \theta = \frac{\alpha_{F2} T_2 - \alpha_{F1} T_1}{e} = \frac{12 \times 10^{-6} \times (25 - 65)}{58.66} = -8.183 \times 10^{-6}/\text{mm}$$

Stresses at the internal support due to wind suction
Unfactored bending moment:

$$M_S = -\frac{w_S L^2}{8(1+k)} = -\frac{(-0.25) \times 4.0^2}{8(1+0.194)} = 0.419\,\text{kN m/m}$$

$$\text{Stresses:} \quad \sigma_{F1} = -\frac{M_S}{e A_{F1}} = -\frac{0.419 \times 10^6}{58.66 \times 584.0} = -12.2\,\text{N/mm}^2 \text{ (compression)}$$

$$\sigma_{F2} = \frac{M_S}{e A_{F2}} = \frac{0.419 \times 10^6}{58.66 \times 469.0} = 15.2\,\text{N/mm}^2 \text{ (tension)}$$

Unfactored shear force close to the support:

$$V = \frac{w_S L}{2}\left[1 + \frac{1}{4(1+k)}\right] = \frac{0.25 \times 4.0}{2}\left(1 + \frac{1}{4(1+0.194)}\right) = 0.605\,\text{kN/m}$$

$$\text{Shear stress:} \quad \tau_C = \frac{V}{e} = \frac{0.605 \times 10^3}{58.66 \times 1000} = 0.0103\,\text{N/mm}^2$$

Unfactored support reaction (for fastener design):

$$F_2 = w_S L\left[1 + \frac{1}{4(1+k)}\right] = -0.25 \times 4.0\left[1 + \frac{1}{4(1+0.194)}\right] = -1.209\,\text{kN/m}$$

Stresses at the internal support due to summer temperature
Unfactored bending moment:

$$M_S = -\frac{3 B_S \theta}{2(1+k)} = -\frac{3 \times 188.0 \times 10^6 \times (-8.183 \times 10^{-6})}{2(1+0.194)} = 1933\,\text{kN m/m}$$

$$\text{Stresses:} \quad \sigma_{F1} = -\frac{M_S}{e A_{F1}} = -\frac{1933 \times 10^3}{58.66 \times 584.0} = -56.4\,\text{N/mm}^2$$

$$\sigma_{F2} = \frac{M_S}{e A_{F2}} = \frac{1933 \times 10^3}{58.66 \times 469.0} = 70.3\,\text{N/mm}^2$$

Unfactored shear force:

$$V = \frac{M_S}{L} = \frac{1933}{4000} = 0.483\,\text{kN/m}$$

$$\text{Shear stress:} \quad \tau_C = \frac{V}{e} = \frac{0.483 \times 10^3}{58.66 \times 1000} = 0.0082\,\text{N/mm}^2$$

Unfactored support reaction:

$$F_2 = \frac{2\,M_S}{L} = \frac{2 \times 1933}{4000} = 0.9665\,\text{kN/m}$$

Load combinations

$$\sum \sigma_{F1} = \gamma_w\, \sigma_{F1.wS} + \gamma_T\, \psi\, \sigma_{F1.TS} \qquad \le \frac{\sigma_{w.support}}{\gamma_M}\ \text{(external face)}$$

$$= 1.0 \times 12.2 + 1.0 \times 0.6 \times 56.4 \le \frac{0.8 \times 120}{1.1}$$

$$= 46.0 \qquad \le 87.3\,\text{N/mm}^2$$

$$\sum \sigma_{F1} = \gamma_w\, \psi\, \sigma_{F1.wS} + \gamma_T\, \sigma_{F1.TS} \qquad \le \frac{\sigma_{w.support}}{\gamma_M}\ \text{(external face)}$$

$$= 1.0 \times 0.6 \times 12.2 + 1.0 \times 56.4 \le \frac{0.8 \times 120}{1.1}$$

$$= 63.7 \qquad \le 87.3\,\text{N/mm}^2$$

Note The factor 0.8 on the right-hand side of the above equations is applied to the wrinkling stress in order to take account of the interaction with the fastener forces, see Section 10.2.3. This value is applicable to two or three screws per panel width of 1 metre. If there are four or more screws per metre, it becomes 0.8 $(11 - n)/8$ where n is the number of screws per metre.

$$\sum \sigma_{F2} = \gamma_w\, \sigma_{F2.wS} + \gamma_T\, \psi\, \sigma_{F2.TS} \qquad \le \frac{f_{ty}}{\gamma_M}$$

$$= 1.0 \times 15.2 + 1.0 \times 0.6 \times 70.3 \le \frac{320}{1.0}$$

$$= 57.4 \qquad \le 320\,\text{N/mm}^2$$

$$\sum \sigma_{F2} = \gamma_w\, \psi\, \sigma_{F2.wS} + \gamma_T\, \sigma_{F2.TS} \qquad \le \frac{f_{ty}}{\sigma_M}$$

$$= 1.0 \times 0.6 \times 15.2 + 1.0 \times 70.3 \le \frac{320}{1.0}$$

$$= 79.4 \qquad \le 320\,\text{N/mm}^2$$

$$\sum \tau_C = \tau_{C.wS} + \gamma_T\, \psi\, \tau_{C.T} \qquad \le \frac{f_{Cv}}{\gamma_M}$$

$$= 1.0 \times 0.0103 + 1.0 \times 0.6 \times 0.0082 \le \frac{0.11}{1.0}$$

$$= 0.0152 \qquad \le 0.11\,\text{N/mm}^2$$

$$\sum \tau_C = \gamma_w\, \psi\, \tau_{C.wD} + \gamma_T\, \tau_{C.T} \qquad \le \frac{f_{Cv}}{\gamma_M}$$

$$= 1.0 \times 0.6 \times 0.0103 + 1.0 \times 0.0082 \le \frac{0.11}{1.0}$$

$$= 0.0144 \qquad \le 0.11\,\text{N/mm}^2$$

Deflection
Under wind suction w_S:

$$w_{max} = \frac{w_S L^4}{48 B_S(1+k)}(0.26 + 2.6k + 2k^2)$$

$$= \frac{-0.25 \times 4000^4 \times 10^{-3}}{48 \times 188.0 \times 10^6(1+0.194)}(0.26 + 2.6 \times 0.194 + 2 \times 0.194^2)$$

$$= -4.99\,\text{mm}$$

Under summer temperature:

$$w_{max} = \frac{\theta L^2}{32} \frac{1.089 + 3.96k}{1+k}$$

$$= -\frac{8.138 \times 10^{-6} \times 4000^2}{32} \frac{1.089 + 3.96 \times 0.194}{1 + 0.194}$$

$$= -6.36\,\text{mm}$$

Load combinations:

$$\sum w_{max} = 0.75 w_{max.wS} + \psi w_{max.TS} = 0.75 \times 4.99 + 0.6 \times 6.36 = 7.56\,\text{mm} = L/529$$

$$\sum w_{max} = 0.75 \psi w_{max.wD} + w_{max.TW} = 0.75 \times 0.6 \times 4.99 + 6.36 = 8.61\,\text{mm} = L/465$$

Design of fasteners for the serviceability limit state
At the intermediate support, under the load combination of wind suction and summer temperature:

$$\sum F_2 = \gamma_w F_{2.wS} + \gamma_T \psi F_{2.TS} \leq \frac{n F_u}{\gamma_M}$$

$$= 1.0 \times 1.209 + 1.0 \times 0.6 \times 0.9665 \leq \frac{2 \times 1.95}{1.0}$$

$$= 1.79 \leq 3.90\,\text{kN}$$

$$\sum F_2 = \gamma_w \psi F_{2.wS} + \gamma_T F_{2.TS} \leq \frac{n F_u}{\gamma_M}$$

$$= 1.0 \times 0.6 \times 1.209 + 1.0 \times 0.9665 \leq \frac{2 \times 1.95}{1.0}$$

$$= 1.69 \leq 3.90\,\text{kN}$$

14.2.7 Comments on the calculations

There are a significant number of load cases and potential failure modes that must be checked in the design of a typical sandwich panels. The calculations are, of course, well ordered and easy to program for a computer.

For the case examined, the critical failure mode is wrinkling of the lower face at the internal support for the serviceability limit state under the load combination of wind pressure and winter temperature. This is a typical outcome.

14.3 TWO-SPAN ROOF PANEL

The structural system is shown in Fig. 14.4.

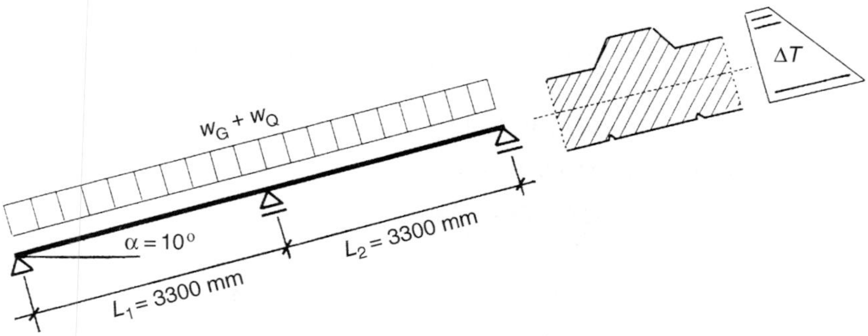

Fig. 14.4 General arrangement of two-span roof panel.

The following points may be noted.

- There are a significant number of alternative load cases for roof panels which are listed in Table 11.4 of Chapter 11. The following three load cases are generally those which govern the design:

  (1) snow and winter temperature with $T_1 = 0°C$
  (2) winter temperature $T_1 = -20°C$ without snow
  (3) wind suction and summer temperature

  The calculation example which follows is primarily concerned with the first of these, namely snow with winter temperature. The calculations for the other load cases follow a similar course.
- In the ultimate limit state, it is conservatively assumed that 'plastic hinges' of zero moment capacity exist over the internal supports. A consequence of this is that the calculation becomes simply that for a single-span panel subject to self-weight plus snow together with a relatively small temperature effect.
- At the serviceability limit state, the combination of wind or snow load and the action of temperature must be considered, bearing in mind the continuity over the internal support. The calculations for this limit state are, therefore, rather more complicated.
- In a roof panel, the influence of creep may be significant and cannot be ignored.

14.3.1 Actions (loads)

The panel is designed for the following actions:

Self-weight: $w_G = 0.166\,kN/m^2$ Perpendicular to the slope of the roof

Snow:	$w_Q = 0.75\,\text{kN/m}^2$	Perpendicular to the slope of the roof	
Wind suction:	w_S = upward force is relieved by self-weight. This case is not considered in detail.		
Temperature:	Winter with snow:	$T_1 = 0°C$	$T_2 = +20°C$
	Winter with no snow:	$T_1 = -20°C$	$T_2 = +20°C$
	Summer:	$T_1 = +65°C$	$T_2 = +25°C$

(colour group 2: medium colours such as sandy yellow or light blue)

14.3.2 Cross-section of the panel

The cross-section of the panel is shown in Fig. 14.5.
The relevant dimensions are as follows:

$t_{nom.1} = 0.60\,\text{mm}$ $\quad t_{d.1} = 0.56\,\text{mm}$ $\quad A_{F1} = 673\,\text{mm}^2/\text{m}$ $\quad I_{F1} = 88.58 \times 10^3\,\text{mm}^4/\text{m}$
$d_{11} = 32.0\,\text{mm}$ $\quad d_{12} = 6.00\,\text{mm}$

$t_{nom.2} = 0.50\,\text{mm}$ $\quad t_{d.2} = 0.46\,\text{mm}$ $\quad A_{F2} = 469\,\text{mm}^2/\text{m}$ $\quad I_{F2} \approx 0$
$d_{21} = 2.45\,\text{mm}$ $\quad d_{22} = 0.20\,\text{mm}$

Nominal depth $\quad d = 60.0\,\text{mm}$

$$e = 60 - \tfrac{1}{2}(t_{nom.1} + t_{nom.2}) + d_{12} - d_{22} = 60 - \tfrac{1}{2}(0.60 + 0.50) + 6.00 - 0.20 = 65.25\,\text{mm}$$

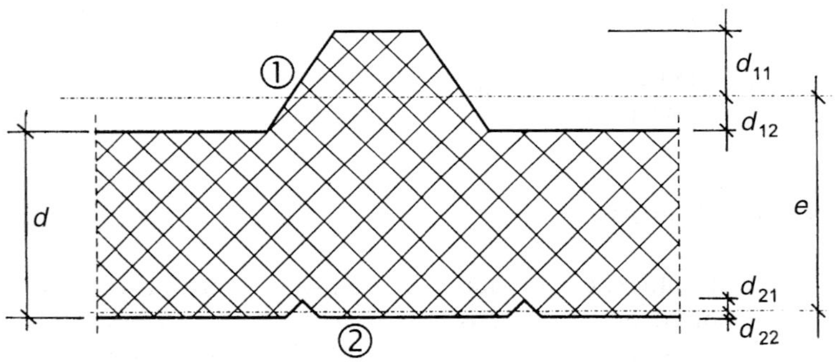

Fig. 14.5 Cross-section of roof panel.

14.3.3 Design values for the panel

The following design values have been determined on the basis of the test procedures described in this book:

Effective shear modulus of the core:	$G_{eff} = 3.02\,\text{N/mm}^2$
Shear strength of the core:	$f_{Cv} = 0.11\,\text{N/mm}^2$
Compressive strength of the core:	$f_{Cc} = 0.10\,\text{N/mm}^2$
Yield strength of the faces:	$f_{ty} = 320\,\text{N/mm}^2$
Wrinkling stress of the face:	f_{Fc} or $\sigma_w = 120\,\text{N/mm}^2$ (inner face only)

Note For this panel, there is no buckling or wrinkling of the outer face prior to yield and it is sufficient to design this face on the basis of $f_{ty} = 320\,N/mm^2$. Depending on the slenderness of the plate elements of the profile, this is not necessarily always the case and it is necessary to verify this assumption by testing or detailed calculation.

The following additional values are required for the calculation:

Elastic modulus of steel: $E_s = 210\,kN/mm^2 = E_{F1} = E_{F2}$

Coefficient of thermal expansion of steel: $\alpha_s = 12 \times 10^{-6}$ per °C $= \alpha_{F1} = \alpha_{F2}$

Creep coefficient (polyurethane – see Table 11.2): permanent load $\phi_t = 7.0$

snow load $\phi_t = 2.4$

14.3.4 Load, material and combination factors

The relevant material factors are given in Table 11.2. The relevant load and combination factors are given in Table 11.4.

14.3.5 Calculation of the initial design parameters

$$B_D = E_{F1} I_{F1} + E_{F2} I_{F2} = 210 \times 88.58 \times 10^3 = 18.602 \times 10^6\,kN\,mm^2/m$$

$$B_S = \frac{E_{F1} A_{F1} E_{F2} A_{F2} e^2}{E_{F1} A_{F1} + E_{F2} A_{F2}} = \frac{210^2 \times 673 \times 469 \times 65.25^2}{210(673 + 469)} = 247.4 \times 10^6\,kN\,mm^2/m$$

$$\frac{B_D}{B_S} = \frac{18.602 \times 10^6}{247.4 \times 10^6} = 0.0752$$

14.3.6 Ultimate limit state

In the ultimate limit state, the individual spans are considered to be simply supported over a span of $L = 3.3$ metres with 'hinges' with zero rest moment at the internal supports. The ultimate limit state is, therefore, defined by wrinkling (for the profile in this example yielding) of the top face of a single span of 3.3 metres.

Only the load case of snow with the relevant winter temperature is considered in detail. The calculation for the other load cases follows a similar pattern but is generally simpler because of the reduced significance of creep.

The equations used are the approximate solutions given in Section 9.4 of this book.

14.3.6.1 Load case: self-weight without creep

Calculation values: $$k_Q = \frac{9.6 B_S}{A_C G_{eff} L^2} = \frac{9.6 \times 247.4 \times 10^9}{65.25 \times 1000 \times 3.02 \times 3300^2} = 1.107$$

$$\beta = \frac{B_D}{B_D + \dfrac{B_S}{1+k_Q}} = \frac{18.602 \times 10^6}{18.602 \times 10^6 + \dfrac{247.4 \times 10^6}{1+1.107}} = 0.137$$

Unfactored bending moments:

$$M_{F1} = \beta \frac{w_G L^2}{8} = 0.137 \frac{0.116 \times 3.30^2}{8} = 0.0216\,\text{kN m/m}$$

$$M_S = (1-\beta)\frac{w_G L^2}{8} = (1-0.137)\frac{0.116 \times 3.30^2}{8} = 0.136\,\text{kN m/m}$$

Stresses:

$$\sigma_{F1.1} = -\frac{M_{F1} d_{11}}{I_{F1}} - \frac{M_S}{e A_{F1}} = -\frac{0.0216 \times 10^6 \times 32.0}{88.58 \times 10^3} - \frac{0.136 \times 10^6}{65.25 \times 673} = -10.9\,\text{N/mm}^2$$

$$\sigma_{F2.2} = \frac{M_S}{e A_{F2}} = \frac{0.136 \times 10^6}{65.25 \times 469} = 4.45\,\text{N/mm}^2$$

Unfactored shear force:

$$V_S = (1-\beta)\frac{w_G L}{2} = (1-0.137)\frac{0.116 \times 3.30}{2} = 0.165\,\text{kN/m}$$

Shear stress: $\tau_C = \dfrac{V_S}{e} = \dfrac{0.165 \times 10^3}{65.25 \times 1000} = 0.002\,53\,\text{N/mm}^2$

At the intermediate support:

Unfactored support reaction: $F_2 = w_G L = 0.116 \times 3.3 = 0.382\,\text{kN/m}$

14.3.6.2 *Load case: snow without creep*

Calculation: k_Q, β as for Section 14.3.6.1: self-weight without creep.

Unfactored bending moments:

$$M_{F1} = \beta \frac{w_Q L^2}{8} = 0.137 \frac{0.75 \times 3.30^2}{8} = 0.140\,\text{kN m/m}$$

$$M_S = (1-\beta)\frac{w_Q L^2}{8} = (1-0.137)\frac{0.75 \times 3.30^2}{8} = 0.881\,\text{kN m/m}$$

Stresses:

$$\sigma_{F1.1} = \frac{M_{F1} d_{11}}{I_{F1}} - \frac{M_S}{e A_{F1}} = -\frac{0.140 \times 10^6 \times 32.0}{88.58 \times 10^3} - \frac{0.881 \times 10^6}{65.25 \times 673} = -70.7\,\text{N/mm}^2$$

$$\sigma_{F2.2} = \frac{M_S}{e A_{F2}} = \frac{0.881 \times 10^6}{65.25 \times 469} = 28.8\,\text{N/mm}^2$$

Unfactored shear force:

$$V_S = (1-\beta)\frac{w_Q L}{2} = (1-0.137)\frac{0.75 \times 3.30}{2} = 1.068\,\text{kN/m}$$

Shear stress: $\tau_C = \dfrac{V_S}{e} = \dfrac{1.068 \times 10^3}{65.25 \times 1000} = 0.017\,16\,\text{N/mm}^2$

At the intermediate support:

Unfactored support reaction: $F_2 = w_Q L = 0.75 \times 3.3 = 2.475\,\text{kN/m}$

14.3.6.3 Load case: winter temperature (with snow; $\Delta T = 20°C$)

Calculation values: $k_T = \dfrac{8.0\,B_S}{A_C\,G_{eff}\,L^2} = \dfrac{8.0 \times 247.4 \times 10^9}{65.25 \times 1000 \times 3.02 \times 3300^2} = 0.922$

$$\beta = \frac{B_D}{B_D + \dfrac{B_S}{1+k_T}} = \frac{18.602 \times 10^6}{18.602 \times 10^6 + \dfrac{247.4 \times 10^6}{1+0.922}} = 0.126$$

Temperature effect: $\theta = \dfrac{\alpha_{F2}\,T_2 - \alpha_{F1}\,T_1}{e} = \dfrac{12 \times 10^{-6} \times (20-0)}{65.25} = 3.678 \times 10^{-6}/\text{mm}$

Unfactored bending moments:

$$M_{F1} = -M_S = B_D\,\theta(1-\beta) = 18.602 \times 10^6 \times 3.678 \times 10^{-6}(1-0.126) = 59.8\,\text{kN mm/m}$$

Stresses:

$$\sigma_{F1.1} = -\frac{M_{F1}\,d_{11}}{I_{F1}} - \frac{M_S}{e\,A_{F1}} = -\frac{59.8 \times 10^3 \times 32.0}{88.58 \times 10^3} + \frac{59.8 \times 10^3}{65.25 \times 673} = -20.2\,\text{N/mm}^2$$

$$\sigma_{F2.2} = \frac{M_S}{e\,A_{F2}} = \frac{-59.8 \times 10^3}{65.25 \times 469} = -1.95\,\text{N/mm}^2$$

In this example, the shear stress due to temperature may be ignored at the ultimate limit state.

14.3.6.4 Load case: creep

The additional bending moments and stresses caused by creep have to be evaluated separately and combined with the values calculated above. It should be noted that, at the ultimate limit state, the stresses caused by snow have a load factor of 1.5 and the stresses caused by self-weight have a load factor of 1.35 whereas those caused by creep only have a load factor of 1.0. Here, the calculation and subsequent combination is carried out in the formally correct way.

Under self-weight: $G_{Ct} = \dfrac{G_C}{1+\phi} = \dfrac{3.02}{1+7.0} = 0.378\,\text{N/mm}^2$

$$k_Q = \frac{9.6 B_S}{A_C\,G_{Ct}\,L^2} = \frac{9.6 \times 247.4 \times 10^9}{65.25 \times 1000 \times 0.378 \times 3300^2} = 8.842$$

$$\beta = \frac{B_D}{B_D + \dfrac{B_S}{1+k_Q}} = \frac{18.602 \times 10^6}{18.602 \times 10^6 + \dfrac{247.4 \times 10^6}{1+8.842}} = 0.425$$

$$M_{F1} = \beta\frac{w_G\,L^2}{8} = 0.425\frac{0.116 \times 3.30^2}{8} = 0.0672\,\text{kN m/m}$$

$$M_S = (1-\beta)\frac{w_G\,L^2}{8} = (1-0.425)\frac{0.116 \times 3.30^2}{8} = 0.0907\,\text{kN m/m}$$

$$\sigma_{F1.1} = -\frac{M_{F1}\,d_{11}}{I_{F1}} - \frac{M_S}{e\,A_{F1}} = -\frac{0.0672 \times 10^6 \times 32.0}{88.58 \times 10^3} - \frac{0.0907 \times 10^6}{65.25 \times 673} = -26.3\,\text{N/mm}^2$$

$$\sigma_{F2.2} = \frac{M_S}{e\,A_{F2}} = \frac{0.0907 \times 10^6}{65.25 \times 469} = 2.96\,\text{N/mm}^2$$

Therefore, additional stresses due to creep: $\Delta_{F1.1.Gc} = -26.3 + 10.9 = -15.4\,\text{N/mm}^2$

$\Delta_{F2.2.Gc} = 2.96 - 4.45 = -1.49\,\text{N/mm}^2$

Under snow: $G_{Ct} = \dfrac{G_C}{1+\phi} = \dfrac{3.02}{1+2.4} = 0.882\,\text{N/mm}^2$

$$k_Q = \frac{9.6\,B_S}{A_C\,G_{Ct}\,L^2} = \frac{9.6 \times 247.4 \times 10^9}{65.25 \times 1000 \times 0.882 \times 3300^2} = 3.790$$

$$\beta = \frac{B_D}{B_D + \dfrac{B_S}{1+k_Q}} = \frac{18.602 \times 10^6}{18.602 \times 10^6 + \dfrac{247.4 \times 10^6}{1+3.790}} = 0.265$$

$$M_{F1} = \beta \frac{w_Q L^2}{8} = 0.265 \frac{0.75 \times 3.30^2}{8} = 0.270\,\text{kN m/m}$$

$$M_S = (1-\beta)\frac{w_G L^2}{8} = (1-0.265)\frac{0.75 \times 3.30^2}{8} = 0.751\,\text{kN m/m}$$

$$\sigma_{F1.1} = -\frac{M_{F1}\,d_{11}}{I_{F1}} - \frac{M_S}{e\,A_{F1}} = -\frac{0.270 \times 10^6 \times 32.0}{88.58 \times 10^3} - \frac{0.751 \times 10^6}{65.25 \times 673} = -114.7\,\text{N/mm}^2$$

$$\sigma_{F2.2} = \frac{M_S}{e\,A_{F2}} = \frac{0.751 \times 10^6}{65.25 \times 469} = 24.5\,\text{N/mm}^2$$

Therefore, additional stresses due to creep: $\Delta_{F1.1.Qc} = -114.7 + 70.7 = -44.0\,\text{N/mm}^2$

$\Delta_{F2.2.Qc} = 24.5 - 28.8 = -4.3\,\text{N/mm}^2$

14.3.6.5 Load combination

$$\sum \sigma_{F1.1} = \gamma_G\,\sigma_{F1.1.G} + \gamma_Q\,\sigma_{F1.1.Q} + \gamma_T\,\sigma_{F1.1.T} + \gamma_c(\Delta\,\sigma_{F1.1.Gc} + \Delta\,\sigma_{F1.1.Qc}) \le \frac{f_{ty}}{\gamma_M}$$

$$= 1.35 \times (-10.9) + 1.5 \times (-70.7) + 1.5 \times (-20.2) + 1.0 \times (-15.4 - 44.0) \le \frac{320}{1.1}$$

$$= 210.5 \le 290.9\,\text{N/mm}^2$$

$$\sum \sigma_{F2.2} = \gamma_G\,\sigma_{F2.2.G} + \gamma_Q\,\sigma_{F2.2.Q} + \gamma_T\,\sigma_{F2.2.T} + \gamma_c(\Delta\,\sigma_{F2.2.Gc} + \Delta\,\sigma_{F2.2.Qc}) \le \frac{f_{ty}}{\gamma_M}$$

$$= 1.35 \times 4.45 + 1.5 \times 28.8 + 1.5 \times 1.95 + 1.0 \times (-1.49 - 4.3)^* \le \frac{320}{1.1}$$

$$= 52.1 \le 290.9\,\text{N/mm}^2$$

* This term is ignored because it is favourable.

$$\sum \tau_C = \gamma_G \tau_{C.G} + \gamma_Q \tau_{C.Q} \qquad \leq \frac{f_{Cv}}{\gamma_M}$$

$$= 1.35 \times 0.00253 + 1.5 \times 0.01716 \leq \frac{0.11}{1.25}$$

$$= 0.0292 \qquad \leq 0.0880\,\text{N/mm}^2$$

Support reaction resistance:

$$F_{R2} = \frac{f_{Cc}}{\gamma_M} L_s = \frac{0.10}{1.25} \times 150 \times 1000 = 12.0 \times 10^3\,\text{N/m} = 12.0\,\text{kN/m}$$

$$\sum F_2 = \gamma_G F_{2.G} + \gamma F_{2.Q} \qquad \leq F_{R2}$$

$$= 1.35 \times 0.382 + 1.5 \times 2.475 \leq 12.0$$

$$= 4.23 \qquad \leq 12.0\,\text{kN/m}$$

14.3.7 Serviceability limit state

The serviceability limit state requires proper consideration of the two-span panel. The following calculations makes use of the relevant design curves in Section 9.4.

14.3.7.1 Load case: self-weight together with snow without creep

Calculation values: $k = \dfrac{B_S}{A_C G_{eff} L^2} = \dfrac{247.4 \times 10^9}{65.25 \times 1000 \times 3.02 \times 3300^2} = 0.115$

$$\frac{B_D}{B_S} = \frac{18.602 \times 10^6}{247.4 \times 10^6} = 0.0752$$

From Fig. 9.14(a) $\quad \beta_{Sq} = 0.18$
From Fig. 9.14(b) $\quad \beta_{0q} = 0.47$

Unfactored bending moments at the internal support:

$$M_{stat} = -\frac{(w_G + w_Q)L^2}{8} = -\frac{(0.116 + 0.75)\,3.3^2}{8} = -1.179\,\text{kN m/m}$$

$$M_{F1} = \frac{\beta_{0q} M_{stat}}{1 + \beta_{Sq}} = \frac{0.47 \times (-1.179)}{1 + 0.18} = -0.470\text{kN m/m}$$

$$M_S = \frac{(1 - \beta_{0q}) M_{stat}}{1 + \beta_{Sq}} = \frac{(1 - 0.47) \times (-1.179)}{1 + 0.18} = -0.530\,\text{kN m/m}$$

Stresses:

$$\sigma_{F1.1} = -\frac{M_{F1} d_{11}}{I_{F1}} - \frac{M_S}{e A_{F1}} = \frac{0.470 \times 10^6 \times 32.0}{88.58 \times 10^3} + \frac{0.530 \times 10^6}{65.25 \times 673} = 181.9\,\text{N/mm}^2$$

$$\sigma_{F2.2} = \frac{M_S}{e A_{F2}} = \frac{-0.530 \times 10^6}{65.25 \times 469} = -17.3\,\text{N/mm}^2$$

Unfactored shear force:

$$V_S \approx 0.625(1-\beta_{0q})(w_G + w_Q)L = 0.625(1-0.47)(0.116+0.75) \times 3.30 = 0.947\,\text{kN/m}$$

Shear stress: $\tau_C = \dfrac{V_S}{e} = \dfrac{0.947 \times 10^3}{65.25 \times 1000} = 0.0145\,\text{N/mm}^2$

At the intermediate support:

Unfactored support reaction:

$$F_2 \leq 1.25(w_G + w_Q)L = 1.25(0.116+0.75)3.3 = 3.57\,\text{kN/m}$$

14.3.7.2 *Load case: winter temperature (with snow; $\Delta T = 20°C$)*

The calculation values are as in Section 14.3.7.1 above.

Therefore, from Fig. 9.14(c) $\quad \beta_{ST} = 1.35$
Therefore, from Fig. 9.14(b) $\quad \beta_{0T} = 0.091$

Unfactored bending moments at the internal support:

$\theta = 3.678 \times 10^{-6}$/mm from Section 14.3.6.3

$$M_{F1} = -\theta B_D \beta_{ST} = -3.678 \times 10^6 \times 18.602 \times 10^6 \times 1.35 = -92.4\,\text{kN mm/m}$$

$$M_S = M_{F1}\left(\frac{1}{\beta_{0T}} - 1\right) = -92.4\left(\frac{1}{0.091} - 1\right) = -922.6\,\text{kN mm/m}$$

Stresses:

$$\sigma_{F1.1} = -\frac{M_{F1} d_{11}}{I_{F1}} - \frac{M_S}{e A_{F1}} = \frac{92.4 \times 10^3 \times 32.0}{88.58 \times 10^3} + \frac{922.6 \times 10^3}{65.25 \times 673} = 54.4\,\text{N/mm}^2$$

$$\sigma_{F2.2} = \frac{M_S}{e A_{F2}} = \frac{-922.6 \times 10^3}{65.25 \times 469} = -30.1\,\text{N/mm}^2$$

Unfactored shear force:

$$V_{ST} = \frac{M_S}{L} = \frac{922.6}{3300} = 0.280\text{kN/m}$$

Shear stress: $\tau_C = \dfrac{V_{ST}}{e} = \dfrac{0.280 \times 10^3}{65.25 \times 1000} = 0.0043\,\text{N/mm}^2$

At the intermediate support:

Unfactored support reaction: $F_2 \leq 2\dfrac{(M_D + M_S)}{L} = 2\dfrac{(0.0924 + 0.9226)}{3.3} = 0.615\,\text{kN/m}$

14.3.7.3 *Load case: creep*

As in Section 14.3.6.4, the additional bending moments and stresses caused by creep have to be evaluated separately and combined with the values calculated above. The procedure is very similar to the calculations already given and is not carried out in detail. The additional stresses due to creep are (all tension):

Self-weight: $\quad \Delta_{F1.1.Gc} = 16.0\,\text{N/mm}^2$
$\quad \Delta_{F2.2.Gc} = 1.92\,\text{N/mm}^2$

Snow: $\Delta_{F1.1.Qc} = 51.8\,N/mm^2$
$\Delta_{F2.2.Qc} = 7.8\,N/mm^2$

14.3.7.4 Load combination

According to Table 11.4, all load factors are 1.0 at the serviceability limit state.

$$\sum \sigma_{F1.1} = \gamma_G \sigma_{F1.1.G} + \gamma_Q \sigma_{F1.1.Q} + \gamma_T \sigma_{F1.1.T} + \gamma_c(\Delta \sigma_{F1.1.Gc} + \Delta \sigma_{F1.1.Qc}) \leq \frac{f_{ty}}{\gamma_M}$$

$$= 1.0 \times [181.9 + 54.4 + (16.0 + 51.8)] \leq \frac{320}{1.0}$$

$$= 304.1 \leq 320\,N/mm^2$$

When considering the maximum compressive stress at the internal face, the contributions of creep are omitted because they cause tension and thus reduce the maximum compression. It is the short-term stresses which are critical.

$$\sum \sigma_{F2.2} = \gamma_G \sigma_{F2.2.G} + \gamma_Q \sigma_{F2.2.T} + \gamma_T \sigma_{F2.2.T} + \gamma_c(\Delta \sigma_{F2.2.Gc} + \Delta \sigma_{F2.2.Qc}) \leq \frac{0.9\,\sigma_w}{\gamma_M}$$

$$= 1.0 \times (-17.3 - 30.1) \leq \frac{0.9 \times 120}{1.1}$$

$$= 47.4 \leq 98.2\,N/mm^2$$

$$\sum \tau_C = \gamma_G \tau_{C.G} + \gamma_Q \tau_{C.Q} \leq \frac{f_{Cv}}{\gamma_M}$$

$$= 1.0 \times 0.0145 + 1.0 \times 0.0043 \leq \frac{0.11}{1.0}$$

$$= 0.0188 \leq 0.11\,N/mm^2$$

Support reaction resistance:

$$F_{R2} = \frac{0.71 f_{Cc}}{\gamma_M} L_s = \frac{0.71 \times 0.10}{1.0} 150 \times 1000 = 10.65 \times 10^3\,N/m = 10.65\,kN/m$$

$$\sum F_2 = \gamma_Q F_{2.(G+Q)}, + \gamma_T F_{2.T} \leq F_{R2}$$

$$= 1.0 \times 3.57 + 1.0 \times 0.615 \leq 10.65$$

$$= 4.19 \leq 10.65\,kN/m$$

14.3.7.5 Deflection

Note that this section of the calculations uses formulae which are used in Germany but which have no formal derivation.

Under self-weight + snow without creep:

Calculation values: $k = \dfrac{B_S}{A_C\,G_{eff}\,L^2} = \dfrac{247.4 \times 10^9}{65.25 \times 1000 \times 3.02 \times 3300^2} = 0.115$

Therefore, from Fig. 9.14(a) $\beta_{Sq} = 0.18$

$$w = \frac{(w_G + w_Q)L^4}{B_D + \dfrac{B_S}{1+15k}}\left[\frac{5}{384} - \frac{1}{139(1+\beta_{Sq})}\right]$$

$$= \frac{(0.116 + 0.75)3.3^4 \times 10^9}{18.6 \times 10^6 + \dfrac{247.4 \times 10^6}{1 + 15 \times 0.115}}\left(\frac{5}{384} - \frac{1}{139(1+0.18)}\right)$$

$$= 6.50\,\text{mm}$$

Under winter temperature ($\Delta T = 20°\text{C}$ with snow on the roof):

Calculation values: $k_T = \dfrac{8.0\,B_S}{A_C G_{eff} L^2} = \dfrac{8.0 \times 247.4 \times 10^9}{65.25 \times 1000 \times 3.02 \times 3300^2} = 0.922$

$$\beta = \frac{B_D}{B_D + \dfrac{B_S}{1+k_T}} = \frac{18.602 \times 10^6}{18.602 \times 10^6 + \dfrac{247.4 \times 10^6}{1+0.922}} = 0.126$$

$$w = \frac{\theta L^2}{32}\,\frac{1 + 2.34 k_T}{1 + 0.59 k_T}\,(1 - \beta)$$

$$= \frac{3.678 \times 10^{-6} \times 3.3^2 \times 10^6}{32}\,\frac{1 + 2.34 \times 0.922}{1 + 0.59 \times 0.922}(1 - 0.126)$$

$$= 2.24\,\text{mm}$$

Additional deflection due to creep:

Load case: self-weight:

$$G_{Ct} = \frac{G_C}{1+\phi} = \frac{3.02}{1+7.0} = 0.378\,\text{N/mm}^2 \text{ (as before)}$$

Calculation values: $k = \dfrac{B_S}{A_C G_{Ct} L^2} = \dfrac{274.4 \times 10^9}{65.25 \times 1000 \times 0.378 \times 3300^2} = 0.921$

Therefore, from Fig. 9.14(a) $\beta_{Sq} = 0.28$

$$w = \frac{w_G L^4}{B_D + \dfrac{B_S}{1+15k}}\left[\frac{5}{384} - \frac{1}{139(1+\beta_{Sq})}\right]$$

$$= \frac{0.116 \times 3.3^4 \times 10^9}{18.6 \times 10^6 + \dfrac{247.4 \times 10^6}{1 + 15 \times 0.921}}\left(\frac{5}{384} - \frac{1}{139(1+0.28)}\right)$$

$$= 2.88\,\text{mm}$$

Additional deflection due to creep: $\Delta w_G = 2.88 - 6.50 \times \dfrac{0.116}{0.866} = 2.01\,\text{mm}$

Load case: snow:

$$G_{Ct} = \frac{G_C}{1+\phi} = \frac{3.02}{1+2.4} = 0.882\,\text{N/mm}^2 \text{ (as before)}$$

Calculation values: $k = \dfrac{B_S}{A_C G_{Ct} L^2} = \dfrac{247.4 \times 10^9}{65.25 \times 1000 \times 0.882 \times 3300^2} = 0.395$

Therefore, from Fig. 9.14(a) $\beta_{Sq} = 0.30$

$$w = \frac{w_Q L^4}{B_D + \dfrac{B_S}{1 + 15k}} \left[\frac{5}{384} - \frac{1}{139(1 + \beta_{Sq})} \right]$$

$$= \frac{0.75 \times 3.3^4 \times 10^9}{18.6 \times 10^6 + \dfrac{247.4 \times 10^6}{1 + 15 \times 0.395}} \left[\frac{5}{384} - \frac{1}{139(1 + 0.30)} \right]$$

$$= 12.26\,\text{mm}$$

Additional deflection due to creep: $\Delta w_Q = 12.26 - 6.50 \times \dfrac{0.75}{0.866} = 6.63\,\text{mm}$

$$w_{max} = w_G + w_Q + w_T + \Delta w_G + \Delta w_Q$$

Maximum deflection: $= 6.50 + 2.24 + 2.01 + 6.63$

$$= 17.4\,\text{mm} = \frac{\text{span}}{190}$$

14.3.8 Design of fasteners

For the design of the fasteners, the load case of wind uplift plus temperature is critical. Here, the dead load (self-weight of the panel) reduces the uplift forces. The calculations for this case are not considered in detail in this example.

Note that special attention should be paid to the uplift forces around the perimeter of the roof.

14.3.9 Comments on the calculations

As with the wall panel considered earlier, the serviceability limit state proves to be more critical than the ultimate limit state. For strength design, it is the stresses at the top of the profiled outer face over the support that govern the design. However, for this roof panel, the deflections are also significant.

Both the stresses and deflections are significantly influenced by creep under long-term loading.

14.4 REFERENCE

14.1 European recommendations for sandwich panels with additional recommendations for mineral wool core material: Part 1: Design. CIB Publication 148 by joint European Convention for Constructional Steelwork Committee TWG 7.9 and International Council for Building Research Studies and Documentation Committee W56. Draft dated Summer 1999. Update published as CIB Publication 257 (2000).

Index